# TELECOMMUNICATIONS RESEARCH TRENDS

# Telecommunications Research Trends

Hans F. Ulrich and Ernst P. Lehrmann
Editors

Nova Science Publishers, Inc.
*New York*

Copyright © 2008 by Nova Science Publishers, Inc.

**All rights reserved.** No part of this book may be reproduced, stored in a retrieval system or transmitted in any form or by any means: electronic, electrostatic, magnetic, tape, mechanical photocopying, recording or otherwise without the written permission of the Publisher.

For permission to use material from this book please contact us:
Telephone 631-231-7269; Fax 631-231-8175
Web Site: http://www.novapublishers.com

### NOTICE TO THE READER

The Publisher has taken reasonable care in the preparation of this book, but makes no expressed or implied warranty of any kind and assumes no responsibility for any errors or omissions. No liability is assumed for incidental or consequential damages in connection with or arising out of information contained in this book. The Publisher shall not be liable for any special, consequential, or exemplary damages resulting, in whole or in part, from the readers' use of, or reliance upon, this material.

Independent verification should be sought for any data, advice or recommendations contained in this book. In addition, no responsibility is assumed by the publisher for any injury and/or damage to persons or property arising from any methods, products, instructions, ideas or otherwise contained in this publication.

This publication is designed to provide accurate and authoritative information with regard to the subject matter covered herein. It is sold with the clear understanding that the Publisher is not engaged in rendering legal or any other professional services. If legal or any other expert assistance is required, the services of a competent person should be sought. FROM A DECLARATION OF PARTICIPANTS JOINTLY ADOPTED BY A COMMITTEE OF THE AMERICAN BAR ASSOCIATION AND A COMMITTEE OF PUBLISHERS.

LIBRARY OF CONGRESS CATALOGING-IN-PUBLICATION DATA

*Telecommunications research trends / Hans F. Ulrich and Ernst P. Lehrmann (editors).*
  p. cm.
 *ISBN 978-1-60456-158-6 (hardcover)*
  1. Telecommunication--Research. 2. Telecommunication--Technological innovations. I. Ulrich, Hans F. II. Lehrmann, Ernst P.
  HE7631.T4467 2008
  384--dc22
                    2007049405

Published by Nova Science Publishers, Inc. ✦ New York

# CONTENTS

| | | |
|---|---|---|
| **Preface** | | vii |
| **Chapter 1** | Factors Influencing Non-Adoption of Telecommunications Services: The Case of Public Wireless LAN in Korea<br>*Moon-Koo Kim and Jong-Hyun Park* | 1 |
| **Chapter 2** | European Union Mobile Telecommunications in the Context of Further Enlargement<br>*Peter Curwen and Jason Whalley* | 13 |
| **Chapter 3** | The Privatization of Intelsat: The Transition from an Integovernmental Organization to Private Equity Ownership<br>*Patricia McCormick* | 45 |
| **Chapter 4** | Contracting for Municipal Broadband Services: A Comparative Study<br>*Carol Ting* | 75 |
| **Chapter 5** | Influencing Infrastructure Performance through Cross-Border Networks of Regulatory Agencies<br>*Jacqueline Horrall* | 91 |
| **Chapter 6** | Value Creation in Acquisitions: Privatization of Turk Telekom<br>*Fırat Demir and Vahap B. Uysal* | 107 |
| **Chapter 7** | Cross Media Ownership: An Analysis of Regulations and Practices in Australia, Hong Kong and Singapore<br>*T.Y. Lau, Katie Look,<br>David Atkin and Carolyn A. Lin* | 127 |
| **Chapter 8** | Nigeria: Reviving a Former Monopoly in a Rapidly Evolving Market<br>*Chuka Onwumechili* | 143 |

| | | |
|---|---|---|
| **Chapter 9** | Strategic Bundling in Telecommunications and its Antitrust Implications for Intermodal Competition<br>*Paul R. Zimmerman* | **159** |
| **Chapter 10** | New Challenges in Raman Amplification for Fiber Communication Systems<br>*P.S. André, A.N. Pinto, A.L.J. Teixeira, B. Neto, S. Stevan Jr., Donato Sperti, F. da Rocha, Micaela Bernardo, J.L. Pinto, Meire Fugihara Ana Rocha and M. Facão* | **177** |
| **Chapter 11** | Fiber Bragg Gratings in High Birefringence Optical Fibers<br>*Rogério N. Nogueira, Ilda Abe and Hypolito J. Kalinowski* | **209** |
| **Index** | | **245** |

# PREFACE

Telecommunication is the transmission of signals over a distance for the purpose of communication. In modern times, this process typically involves the sending of electromagnetic waves by electronic transmitters, but in earlier times telecommunication may have involved the use of smoke signals, drums or semaphore or heliograph. Today, telecommunication is widespread and devices that assist the process, such as the television, radio and telephone, are common in many parts of the world. There are also many networks that connect these devices, including computer networks, public telephone networks, radio networks and television networks. Computer communication across the Internet is one of many examples of telecommunication. This new book presents the latest research from around the world.

Chapter 1 - Public wireless LAN has been stagnated as it fails to overcome chasm phase and facing crisis of service switching in spite of its flourishing expectation at the beginning time. More over, introduction of HSDPA and WiBro (Mobile WiMAX) representing mobile broadband internet service intimidate future market of Wireless LAN. However, in other aspects, bundling service of Wireless LAN with other internet services is expected to provide a new opportunity of growth as wireless mesh network is deployed based on the technology evolution.

As introduction of new telecommunication service induces great deal of investment and results tremendous impact over national economy, secure guarantee of market is very important point for operators. This paper aims to analyze the non-adoption factors for Public Wireless LAN in Korea, and provide useful implication for newly introduced telecommunication service. Especially analyzed factors are perceived service characteristics of public wireless LAN, individual characteristics, and perceived lack of service utility to figure out reason not to adopt public wireless LAN.

Chapter 2 - On 1 May 2004, the European Union (EU) witnessed its single largest expansion when ten countries joined. The accession of these ten countries – Cyprus (South), Czech Republic, Estonia, Hungary, Latvia, Lithuania, Malta, Poland, Slovakia and Slovenia was followed by that of two further countries, Bulgaria and Romania, on 1 January 2007. These countries may well not be the last to join the EU, but their accession represents a useful point at which to assess just how big an effect EU expansion has had upon its industrial structure and the strategies of its companies, in this case with respect to mobile telecommunications.

The purpose of this paper is to explore two particular issues in the context of the mobile telecommunications market: Firstly, whether as a consequence of EU expansion, the accession countries will indeed be the recipients of inward investment from those companies that have, until now, largely ignored them; and secondly, whether those companies already active in the accession countries will seek to reinforce their existing positions.

To this end, the paper is structured as follows. The initial section charts the evolution of telecommunications regulation within the EU and addresses the accession process. In the main section, the ownership of mobile communication licences across the enlarged EU is described. In addition, the geographical footprint of operators is established, with a distinction being made between those operators that have invested in the accession countries and those that have not. In the second main section the focus shifts onto the accession countries, with the analysis being driven by the issues outlined above. Conclusions are drawn in the final section.

Chapter 3 - This chapter examines the privatization of the world's most prominent international satellite organization, namely the International Telecommunications Satellite Organization (INTELSAT), which underwent a profound restructuring in November 2000 when it was separated into two entities. INTELSAT's intergovernmental treaty arrangement was fundamentally altered in March 2000 with the enactment of U.S. legislation, the Open-market Reorganization for the Betterment of International Telecommunications Act or ORBIT Act, which set forth specific criteria relating to the privatization of INTELSAT, which transpired on 18 July 2001. Simultaneously, a continuing intergovernmental organization, retaining the name the International Telecommunications Satellite Organization with the acronym 'ITSO', was designated to serve for a minimum period of twelve years in a supervisory and monitoring capacity, ensuring that the private company meet its public service and lifetime connectivity obligations to those developing countries largely dependent on the services and pricing mechanisms of Intelsat, Ltd.

This research examines the transformation of INTELSAT from an intergovernmental organization, owned by governments and by extension, citizens, to its current ownership by a private equity firm, within the larger context of the privatization of the global commons. This research analyzes the complicated consequences of the notable structural and ownership alterations of INTELSAT as well as the influence exerted by Lockheed Martin Corporation in the restructuring process. Methodologically, this research employs institutional ethnography in conjunction with the case study. In the context of international regime change and increased cooperation between national governments and the private sector, this work critically examines the process of privatizing INTELSAT and post-privatization developments as well the ability of ITSO to effectively oversee its charge.

Chapter 4 - In the past few years, a number of municipal governments have launched citywide broadband networks to promote economic development, digital inclusion and economization of city services. Such projects often face the challenges of building a sustainable business model and dealing with long-term transaction cost issues arising from information problems, market power and lock-in.

Building on the literature of transaction cost theory and history of public utility networks, this chapter examines franchise agreements of major municipal broadband projects and analyzes local governments' approaches to implementing their goals and addressing transaction cost issues with deployment of broadband networks. The study identifies significant variation in most of these major aspects and suggests subjects for long-term research.

Chapter 5 - Increasingly, networks of sectoral regulatory agencies in telecommunications, energy, transportation, and water have been providing regional public goods (RPGs). Developing countries especially stand to benefit from shared resources permitted by the provision of RPGs. RPGs are typically facilitated by regional networks and include data sharing for benchmarking, best practice techniques, capacity building and training, development studies, and the facilitation of events and meetings. The development of transnational networks and the RPGs they provide appears to be a strong answer that would encourage regional infrastructure development. The effects would be most noticeable in developing countries that face capacity, resource and financial and other constraints to national infrastructure development. The success of cross-border provision of public goods, however, may prove to be challenging, because of issues that arise with the provision of public goods. These problems include adverse selection, moral hazard, the prisoners' dilemma, and free-rider problems. Generally, any kind of problems having to do with collective action agreement, such as agreements on data collection standards, cross-country conflict of regulatory policies and legislation form noticeable obstacles to successfully providing cross-border public goods. Improving the provision of cross-border public goods may only be achieved by tackling provision problems that are not only financially related, but from the production end as well.

Chapter 6 - This paper examines value creation and distribution in acquisitions in the context of Turk Telekom privatization. The authors find that operational synergies, in addition to traditional sources of synergies, play an important role in value creation. Specifically, re-allocation of resources from inefficient business units to efficient ones contributes to value creation. The authors also document that the privatization method as well as the post-privatization regulatory framework are significant determinants of the success or failure of privatization programs in developing countries especially with regard to increasing competition, efficiency and value transfer to the public.

Chapter 7 - This study compares the regulations and practices of cross media ownership in Australia, Hong Kong, and Singapore, utilizing a conceptual framework to explore relationships between the government, service providers and consumers. Government regulatory policies are then analyzed in the context of recent technological trends pushing media convergence. Study findings suggest that market size does not affect policy-making and that private ownership is the key determinant of policy outcomes. The relative merit of various regulatory approaches (i.e. "government- guided" vs. "market-oriented") are discussed, alongside implications of cross media ownership changes wrought by technology.

Chapter 8 - Nigeria's telephone market has changed remarkably in the last few years with NITEL, the erstwhile monopoly, rapidly losing market share to vibrant competitors. Prior to 1992, NITEL was the sole provider of telephone services to Nigeria. However, NITEL was unsuccessful and regulatory changes in 1992 led to policies allowing the entrance of private providers into the market.

A decade later, the market reached a watershed when the regulator, the Nigerian Communications Commission (NCC), auctioned GSM licenses that dramatically changed the market. The new licenses quickly re-shaped the market environment by generating rapid increases in customer subscription and introduction of value added services. NITEL, clearly unable to compete, attempted to stem the rising tide by creating interconnection bottlenecks.

The attempts failed and NITEL's market share went into a free fall. The government later sold NITEL to private investors – Transcorp. Ltd. – but there remains a tough road ahead for the erstwhile monopoly.

This chapter addresses the issues identified above by reviewing the Nigerian telephone market before 1992 and discussing what followed market liberalization. A key focus is an analysis of NITEL's options in a market, which continues to evolve.

Chapter 9 - The Telecommunications Act of 1996 and subsequent regulatory actions sought to enhance the degree of competition in the local and long-distance wireline exchange markets by, among other measures, opening the local networks of monopoly incumbent local exchange carriers (in particular, those of the Regional Bell Operating Companies or "RBOCs") to competitive entry and implementing wireless local number portability to facilitate wireless substitution. However, the recent wave of merger activity in the U.S. telecommunications industry (as well as the recent revocation of the Act's unbundling provisions) has led to rapid consolidation within both the wireline and wireless segments of the industry. At the same time, the capacity of independent wireless and Internet-based telecommunications providers to function as sources of intermodal competition against the RBOCs is often championed by policy makers and industry analysts. This chapter examines the validity of this latter contention in the context of several important characteristics of the current telecommunications industry including (but not limited to) the complementarity of wireline and wireless access, the increased tendency for subscribers to take bundled telecommunications services, and the ownership of the largest national facilities-based wireless carriers by the largest regional wireline firms. Particular attention is paid to the interaction between these factors and the use of strategic wireline/wireless bundling by the RBOCs to affect the development of intermodal competition (and thus consumer welfare). It is argued that RBOC wireline/wireless bundling may confer various value-added services and other benefits (*e.g.*, lower prices) to consumers in the short-run as the RBOCs employ these bundles to retain their wireline customer base. However, in the long-run, these same benefits may serve to disadvantage independent wireless (and Internet) carriers who cannot offer the same bundled offerings. To the extent that this results in the latter firms being forced to exit the market, wireline/wireless bundling may serve as means for the RBOCs to strategically retain their market power in the local wireline exchange access market while leveraging their wireline market power into the in-region wireless market. As such, the potential for independent wireless or Internet-based carriers to serve as viable intermodal competitors in the long-run appears tenuous at best. Finally, it is posited that the most viable source of intermodal competition in the local residential exchange market will stem from the entry of cable operators into the voice telephony market. As such, the long-run market structure of the telecommunications industry may evolve into a duopoly characterized by RBOCs and cable providers competing in bundles. Whether or not such a market structure is sufficient to induce "aggressive" competition between these two carriers is uncertain, which in turn highlights the need for careful antitrust and regulatory oversight in the future in order to protect the interests of telecommunications subscribers.

Chapter 10 - Raman fiber amplifiers (RFA) are among the most promising technologies in lightwave systems. In recent years, Raman optical fiber amplifiers have been widely investigated for their advantageous features, namely the transmission fiber can be itself used as the gain media reducing the overall noise figure and creating a lossless transmission media.

The introduction of RFA based on low cost technology will allow the consolidation of this amplification technique and its use in future optical networks.

This paper reviews the challenges, achievements, and perspectives of Raman amplification in optical communication systems. In Raman amplified systems, the signal amplification is based on stimulated Raman scattering, thus the peak of the gain is shifted by approximately 13.2 THz with respect to the pump signal frequency. The possibility of combining many pumps centered on different wavelengths brings a flat gain in an ultra wide bandwidth.

An initial physical description of the phenomenon is presented as well as the mathematical formalism used to simulate the effect on optical fibers.

The review follows with one section describing the challenging developments in this topic, such as using low cost pump lasers, in-fiber lasing, recurring to fiber Bragg grating cavities or broadband incoherent pump sources and Raman amplification applied to coarse wavelength multiplexed networks. Also, one of the major issues on Raman amplifier design, which is the determination of pump powers in order to realize a specific gain will be discussed. In terms of optimization, several solutions have been published recently, however, some of them request extremely large computation time for every interaction, what precludes it from finding an optimum solution or solve the semi-analytical rate equation under strong simplifying assumptions, which results in substantial errors. An exhaustive study of the optimization techniques will be presented.

This paper allows the reader to travel from the description of the phenomenon to the results (experimental and numerical) that emphasize the potential applications of this technology.

Chapter 11 - Fiber Bragg gratings (FBG) are a key element in optical communication devices and in fiber sensors. This is mainly due to its intrinsic characteristics, which include low insertion loss, passive operation and immunity to electromagnetic interferences. Basically a FBG is a periodic modulation of the core refractive index formed by exposure of a photosensitive fiber to a spatial pattern of ultraviolet light in the region of 244–248 nm. The lengths of FBGs are normally within the region of 1–20 mm. Usually a FBG operates as a narrow reflection filter, where the central wavelength is directly proportional to the periodicity of the spatial modulation and to the effective refractive index of the fiber. The production technology of these devices is now in a mature state, which enables the design of gratings with custom-made transfer functions, crucial for all-optical processing. Recently, some work has been done in the application of FBG written in highly birefringent fibers (HiBi). Due to the birefringence, the effective refractive index of the fiber will be different for the two transversal modes of propagation. Therefore, the reflection spectrum of a FBG will be different for each polarization. This unique property can be used for advanced optical processing or advanced fiber sensing.

The chapter will describe in detail this unique device. The chapter will also analyze the device and demonstrate different applications that take advantage of its properties, like multiparameter sensors, devices for optical communications or in the optimization of certain architectures in optics communications systems.

*Chapter 1*

# FACTORS INFLUENCING NON-ADOPTION OF TELECOMMUNICATIONS SERVICES: THE CASE OF PUBLIC WIRELESS LAN IN KOREA

### *Moon-Koo Kim and Jong-Hyun Park*
Electronics and Telecommunications Research Institute, Korea

## ABSTRACT

Public wireless LAN has been stagnated as it fails to overcome chasm phase and facing crisis of service switching in spite of its flourishing expectation at the beginning time. More over, introduction of HSDPA and WiBro (Mobile WiMAX) representing mobile broadband internet service intimidate future market of Wireless LAN. However, in other aspects, bundling service of Wireless LAN with other internet services is expected to provide a new opportunity of growth as wireless mesh network is deployed based on the technology evolution.

As introduction of new telecommunication service induces great deal of investment and results tremendous impact over national economy, secure guarantee of market is very important point for operators. This paper aims to analyze the non-adoption factors for Public Wireless LAN in Korea, and provide useful implication for newly introduced telecommunication service. Especially analyzed factors are perceived service characteristics of public wireless LAN, individual characteristics, and perceived lack of service utility to figure out reason not to adopt public wireless LAN.

## I. INTRODUCTION

In Korea, public wireless LAN has received much attention from the industry since its market debut in 2002, when it was touted as one of the most promising solutions to lead FMC (Fixed-Mobile Convergence) and bring on new growth spurts for the local telecom sector. The demand for public wireless LAN, however, has been stagnating since quite a while, as though caught in a market chasm. Public wireless LAN services are provided chiefly through a few carriers. With hot spots created in urban centers, providers have engaged intense marketing efforts to expand their subscriber base. The number of subscribers sharply surged during the initial period following the roll-out. The rate of growth, however, gradually slowed down. From the end of 2005, the number of subscribers either declined or stayed static. Whilst attempts to restimulate the demand for public wireless LAN are currently underway, notably, by upgrading transmission speed, the outlook is far from rosy, as new mobile broadband internet technologies rolled out in the intervening years, such as HSDPA and WiBro (Mobile WiMAX), are poised to eat away at a significant portion, if not the entirety, of its demand base.

Thanks to strides made in telecommunications technology since the late 1990s, a host of innovative services have been bought to the market. However, in spite of massive investments by operators, most of these new services saw demand stagnate after the early market or were replaced by other more evolved services and retired from the market[3]. Market failures by innovative telecom services, while primarily business failures of individual companies, can also have a negative consequence for the overall economy and the welfare of consumers.

Much of the existing literature on telecommunications services is concerned with identifying factors determining consumers' adoption of an innovative service. Few studies were devoted to factors influencing the opposite response from consumers, in other words, non-adoption of an innovative service. The great majority of prior studies focused on what incites consumers to adopt new telecommunications services, but not why certain services are rejected by consumers or are unable to cross the market chasm. By analyzing the phenomenon of demand stagnation or non-adoption, one can determine precise causes for consumers' rejection of certain services from a user perspective. Understanding these factors of non-adoption is essential for developing services that are not just based on technology pull, but address real needs of their users and are adapted to consumers' adoption behavior. Developing user-oriented services, furthermore, can help reduce market failures, which entail particularly huge financial losses in the telecom sector, as new telecommunications services tend to involve staggering amounts of capital investment.

This study is an attempt to determine factors influencing the non-adoption of public wireless LAN in Korea, from a user perspective. Our findings can assist providers of public wireless LAN service with restimulating demand for this service, helping them turn the challenge they are currently facing into an opportunity. Our aim is both to develop a theoretical understanding of the non-adoption in telecom services and draw practical implications that are concretely useful for service providers. The rest of this paper is organized as follows: Section II discusses the current status of public wireless LAN in Korea and the outlook, as well as the potential of wireless mesh-based wireless LAN. In Section III we analyze the influence factors for non-adoption of public wireless LAN, using the results of a consumer survey. Non-adopters are classified into several subgroups according to criteria including demographic and lifestyle characteristics, telecom service usage characteristics and factors hindering adoption or non-adoption factors indicated by them. Finally, in Section IV, we re-summarize our results and present their implications, pointing out strategic directions to take when attempting to broaden consumers' take-up of a new telecom service.

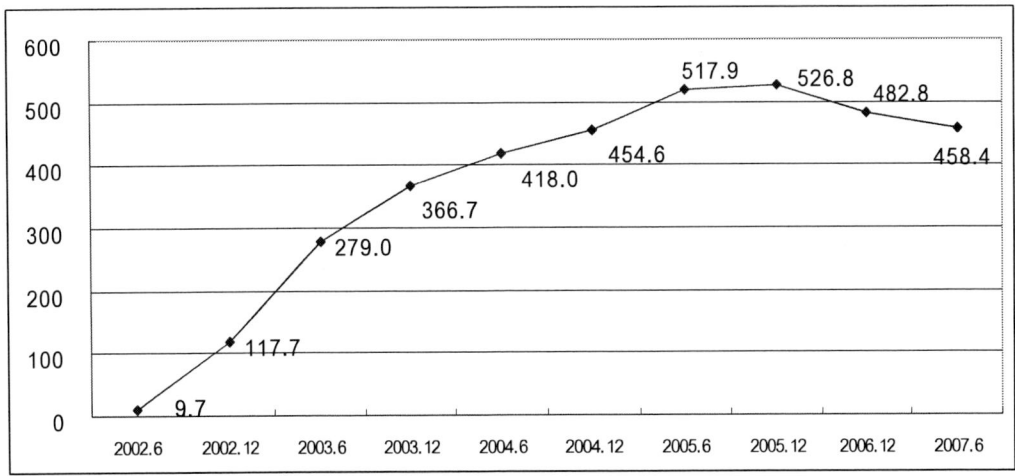

Unit: 1,000 subscribers.
Source: Korea Ministry of Information and Communication website.

Figure 1. Public Wireless LAN Subscribers in Korea.

## II. THE CURRENT STATUS AND OUTLOOK OF PUBLIC WIRELESS LAN IN KOREA

In Korea, public wireless LAN, as it permits access to fixed broadband internet both from indoors and outdoors, was welcomed as a technology capable of accelerating the process of convergence between fixed and mobile communications. Public wireless LAN was commercialized in 2002, by Korea Telecom and Hanaro Telecom. At the time of its roll-out, the demand for public wireless LAN was projected to grow at an average annual rate of over 58.3%, securing a subscriber base of 2 million by 2007[7].

Providers set up wireless LAN hot spots in urban centers and launched active marketing efforts to expand their subscriber base. However, contrary to initial expectations, as can be seen in the trend graph in (Figure 1) below, the number of public wireless LAN subscribers has declined after reaching a peak of 0.52 million in late 2005.

Public wireless LAN services are currently provided not only at 2.4GHz band, but also at 5GHz band. Providers are investing in improving transmission speeds, mobility and indoor access, and designing new business models to bundle the service with multimedia, to rekindle demand. There have been, however, some major changes in the competition landscape for public wireless LAN, in recent years. Mobile broadband services like HSDPA (High Speed Downlink Packet Access) and WiBro, rolled out in the first half of 2006, offer similar functions to public wireless LAN, and are poised to substitute some if not most of the demand for the latter[6].

The outlook, nevertheless, is not altogether bleak for public wireless LAN. Wireless mesh networks, enabling broadband internet access via AP-to-AP communication, are shortly due to be added to public wireless LAN infrastructure. Dispensing with wiring between existing wireless LAN access points, wireless mesh networks are easy to set up in most urban areas, including subways, university campuses, parks and downtown centers, and can dramatically extend the service coverage. Furthermore, this self-organizing network supporting dynamic path selection makes it possible for providers to more flexibly respond to changes in service environments; which also contributes to the stability of service. The cost is minimal, as existing wireless LAN APs can be re-used. Mesh networking also allows continuous evolution of services, as the open architecture enables easy integration of new and enhanced wireless LAN technologies as they become available. In sum, the new networking technology opens up the way for public wireless LAN to evolve into a wireless broadband access solution, combining a high transmission speed with an optimal degree of mobility. It also opens up a whole-new avenue of possibilities in terms of business models; fixed-wireless convergence services, affordable telephone services through VoIP (Voice of IP), services bundling telematics and IPTV functions, home networking linked to ubiquitous city based network services are just a few[1]. In Korea, wireless LAN-based mesh networks are under construction in several university campuses. Outside Korea, mesh networking is used in Taiwan for Taipei's city network and university networks.

Wireless LAN is also slated to be coupled with HSDPA and WiBro to improve access from dead spots and indoor environments. Both mesh networking and combination with HSDPA and WiBro mark an important point of transition for public wireless LAN.

## III. FACTORS INFLUENCING NON-ADOPTION OF PUBLIC WIRELESS LAN

In this section, we perform an empirical analysis of factors influencing consumers' non-adoption of public wireless LAN, using data obtained from a survey of Korean consumers. The section will focus on the individual characteristics and service characteristics influencing consumers' intention not to adopt public wireless LAN, and their practical implications.

### 1. Research Model and Variables

Perceived service characteristics, individual characteristics and perceived lack of service utility (see (Figure 2)) were selected as influence factors for non-adoption of public wireless LAN, by reviewing the existing literature. The impact of influence factors was measured using a structural equation model. The research model sets the perceived lack of service utility as the primary factor determining a consumer's decision not to use public wireless LAN, and perceived service characteristics and individual characteristics as the secondary influence factors.

Perceived service characteristics, the independent variables, are given five detailed variables, including price, limitation of service coverage, risk (especially security risk), low quality (service interruptions, etc.), and availability of alternative services (roll-out of WiBro or HSDPA). Two detailed variables were created under individual characteristics: innovation acceptance and proficiency with IT (Information Technology).

Perceived service characteristics and individual characteristics are each assigned three variables, and 21 total detailed items were constructed. All variables were measured using a 7-point Likert scale. Perceived lack of service utility, the mediator, and intention not to adopt the service, the independent variable, were also given three questionnaire items each, which were measured using a 7-point Likert scale.

## 2. Data Collection

An external market research organization surveyed 650 Koreans aged 15-49, residing in urban areas, who have knowledge of public wireless LAN or are at least aware of its existence, but are not currently using it. The respondents were selected through stratified sampling, using the age quota based on 2,000 year national census data. The survey was conducted through one-on-one interviews, and a structured questionnaire was used. After discarding incomplete and otherwise invalid responses, 629 responses were retained. The demographic characteristics and characteristics related to the use of telecom services of the samples closely coincided with those observed in the general population.

SPSS 11.0 and the AMOS 4.0 package were used for basic statistical analysis, factor analysis, reliability testing and structural equation modeling.

## 3. Validity and Reliability

To test the construct validity of the measures, we performed an exploratory factory analysis on independent variables, using a principal components method with VARIMAX rotation. Further, to verify internal consistency, we tested the variables for reliability. The factor analysis used a factor loading of 0.6 and above and Cronbach's α, the coefficient of reliability, was set to 0.7 and above[5].

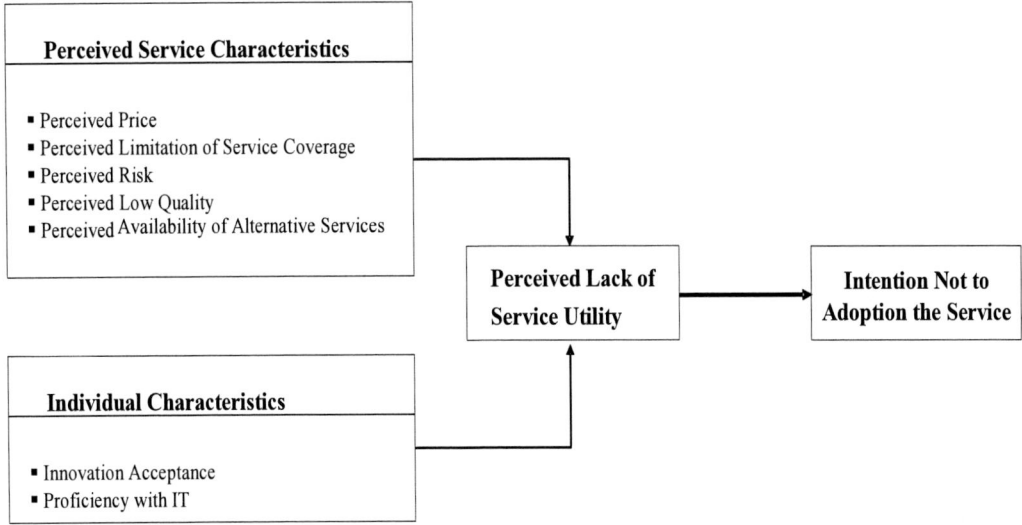

Figure 2. Research Model.

**Table 1. Results of Factor Analysis and Reliability Testing on Independent Varaibles**

|  | Perceived Price | Perceived Limitation of Service Coverage | Perceived Risk | Perceived Low Quality | Perceived Availability of Alternative Services | Innovation Acceptance | Proficiency with IT |
|---|---|---|---|---|---|---|---|
| Variable 1 | 0.864 | 0.867 | 0.878 | 0.908 | 0.842 | 0.899 | 0.901 |
| Variable 2 | 0.815 | 0.823 | 0.836 | 0.881 | 0.827 | 0.868 | 0.832 |
| Variable 3 | 0.810 | 0.766 | 0.748 | 0.844 | 0.701 | 0.859 | 0.761 |
| Factor Loading | 11.285 | 8.997 | 6.853 | 23.161 | 5.418 | 13.545 | 6.215 |
| Eigen Value | 2.370 | 1.889 | 1.439 | 4.864 | 1.138 | 2.845 | 1.305 |
| Coefficient of Reliability | 0.841 | 0.838 | 0.806 | 0.878 | 0.774 | 0.859 | 0.793 |

The factor analysis performed on seven independent variables revealed seven factors, as intended (see Table 1). The explanatory power of the seven factors was 75.5%, and their eigen value was 1 or greater, indicating a satisfactory degree of construct validity. The coefficient of reliability was equal to or greater than 0.7, confirming the existence of internal consistency.

The confirmatory factor analysis, performed to test the convergent validity between the factors, revealed that all indicators of goodness of fit were above their recommended thresholds.

## 4. RESULTS

The results of goodness of fit testing are given in Table 2 below. The research model met or exceeded recommended thresholds of all indicators including GFI, AGFI, RMR, NFI, NNFI, CFI and RMSEA; a confirmation that the model is adapted for structural equation modeling.[1]

**Table 2. Goodness of Fit Indicators**

| Indicator | $\chi^2$ | Degree of Freedom | GFI | AGFI | RMR | NFI | NNFI | CFI | RMSEA |
|---|---|---|---|---|---|---|---|---|---|
| Research Model | 835.003*** | 280 | 0.918 | 0.889 | 0.071 | 0.909 | 0.921 | 0.937 | 0.056 |

*: <0.1, **: <0.05, ***:<0.01

---

[1] Goodness of fit indicators and their respective recommended ranges of acceptability used in this study are as follows[4]: GFI(≥0.90), AGFI(≥0.80), RMR(≤0.08), NFI(≥0.90), $\chi^2$ (≥α=0.05; we did not assign too much importance to the significance level, as the vaue of $\chi^2$ is known to abnormally increase, when the number of respondents exceeds 200).

The results of the structural equation modeling, summarized in Table 3, were as follows:

The perceived lack of service utility appeared to have a positive influence on consumers' intent not to adopt public wireless LAN. What this suggests is that consumers, as long as they do not find public wireless LAN useful, would not consider adopting it. This result corroborates the known relationship between the perceived utility of an IT service and consumers' intention to use the service under a technology acceptance model (TAM).

Perceived service characteristics, perceived price, perceived limitation of service coverage and perceived risk proved to have a significant positive impact on the perceived lack of utility of public wireless LAN. As the price partakes of the economic value of a service, the price of public wireless LAN service appears to be closely related to consumers' perception of its utility. This result, further, suggests that the coverage of public wireless LAN, limited to hot spots, and security and other risks involved in its use are hindrances to the adoption of this service. On the other hand, low quality and availability of alternative services like HSDPA or WiBro did not appear to affect the perceived lack of utility regarding public wireless LAN.

Also, we did not find that individual characteristics of consumers such as innovation acceptance and proficiency with IT had a statistically significant relationship with the intention not to adopt public wireless LAN. This indicates that there is no substantial difference between innovators and laggards in their perception of utility of public wireless LAN, and that the level of proficiency in IT use is not a real factor for this relatively easy-to-use medium.

**Table 3. Results of Structural Equation Model**

| Path | Estimate | S.E. | C.R. | P-value |
|---|---|---|---|---|
| Perceived Price → Perceived Lack of Service Utility | 0.149 | 0.076 | 1.963 | 0.050* |
| Perceived Limitation in Service Coverage → Perceived Lack of Service Utility | 0.230 | 0.075 | 3.074 | 0.002*** |
| Perceived Risk → Perceived Lack of Service Utility | 0.258 | 0.064 | 4.006 | 0.000*** |
| Perceived Low Quality → Perceived Lack of Service Utility | 0.040 | 0.064 | 0.631 | 0.528 |
| Perceived Availability of Alternative services → Perceived Lack of Service Utility | 0.090 | 0.087 | 1.038 | 0.299 |
| Innovation Acceptance → Perceived Lack of Service Utility | -0.041 | 0.049 | -0.840 | 0.401 |
| Proficiency with IT → Perceived Lack of Service Utility | -0.031 | 0.064 | -0.492 | 0.623 |
| Perceived Lack of Service Utility → Intention Not to Adopt the Service | 0.565 | 0.041 | 13.934 | 0.000*** |

*: <0.1, **: <0.05, ***: <0.01.

**Table 4. Results of Segmentation of Non-adopters of Public Wireless LAN**

| Subgroup | Future Adopters | Possible Future Adopters | Adopters of Alternatives | Absolute Non-adopters |
|---|---|---|---|---|
| Percentage Share | 3.8% | 5.2% | 40.1% | 50.9% |
| Demographic Characteristics | Aged 20 to 30 Students, office workers, professionals, Earners of 3 to 4 million won in monthly income, Reside in Seoul metropolitan area | Aged 20 to mid-30s Students, office workers, self-employed workers' Earners of 2 to 3 million won in monthly income, Reside in Seoul and other large cities | Aged 10 to early 30s Professionals, students, office workers, self-employed workers, Earners of 2 to 4 million won in monthly income, Reside in Seoul metropolitan area or other large cities | Aged late 30s to 40 Bluecollar workers, housewives, Earners of 1 to 2 million won, and 5 million won and more in monthly income, Reside in small to medium cities |
| Lifestyle | Active individuals, Rational consumers Highly innovative, Highly proficient in technology use | Knowledge and information-seeking individuals, Rational consumers, Medium degree of innovativeness, Medium level of technology proficiency | High-tech buffs, Fashion-conscious individuals Knowledge and information-seeking, Highly innovative, Medium or above level of technology proficiency | Conservative consumers, Low deree of innovativeness, Low level of technology proficiency |
| Telecom Service Use Characteristics | Extensive use of fixed-line internet | Extensive use of fixed-line internet | High rate of use of mobile phone-based wireless internet | Spend generally little time on communications media |
| Hindrance Factors for Public Wireless LAN Adoption | Price, limited coverage, perception of risk | Price, limited coverage, perception of risk | Price, limited coverage, perception of risk, likelihood of future availability of alternative services | Price, limited coverage, perception of risk, low quality |

## 5. ADDITIONAL ANALYSIS: SEGMENTATION OF NON-ADPOTERS OF PUBLIC WIRELESS LAN

We segmented non-adopters of public wireless LAN into smaller subgroups, using the survey responses on additional questionnaire items. The four subgroups are people with the intention to use public wireless LAN sometime in the future (future adopters), people with intention to use public wireless LAN on the condition that significant improvements are made to the service (possible future adopters), those with the intention to use alternatives to public wireless LAN, such as HSDPA and WiBro (adopters of alternatives), and those with no intention to adopt public wireless LAN now or ever (absolute non-adopters). The four groups each account for 3.8%, 5.2%, 40.1% and 50.9% of total respondents. The respective characteristics of the four sub-groups are as follows:

- Future Adopters: This group of consumer are not currently subscribed to a public wireless LAN service, but have the intention to use it sometime in the future. Members of this sub-group representing a small minority of 3.8%, are in their mid-20s to early 30s. College students, office workers and professionals accounted for a comparatively large share. Many of them resided in the metropolitan area, and brought in 3 to 4 million won (medium level household income in Korea) in monthly income. They appear to be active individuals and rational consumers with a high degree of innovativeness and proficiency in technology use. They, furthermore, tend to spend many hours on the internet, via fixed line access, and cited price, limited coverage and risk as reasons that may lead potential users of public wireless LAN not to adopt it.
- Possible Future Adopters: The second smallest sub-group, representing 5.2% of total respondents, is made up of people who are currently not subscribed to a public wireless LAN service, but may consider using it in the future. Members of this sub-group were similar to future adopters in their general characteristics. With a median age of 25-35, this group had a high percentage of students, office workers and self-employed workers. Many of them were residents of large cities, earning 2 to 3 million won (low level household income in Korea) monthly. They appear to be knowledge and information-seeking individuals and rational consumers with a medium level of innovativeness and proficiency in technology use. They spend long hours on the internet and cited price, limited coverage and risk as principal reasons for not adopting public wireless LAN.
- Adopters of Alternatives: Members of this group are not currently subscribed to a public wireless LAN service and are inclined to use one of its alternatives like HSDPA or WiBro. 40.1% of all respondents fell into this sub-group. Adopters of alternatives have a profile quite distinct from the two previous sub-groups in that they are individuals in their late teens to early 30s. Professionals, students, office workers, and self-employed workers represented a comparatively large percentage of this group. Many of them resided in the Seoul metropolitan area with their monthy income ranging from 2 to 4 million won (low and medium level household income in Korea). This group of people with varied lifestyles was characterized by their common interest in high-tech gadgets and fashion-consciousness. They were avid

consumers of information with an above-average level of technology use proficiency. Heavy users of the internet, they accessed the web both via fixed-line and wireless connection, and picked price, limited service coverage, risk and likelihood of future availability of alternative services as the chief factors influencing their decision not to use public wireless LAN.
- Absolute Non-adopters: This segment grouping together consumers expressing the lack of intent to use public wireless LAN currently as well as in the future, was the largest of all four subgroups, accounting for 50.9% of total respondents. Consumers belonging to this subgroup stood out from the rest of the respondents in many aspects. They were mostly blue-collar workers or housewives in their late 30s to 40s. A majority of them lived in small and medium-sized cities outside the capital area and fell into the income range of either 1 to 2 million won or 5 million won and above (they belonged to either the bottom end or top end of the income distribution). Their lifestyle was generally conservative, with a low level of innovative tendency. Proficiency in technology use also was rather low. The use of telecom services was overall quite modest among consumers of this group. They attributed their lack of intent to use public wireless LAN to price, limited service coverage, risk and low quality.

## IV. Conclusion and Implications

This study has been an attempt to identify factors influencing the non-adoption of public wireless LAN, an innovative communications service currently experiencing a chasm-like stagnation in demand growth, after generating much excitement in the technology world at its initial launch. The relationship between consumers' intention not to use public wireless LAN and factors such as perceived service characteristics, individual characteristics and perceived lack of service utility, was empirically analyzed through structural equation modeling. As an additional step, we segmented non-adopters of public wireless LAN into four subgroups, and described their demographic and lifestyle characteristics and characteristics related to telecom service usage and the details of their attitude toward the (non-)adoption of this service. The summary of our results and their implications for marketers of new communications service, are as follows:

- The most decisive factor influencing the non-adoption of public wireless LAN is the low degree of perceived utility. Therefore, for public wireless LAN as well as other new communications services to successfully elicit demand, it is crucial to increase their perceived service utility. In order to enhance the level of perceived service utility, the marketing strategy must aim at creating a perception that the service offers practical benefits or relative advantages over other alternatives. Relative advantages are known to most directly affect a consumer's intent to purchase a product or service. Also important is to differentiate the service from existing alternatives or target a different level of service[2].
- The price of service is another factor hindering the growth of the public wireless LAN market. Value barriers are the single most important determinant of consumers'

resistance to innovation and adoption barriers. Value barriers appear when an innovation fails to offer a strong performance-to-price value. Whilst the basic monthly charge for public wireless LAN, when purchased bundled with broadband access service, is currently only at about one million won (appx. 11 USD), it can easy cost much more depending on the frequency or length of use. Reducing the price ahead of the introduction of service bundles (especially bundles with HSDPA or WiBro), lifestyle-tailored discount programs and hybrid pricing combining a flat fee and metered rates are some of the options providers can consider to improve the perceived price competitiveness of public wireless LAN.

- Difficulties associated with usage (e.g. limited coverage) or other impediments to the use (e.g. security and other risks) can directly contribute to stunting the growth of demand for a new service. Inconvenience and risk can be fatal blows to the prospect of mass adoption of an innovative service, serving respectively as a usage and a risk barrier, two of the classic adoption barriers. It is, therefore, of paramount importance to resolve these shortcomings by improving the performance of the service as well as the convenience of its use.

- Individual characteristics of consumers appeared not to have a significant impact on the expansion of the public wireless LAN market. This result may be attributable to the high degree of technology-friendliness among consumers. Once an innovative service proves to satisfy existing needs of consumers or attains a high level of performance, its diffusion tends to quickly accelerate, generally unaffected by the individual lifestyles of users. The rapid expansion undergone by mobile communication market, its broadband internet access and mobile phone-based wireless internet markets is the case in point.

- Based strictly on the adopter-to-non-adopter ratio of consumers, the outlook for public wireless LAN appears less than certain. The percentage of those expressing intent to adopt public wireless LAN and those who indicated that they may consider adopting it were made up only small percentages of total respondents. To turn around this situation, providers of public wireless LAN services should try to abandon their current standalone service model, centered on hot spots, in favor of one that forms a complementary relationship to other services. The two best strategies would be to evolve towards a mesh-based network (serving college campuses, company buildings or small and medium-sized cities) and linking public wireless LAN with HSDPA or WiBro (ensuring access in indoor environments and dead spots, and deployment in high-traffic areas).

## REFERENCES

[1] Kim, M. K., Park, J. H. & Lee, S. M. (2006) Characteristics of Wireless LAN-based Mesh Networks: Current Status and Outlook, ETRI.
[2] Kim, S. H. (2004) *High-tech Marketing*, Seoul: Pakyoungsa.
[3] Ahn, J. H., Kim, M. S. & Lee, D. J. (2005) Learning from the Failure: Experiences in the Korean Telecommunications Markets. *Technovation, vol. 25, no. 1*, pp. 69-82.

[4] Bentler, P. M. (1990) Multivariate Analysis with Latent Variables: Causal Modeling. *Annual Review of Psychology, Vol 31*, pp.419-456.
[5] Hair, J. F. Jr., Anderson, R. E., Tatham, R. L. & Black, W. C. (1998) *Multivariate Data Analysis*. 5th Edition, Upper Saddle River, New Jersey: Prentice Hall.
[6] Kim, M.K. & Jee, K. Y. (2006) Characteristics of Individuals Influencing Adoption Intentions for Portable Internet Service. *ETRI Journal, vol.28, no.1,* pp.67-76.
[7] ETRI (2003) The Prospect for the IT Industry.
[8] http://www.mic.go.kr (Korea Ministry of Information and Communication website)

Chapter 2

# EUROPEAN UNION MOBILE TELECOMMUNICATIONS IN THE CONTEXT OF FURTHER ENLARGEMENT

*Peter Curwen and Jason Whalley*
University of Strathclyde, Glasgow, Scotland

## ABSTRACT

On 1 May 2004, the European Union (EU) witnessed its single largest expansion when ten countries joined. The accession of these ten countries – Cyprus (South), Czech Republic, Estonia, Hungary, Latvia, Lithuania, Malta, Poland, Slovakia and Slovenia was followed by that of two further countries, Bulgaria and Romania, on 1 January 2007. These countries may well not be the last to join the EU, but their accession represents a useful point at which to assess just how big an effect EU expansion has had upon its industrial structure and the strategies of its companies, in this case with respect to mobile telecommunications.

The purpose of this paper is to explore two particular issues in the context of the mobile telecommunications market: Firstly, whether as a consequence of EU expansion, the accession countries will indeed be the recipients of inward investment from those companies that have, until now, largely ignored them; and secondly, whether those companies already active in the accession countries will seek to reinforce their existing positions.

To this end, the paper is structured as follows. The initial section charts the evolution of telecommunications regulation within the EU and addresses the accession process. In the main section, the ownership of mobile communication licences across the enlarged EU is described. In addition, the geographical footprint of operators is established, with a distinction being made between those operators that have invested in the accession countries and those that have not. In the second main section the focus shifts onto the accession countries, with the analysis being driven by the issues outlined above. Conclusions are drawn in the final section.

## INTRODUCTION

On 1 May 2004, the European Union (EU) witnessed its single largest expansion when ten countries joined. The accession of these ten countries – Cyprus (South), Czech Republic, Estonia, Hungary, Latvia, Lithuania, Malta, Poland, Slovakia and Slovenia – irrevocably changed the EU [Rachman, 2001, Cottrell, 2003]. Although the absorption of so many new Member States at the same time inevitably created problems [Dinan, 2006; Peet, 2007; Sedelmeier and Young, 2006], the EU was already committed to welcoming two additional members into the fold on 1 January 2007, namely Bulgaria and Romania. These countries may well not be the last to join the EU – there is still a lengthy queue of prospective entrants – but their accession represents a useful point at which to assess just how big an effect EU expansion has had upon its industrial structure and the strategies of its companies, in this case with respect to mobile telecommunications.

The purpose of this paper is to explore two particular issues in the context of the mobile telecommunications market: Firstly, whether as a consequence of EU expansion, the accession countries will indeed be the recipients of inward investment from those companies that have, until now, largely ignored them; and secondly, whether those companies already active in the accession countries will seek to reinforce their existing positions.

To this end, the paper is structured as follows. The initial section charts the evolution of telecommunications regulation within the EU and addresses the accession process. In the main section that follows, the ownership of mobile communication licences across the enlarged EU will be described. In addition, the geographical footprint of operators will be established, with a distinction being made between those operators that have invested in the accession countries and those that have not. In the second main section the focus shifts onto the accession countries, with the analysis being driven by the issues outlined above. Conclusions will be drawn in the final section.

## ISSUES PERTAINING TO ACCESSION

Until the mid-1980s the EU had no role in the regulation of the telecommunications industry [Thatcher, 2001: 562]. Since then, however, the role of the EU has been transformed to such an extent that it is now a key actor in the industry's regulation and development. Drawing on Thatcher (2001) and Walden (2005), among others it is possible to divide this transformation into three phases. The first phase is associated with the 1987 Green Paper that sought to create a telecommunications single market across the EU. This set out three principles to guide the subsequent creation of the regulatory framework: liberalisation, harmonisation and the application of competition rules (*Ibid.*: 110) [1].

The second phase commenced in late 1994 when the EU committed itself to full liberalisation by 1 January 1998 [2]. To this end, Directives were issued covering those parts of the telecommunications industry hitherto excluded from liberalisation by the EU, namely satellite services, mobile communications, voice telephony and infrastructure [Thatcher, 2001: 568]. In addition, the EU also issued Directives covering licensing, numbering, interconnection and universal service. The first three of these are central to the development and functioning of competitive telecommunication markets, while the latter focuses on the provision of telecommunication services to all citizens throughout the EU [3]. According to Walden (2005: 112), the third phase emerged from the 1999 review of communications that sought to address external pressures such as conformity with the rules of the World Trade Organisation, diverging practices between Member States [4] and a desire to simplify regulation to reflect the widening prevalence of competition within the industry. The EU proposed six Directives in July 2000 for implementation by July 2003 [5]. The new framework took into account convergence, and distinguished between ex-ante regulation and competition law [6].

When countries join the EU they are required to adopt its existing laws and institutions [7]. As a consequence, the accession countries joining in 2004 and 2007 were required to incorporate the considerable and wide-ranging body of EU telecommunications regulation outlined above into national legislation. This is by no means an insignificant task. Not only is the body of telecommunications regulation substantial [8], but it is just one of thirty-one areas where accession negotiations are undertaken [9]. Having said this, benefits begin to accrue to accession countries before they become actual members. Grabbe and Hughes (1998: 24) note how the 2004 accession countries benefited from foreign direct investment, preferential access to the EU and increased competition before they became full members of the EU. While foreign investors are likely to be attracted to accession countries for a variety of reasons, they are also likely to be reassured by the country's negotiation of membership as this necessitates a functioning market economy, efficient administration and the rule of law [10]. Hitran (2006) illustrates how Romania adopted the 2003 framework and liberalised its telecommunications market before it formally joined the EU, accruing substantial benefits in the process. Thus, the date at which the EU agrees to begin accession negotiations is as important as the actual accession date.

## MOBILE LICENCE DISTRIBUTION ACROSS THE ENLARGED EU

At the heart of the analysis of the implications of EU expansion on the strategies of mobile communications companies is the information contained in Table 1. This table depicts networks using second-generation (2G) technology – known generically in the EU as the Global System for Mobile (GSM) – and third-generation (3G) technology – known in the EU as the Universal Mobile Telecommunications System (UMTS) [11]. The table encompasses mobile licence ownership across the twenty-seven countries that were Member States of the EU on 1 January 2007. In essence, 2G represents a digital technology whose main purpose is to carry voice telephony while also accommodating low-speed data transfer as exemplified by the short message service (SMS), while 3G is capable of much higher speeds of data transfer suitable for large data files and still and video photography. UMTS normally requires the licensing of new spectrum, but there are also two intermediate technologies, known as the General Packet Radio Service (GPRS) and Enhanced Data rates for GSM Evolution (EDGE) that can operate at higher speeds than GSM while using the same spectrum, as well as a recently launched technology, High-Speed Downlink Packet Access (HSDPA), capable of even higher speeds than UMTS over the same spectrum band.

Table 1 builds on Whalley & Curwen (2003) in several ways. In the first place, the table differentiates between the two bandwidths used for what is generically called GSM, namely

GSM900 and GSM1800 (PCNs). Secondly, the table identifies when each mobile service was launched. Thirdly, the table details the number of subscribers that each company had as of 1 January 2007. By detailing the number of subscribers that a company has in each country, the table begins to differentiate between a simple presence in a country where the company is not a significant player and a presence where the company is actually (one of) the largest in the market in terms of number of subscribers. By combining the service launch date and the number of subscribers, the table also provides an impression as to how fast the market is growing and how successful the mobile operator has been in gaining subscribers.

Using Table 1 as our starting point, it is possible to make a series of preliminary observations about the mobile market of the enlarged EU in general and the mobile markets of accession countries in particular as of 1 January 2007. In the first place, 2G licences had been issued in all Member States. Although 137 licences had been issued – comprising 67 GSM900 and 70 GSM1800 licences (of which 54 had been issued to GSM900 licensees) – the number of licensees in each Member State varied between two and four. Cyprus (South), Malta and Slovakia had issued only two 2G licences (and, in addition, Slovenia had only two operational licensees), while the Netherlands, which uniquely had originally issued five, had seen the number fall to four as a result of a takeover. Clearly, therefore, the optimum number of 2G licences for an EU Member State, regardless of its size, appears currently to be either three or four, although as noted below there is considerable consolidation underway which is likely to lead to a reduction in the number of licensees by the year 2010.

Secondly, all EU Member States had issued 3G licences – in some cases such as Lithuania, fairly recently. In practice, somewhat more 3G licences than 2G licences had been put on offer, but because not all of the licences on offer had actually been awarded while others had been returned, revoked or sold on, the number of operators with 3G licences was only slightly larger – 87 compared to 83 – than the number of 2G licensees, which was somewhat surprising as many governments had indicated their desire to use the 3G licensing process as a way to increase the number of operators, and therefore the amount of competition, in the market [Whalley & Curwen, 2006]. However, it is – not surprisingly – the case that the 2G and 3G licensee groups heavily overlap. Altogether, 76 operators held both licences while there were 11 3G-only operators and a mere seven 2G operators without (so far) a 3G licence.

Despite the recent issuance of 3G licences in some cases, every Member State had witnessed the launch of 3G services prior to the end of 2006. By that point, 78 launches had taken place across the EU – that is, the vast majority of the then licensees had launched – although a number of the original licensees were no longer in that capacity [Whalley and Curwen, 2006] so the proportion was noticeably smaller when expressed in relation to the original number of licensees. In terms of individual Member States, only one – the UK – still had five operational licensees while none remained with only one.

Thirdly, the situation with respect to HSDPA (sometimes referred to as 3.5G) is worthy of note. Prior to 2005, almost nothing was expected from this technology given the emphasis placed upon UMTS, and although there was the very occasional launch during 2005 such as that of T-Mobile Austria, 2006 witnessed a positive tidal wave of launches resulting in a total of 56 being reached by 1 January 2007. HSDPA is essentially a software overlay – that is, it utilises existing networks and does not require new spectrum – and hence, crucially, does not necessitate a new licence which accounts for its rapid adoption compared to UMTS. Virtually every surviving EU network will be HSDPA-enabled by the end of 2007, and most will have gravitated from an initial maximum data transfer speed of 1.8 Mbps (megabits per second) to a maximum of 3.6 Mbps or even occasionally 7.2 Mbps.

## MARKET STRUCTURE

Across the enlarged EU in 2004, the incumbent fixed-wire operator owned the largest mobile operator as measured in terms of subscribers in nineteen of the twenty-five Member States: Austria, Belgium, Cyprus (South), Czech Republic, Denmark, Estonia, Finland, France, Germany, Greece, Hungary, Italy, Latvia (minority stake), Luxembourg, Netherlands, Portugal, Slovenia, Spain and Sweden. By 1 January 2007, the Czech Republic had disappeared from the list and both of the two new entrants qualified, albeit via a minority stake in Romania, so it was now a case of twenty-one among twenty-seven. This observation can also be expressed informatively in a slightly different fashion; that is, in just two of the fifteen 'old' Member States of the EU the incumbent operator did *not* own the largest mobile operator either in 2004 or subsequently. The exceptions were Ireland and the UK. In the case of Ireland, Eircom, the incumbent fixed-wire operator, divested its mobile subsidiary, Eircell, in May 2001 and Vodafone subsequently acquired Eircell for €4.5 billion in December 2001. BT also divested its mobile arm, known at the time as mmO$_2$ (which has recently become part of Telefónica). In November 2001, BT spun off mmO$_2$ in order to ease the financial problems that it was facing in the aftermath of acquiring 3G licences and buying out its partners in its British, Irish, Dutch and German mobile businesses. Moreover, mmO$_2$ was not at that time the largest mobile operator in the UK and had not been for many years. This accolade had alternated between Orange, a subsidiary of France Télécom, and Vodafone. As of December 2003, all four GSM network operators had at least 13 million subscribers and only 897,000 subscribers separated the largest company, Vodafone, from the smallest, mmO$_2$. No other Member State has ever had anything like such consistent equality between so many operators although Lithuania in 2006 produced virtual equality between three much smaller operators.

In 2004 this left four Member States – all accession countries – where the incumbent fixed-wire operator did not own the largest mobile operator. In all of these – Lithuania, Malta, Poland and Slovakia – the largest mobile operator was partially owned by foreign investors. In Lithuania, the incumbent fixed-wire operator did not own a stake in any mobile operator, while in the case of the other three countries the incumbent owned a stake in the second-largest mobile operator. Subsequently, Ceský Telecom sold its mobile subsidiary, EuroTel Praha, to what is now Telefónica O$_2$ Czech Republic. Meanwhile, in Bulgaria, BTC eventually got around to launching a GSM network in November 2004 and, in Romania, RomTelecom continues to own 30 per cent of CosmOTE Romania.

Table 1. European Union mobile networks, 1 January 2007. (000s of subscribers)

| Country | GSM[1] | | | PCNs[1] | | | UMTS[2,3] | | |
|---|---|---|---|---|---|---|---|---|---|
| Austria | - | - | - | - | - | - | Hutchison 3G* | 04/03 | 409 |
|  | mobilkom | 12/93 | 3,630 | mobilkom | - | - | Mobilkom* | 04/03 | - |
|  | - | - | - | ONE | 10/98 | 2,070 | ONE* | 12/03 | - |
|  | T-Mobile | 10/96 | 3,180 | T-Mobile[6] | 05/00 | - | T-Mobile*[7] | 01/04 | - |
| Belgium | - | - | - | Base | 06/99 | 2,358 | KPN Mobile 3G | - | - |
|  | Mobistar | 08/96 | 3,139 | Mobistar | 11/00 | - | Mobistar* | 08/06 | - |
|  | Proximus | 01/94 | 4,247 | Proximus | - | - | Proximus* | 04/04 | - |
| Bulgaria | BTC | 11/05 | 692 | - | - | - | - | - | - |
|  | GloBul | 09/01 | 3,271 | - | - | - | GloBul* | 10/06 | - |
|  | MobilTel | 09/95 | 4,270 | - | - | - | MobilTel* | 03/06 | - |
| Cyprus (South) | CyTA | 04/95 | 784 | - | - | - | CyTA | 04/06 | - |
|  | Investcom | 10/03 | 77 | - | - | - | Investcom | 12/04 | - |
| Czech Republic | Telefónica $O_2$ | 07/96 | 4,865 | Telefónica $O_2$ | 07/00 | - | Telefónica $O_2$* | 12/05 | - |
|  | T-Mobile | 09/96 | 5,049 | T-Mobile | 07/00 | - | T-Mobile | 12/06 | - |
|  | Vodafone | 03/00 | 2,413 | Vodafone | 01/00 | - | - | - | - |
| Denmark | - | - | - | - | - | - | Hi3G Denmark* | 11/03 | 215 |
|  | Sonofon | 07/92 | 1,398 | Sonofon | 11/00 | - | Sonofon* | 06/06 | - |
|  | TDC | 07/92 | 2,403 | TDC | 11/00 | - | TDC | 10/05 | - |
|  | TeliaSonera | 06/97 | 1,123 | TeliaSonera | 03/98 | - | - | - | - |
| Estonia | EMT | 01/95 | 759 | EMT | - | - | EMT* | 10/05 | - |
|  | Radiolinja Eesti | 01/95 | 295 | Radiolinja Eesti | - | - | Radiolinja Eesti* | 07/06 | - |
|  | - | - | - | - | - | - | Renberg Invest | - | - |
|  | Tele2 | 09/96 | 536 | - | - | - | Tele2* | 11/06 | - |
| Finland | Radiolinja | 12/91 | 2,194 | Radiolinja | 09/98 | - | Radiolinja* | 11/04 | - |
|  | Suomen 2G | 01/00 | 983 | Suomen 2G | 04/97 | - | Suomen 2G* | 12/05 | - |
|  | TeliaSonera | 06/92 | 2,407 | TeliaSonera | - | - | TeliaSonera | 10/04 | - |

| Country | Operator | Date | Subs | Operator | Date | Value | Operator | Date | Value |
|---|---|---|---|---|---|---|---|---|---|
| France | Bouygues Tél | 05/96 | 8,587 | Bouygues Tél | - | - | Bouygues Tél | - | - |
|  | Orange | 07/92 | 24,109 | Orange | - | - | Orange* | 02/04 | - |
|  | SFR | 07/92 | 15,200 | SFR | - | - | SFR* | 06/04 | 2,693 |
| Germany | - | - | - | E-Plus | 05/94 | 12,654 | E-Plus | 06/04 | - |
|  | T-Mobile | 07/92 | 31,398 | O$_2$ | 10/98 | 11,025 | O$_2$* | 06/04 | - |
|  | Vodafone | 06/92 | 30,622 | T-Mobile | - | - | T-Mobile* | 01/04 | - |
|  |  |  |  | Vodafone | - | - | Vodafone* | 01/04 | - |
| Greece | CosmOTE | 04/98 | 5,217 | CosmOTE | - | - | CosmOTE* | 06/04 | - |
|  |  |  |  | Q-Telecom | 06/02 | 1,072 |  | - | - |
|  | TIM | 06/93 | 2,832 | TIM | 07/01 | - | TIM* | 01/04 | - |
|  | Vodafone | 07/93 | 4,960 | Vodafone | 07/01 | - | Vodafone | 08/04 | - |
| Hungary | Pannon | 03/94 | 3,153 | Pannon | 11/00 | - | Pannon | 10/05 | - |
|  | T-Mobile | 03/94 | 4,431 | T-Mobile | 11/00 | - | T-Mobile* | 08/05 | - |
|  | Vodafone | 11/99 | 2,134 | Vodafone | 12/99 | - | Vodafone | 12/05 | - |
| Ireland | Meteor | 02/01 | 803 | Meteor | - | - | H3G* | 07/05 | 87 |
|  | O$_2$ | 03/97 | 1,632 | O$_2$ | - | - | O$_2$ | 03/05 | - |
|  | Vodafone | 06/93 | 2,178 | Vodafone | - | - | Vodafone* | 07/04 | - |
| Italy | - | - | - | - | - | - | H3G* | 03/03 | 7,080 |
|  |  |  |  |  |  |  | IPSE 2000 | - | - |
|  | TIM | 04/95 | 34,245 | TIM | - | - | TIM* | 05/04 | - |
|  |  |  |  | Wind | 03/99 | 14,634 | Wind | 10/04 | - |
|  | Vodafone | 09/95 | 26,175 | Vodafone | - | - | Vodafone* | 02/04 | - |
| Latvia[4] | - | - | - | Bité | 05/05 | 204 | Bité* | 09/06 | - |
|  | LMT | 01/95 | 803 | LMT | - | - | LMT* | 12/04 | - |
|  | Tele2 | 03/97 | 1,025 |  |  |  | Tele2 | 12/05 | - |
| Lithuania | Bité | 08/95 | 1,839 | Bité | 01/99 | - | Bité* | 04/06 | - |
|  | Omnitel | 03/95 | 2,074 | Omnitel | 01/99 | - | Omnitel* | 05/06 | - |
|  | Tele2 | 12/99 | 1,708 | Tele2 | 06/00 | - | Tele2 | - | - |
| Luxembourg | EPT | 06/93 | 303 | EPT | 03/99 | - | EPT | 06/03 | 14 |
|  | LuXcom | 05/05 | 102 |  | - | - | LuXcom | 05/05 | 35 |
|  |  |  |  |  |  |  | Orange | - | - |
|  | Tele2 | 05/98 | 224 | Tele2 | 03/99 | - | Tele2 | 07/04 | 11 |

**Table 1. Continued**

| Country | GSM[1] | | | PCNs[1] | | | UMTS[2,3] | | |
|---|---|---|---|---|---|---|---|---|---|
| Malta | - | - | - | Go Mobile | 12/00 | 160 | Go Mobile* | 09/06 | - |
| | Vodafone | 09/97 | 188 | - | - | - | Vodafone* | 08/06 | - |
| | - | - | - | - | - | - | 3G Telecoms | - | - |
| Netherlands | KPN | 07/94 | 8,642 | KPN (Telfort) | 10/98 | - | KPN* | 07/04 | - |
| | - | - | - | Orange | 01/99 | 2,115 | Orange | 11/06 | - |
| | - | - | - | T-Mobile | 02/99 | 2,552 | T-Mobile* | 01/06 | - |
| | Vodafone | 09/95 | 3,817 | Vodafone | - | - | Vodafone* | 02/04 | - |
| Poland | Centertel | 07/99 | 12,521 | Centertel | 08/97 | - | Centertel* | 01/06 | - |
| | - | - | - | - | - | - | Netia P4 | - | - |
| | Polkomtel | 02/96 | 11,855 | Polkomtel | 09/99 | - | Polkomtel* | 09/04 | - |
| | PTC | 02/96 | - | PTC | 08/99 | - | PTC* | 08/04 | - |
| Portugal | Optimus | 08/98 | 2,023 | Optimus | 08/98 | - | Optimus* | 06/04 | - |
| | TMN | 10/92 | 5,700 | TMN | - | - | TMN* | 04/04 | - |
| | Vodafone | 10/92 | 4,618 | Vodafone | - | - | Vodafone* | 02/04 | - |
| Romania[5] | - | - | - | CosmOTE | 04/00 | 1,226 | - | - | - |
| | Orange | 06/97 | 7,760 | - | - | - | Orange | 06/06 | 383 |
| | Vodafone | 03/97 | 7,419 | - | - | - | Vodafone* | 04/05 | 298 |
| Slovakia | Orange | 01/97 | 2,690 | Orange | - | - | Orange* | 03/06 | - |
| | T-Mobile | 02/97 | 2,201 | - | - | - | T-Mobile* | 01/06 | - |
| Slovenia | Mobitel | 06/96 | 1,374 | Mobitel | - | - | Mobitel* | 12/03 | - |
| | Si.mobil | 03/99 | 421 | Si.mobil | - | - | Si.mobil | - | - |
| | - | - | - | Tus mobil | 10/07 | - | | | |
| | - | - | - | - | - | - | T-2 | - | - |
| Spain | - | - | - | Orange | 04/99 | 10,692 | Orange* | 10/04 | 422 |
| | Telefónica | 07/95 | 21,446 | Telefónica | 07/98 | - | Telefónica* | 02/04 | - |
| | Vodafone | 10/95 | 14,464 | Vodafone | 07/98 | - | Vodafone* | 02/04 | - |
| | - | - | - | - | - | - | Xfera | 12/06 | 24 |

| Sweden | | | | Hi3G* | 04/03 | 518 |
|---|---|---|---|---|---|---|
| | Spring Mobile | 02/04 | - | - | - | - |
| | Tele2 | 09/92 | 3,508 | Tele2 | 06/04 | - |
| | Telenor | 09/92 | 1,733 | Telenor | 01/04 | - |
| | TeliaSonera | 11/92 | 4,603 | TeliaSonera | 03/04 | - |
| UK | O₂ | 01/94 | 17,633 | O₂ | 09/04 | - |
| | - | - | - | Orange* | 07/04 | 931 |
| | Orange | 04/94 | 14,404 | T-Mobile* | 02/04 | - |
| | T-Mobile | 09/93 | 16,905 | Vodafone* | 04/04 | - |
| | Vodafone | 07/92 | 16,939 | 3 UK* | 03/03 | 3,863 |

[1] The entries consist of name of operator, the month when its service was first launched and the number of subscribers in thousands on the first of January 2007. Where an operator provides both GSM (900 MHz band) and PCN (1800 MHz band) services, the subscriber data are generally provided for both services together in the GSM column. Subscriber data often differ depending upon the source, but such differences tend not to be significant in the context of EU countries. There is, however, some controversy over the counting of 'inactive' customers which may be done differently by individual operators. For example, in Germany Vodafone and E-plus release both registered and active subscriber numbers while T-Mobile only releases registered subscribers and O₂ only active subscribers.

[2] UMTS is the name used for W-CDMA technology in the EU. It is also known as 3GSM. The licensees on 1 January 2007 are listed but not all have yet launched. The term 'launch' in the context of UMTS can mean many things, but normally refers to the launch of a service for corporate customers via data cards inserted in laptops. A consumer service via handsets – sometimes referred to as a 'commercial' launch – usually follows months later but may be simultaneous or even come first. Where subscriber data are known they are included in this column and the total deducted from the total number of subscribers to derive the total for GSM.

[3] Operators marked with an asterisk have launched High-Speed Downlink Access (HSDPA).

[4] Zapp Mobile launched cdma2000 1xEV-DO (approximating to somewhere between UMTS and HSDPA) in October 2004. It had 370,000 subscribers.

[5] Telekom Baltija operated a cdma2000 1xRTT network with roughly 30,000 subscribers.

[6] Formerly known as tele.ring.

[7] Including the former tele.ring.

Source: Details obtained from a wide variety of regulators' websites, company websites and media and Internet websites.

## Table 2. Market position, 1 January 2007

| | Deutsche Telekom | France Télécom | KPN | mobilkom | OTE | TDC | Telefónica incl. $O_2$ | Telenor[1] | Tele2[2] | Telia Sonera | Vodafone[3] |
|---|---|---|---|---|---|---|---|---|---|---|---|
| Austria | 2 | 3 | | 1 | | | | 3 | | | |
| Belgium | | 2 | 3 | | | | | | | | |
| Bulgaria | | | | 1 | 2 | | | | | | |
| Cyprus (S) | | | | | | | | | | | |
| Czech Rep. | 1 | | | | | | | | | | |
| Denmark | | | | | | 1 | 2 | 2 | | 3 | 3 |
| Estonia | | | | | | | | | 2 | 1 | |
| Finland | | | | | | | | | | 1 | |
| France | | 1 | | | | | | | | | 2 |
| Germany | 1 | | 3 | | | | 4 | | | | 2 |
| Greece | | | | | 1 | | | | | | 2 |
| Hungary | 1 | | | | | | 2 | 2 | | | 3 |
| Ireland | | | | | | | 2 | | | | 1 |
| Italy | | | | | | | | | | | 2 |
| Latvia | | | | | | 3 | | | 1 | 3 | |
| Lithuania | | | | | | 2 | | | 3 | 1 | |
| Luxembourg | | | | | | | | | 2 | | |
| Malta | | | | | | | | | | | 1 |
| Netherlands | 3 | 4 | 1 | | | | | | | | 2 |
| Poland | 3 | 1 | | | | 2 | | | | | 2 |
| Portugal | | 3 | | | | | 1 | | | | 2 |
| Romania | | 1 | | | 3 | | | | | | 2 |
| Slovakia | 2 | 1 | | | | | | | | | |
| Slovenia | | | | 2 | | | | | | | |
| Spain | | 3 | | | | | 1 | | | 4 | 2 |
| Sweden | | | | | | | | 3 | 2 | 1 | |
| UK | 2 | 4 | | | | | 3 | | | | 1 |

[1] Telenor is the largest mobile operator in its home market of Norway, a non-EU Member State.
[2] Also present through MVNO arrangements in Austria, Denmark, France and The Netherlands.
[3] Present through a Partner Network Agreement in Austria, Belgium, Bulgaria, Cyprus (South), Denmark, Estonia, Finland, Latvia, Lithuania, Luxembourg, Slovenia and Sweden.

Source: Calculated by authors from data in Table 1.

**Table 3. Mobile market concentration, 1 January 2007**

| Country | Total subscribers (000s) | % market share: largest operator | % market share: largest 2 operators | % market share: largest 3 operators |
|---|---|---|---|---|
| Austria | 9,289,000 | 39.0 | 73.1 | 95.3 |
| Belgium | 9,744,000 | 43.6 | 75.8 | 100 |
| Bulgaria | 8,233,000 | 51.9 | 91.6 | 100 |
| Cyprus (South) | 861,000 | 91.2 | 100 | n/a |
| Czech Republic | 12,327,000 | 41.0 | 80.4 | 100 |
| Denmark | 5,139.000 | 46.8 | 74.0 | 95.8 |
| Estonia | 1,590,000 | 47.7 | 81.4 | 100 |
| Finland | 5,584,000 | 43.1 | 82.4 | 100 |
| France | 50,589,000 | 47.7 | 83.0 | 100 |
| Germany | 85,699,000 | 36.6 | 72.4 | 87.2 |
| Greece | 14,081,000 | 37.0 | 72.3 | 92.4 |
| Hungary | 9,718,000 | 45.6 | 78.0 | 100 |
| Ireland | 4,701,000 | 46.3 | 81.1 | 98.1 |
| Italy | 82,134,000 | 41.7 | 73.6 | 91.4 |
| Latvia | 2,032,000 | 50.4 | 90.0 | 100 |
| Lithuania | 5,621,000 | 36.9 | 69.6 | 100 |
| Luxembourg | 689,000 | 46.0 | 80.1 | 100 |
| Malta | 348,000 | 52.0 | 100 | n/a |
| Netherlands | 17,126,000 | 50.5 | 72.7 | 87.6 |
| Poland | 36,334,000 | 33.7 | 67.4 | 100 |
| Portugal | 12,341,000 | 46.2 | 83.6 | 100 |
| Romania | 17,371,000 | 46.3 | 90.7 | 97.8 |
| Slovakia | 4,891,000 | 55.0 | 100 | n/a |
| Slovenia | 1,795,000 | 76.5 | 100 | n/a |
| Spain | 47,048,000 | 45.6 | 76.4 | 100 |
| Sweden | 10,362,000 | 44.4 | 78.3 | 95.0 |
| UK | 70,675,000 | 24.9 | 48.9 | 72.8 |

Source: Calculated by authors from data in Table 1.

Related to the above is the observation that those mobile operators with multiple licences across the EU were usually the second- or third-largest operators in the market. It is possible to determine the market position of operators in EU Member States if we assume that an operator is present provided it has a stake in any operator outside its home market. The precise size of most of these stakes is set out in the discussion that follows, and it may be observed that some are controlling stakes – which is not necessarily the same thing as majority stakes – while others are not, and that in certain cases there is multiple ownership of the same operator by companies listed in the table. With two exceptions – Tele2 and Telefónica subsidiary $O_2$ (the former mmO$_2$) – each company identified above was the largest operator in its home market denoted in bold. Tele2 was the second-largest operator in Sweden after TeliaSonera while $O_2$ was the third-largest (albeit not by much) of the four second-generation network operators in the UK.

If we had turned our attention to those mobile operators that were the largest operators in a foreign country just three years previously, then a common trait was that those foreign markets where they were the largest were comparatively small. However, this was no longer the case at the end of 2006. It remained true, for example, that TeliaSonera was the largest mobile operator in two of the three Baltic States which were numbered among the smaller EU markets, while Vodafone was dominant in Ireland and Malta. In contrast, however, Deutsche Telekom was the largest operator in the Czech Republic and Hungary, and France Télécom in Poland and Romania.

## MARKET CONCENTRATION

Drawing on the subscriber information contained in Table 1, it is possible to calculate the percentage of the mobile market controlled by the largest (two/three) mobile operator(s) as of 31 December 2006. As can be observed from Table 3 above, it was commonly the case – basically reflecting the advantage of GSM incumbency – that the largest mobile operator controlled at least 40 per cent (but generally less than 50 per cent) of the market although, clearly, the figure always exceeded 50 per cent in the case of duopolies. Nevertheless, in five countries – Austria, Germany, Greece, Lithuania and Poland – the largest operator accounted for only between 30 per cent and 40 per cent of all subscribers while in the UK the figure was a mere one-quarter, indicating fierce competition.

One explanatory factor is clearly the existence of a fourth operator, although only in Germany, the Netherlands and the UK do these control more than 10 per cent of the market, and it has to be said that even attaining 10 million subscribes in the largest markets may not prove to be economic in the medium term.

If the calculation is extended to include the second-largest mobile operator in each market, then in a majority of Member States the mobile market was, to all intents and purposes, a duopoly with the percentage standing in excess of 80 per cent. Elsewhere, the two largest mobile operators still tended to control over 70 per cent of the market, with the only exceptions being Lithuania and Poland (both narrowly) and the UK. Where three or more mobile operators had been licensed, a considerable gap often existed between the number of mobile subscribers controlled by the second-largest operator and the number of subscribers controlled by the third-largest operator. In eleven Member States, the subscriber base of the third-largest mobile operator was more than half the size of that of the second-largest, but in a further eight cases it was less than half the size – notably so in the case of recent entrants Bulgaria and Romania. In the Czech Republic and Ireland the third-largest was almost exactly half the size of the second-largest, but the real anomaly was the UK with four almost equal-sized operators plus a respectably-sized 3G operator (3 UK).

It may also be noted that in the majority of Member States the most recent mobile operator to launch its service was also the one with the fewest subscribers. Although this was true in fifteen cases, it is surprising that only four of these could be found among the accession countries. In other words, the date when a mobile operator launched its services was more important in the EU15 than in the accession countries. For five Member States – Austria, Finland, Latvia, Slovenia and the UK – the gap between the smallest and the next-largest operator was less than one million subscribers, while for the remaining ten countries the gap was greater, sometimes considerably greater, than one million. For example, in Italy there was a gap of almost ten million subscribers between Wind, the last of the three 2G operators to launch its service, and Vodafone, the second-largest operator in the market. Gaps of more than two million subscribers could also be found in France (8.1 million subscribers), Germany (2.6 million), the Czech Republic (2.3 million) and Greece (2.1 million). In Belgium, Hungary, Ireland, Portugal and Spain the gap was less than one million subscribers.

## MOBILE COMMUNICATION MARKETS IN ACCESSON COUNTRIES

The first issue to address at this point is the extent to which mobile operators were EU-centric in respect of their geographical footprints, distinguishing between operators with a heavy presence in the EU-15 and those with a presence in the 12 accession countries.

Table 4 is drawn up so as to include those operators with licences in at least two accession countries. This is a modest enough total, but reflects the fact that only one operator, Vodafone, was present in more than four of the twelve. Even here, however, there is a need to distinguish carefully between operators with licences and those companies operating under other arrangements. For example, it is possible for an operator to act as a mobile virtual network operator (MVNO) by leasing spare capacity on an incumbent's network [Shin & Bartolacci, 2007]. Technically, the definition of an MVNO requires an operator to own its own switches and sell under its own brand, although there are also less rigorous ways to operate such as an enhanced service provider or simply as a reseller of another operator's branded service. The primary advocate of the MVNO approach has been Tele2 [Curwen & Whalley, 2007] although, as Table 4 shows, it has tended to prefer direct investment in networks in accession countries while operating as a MVNO in more established markets. For its part, Vodafone has preferred to negotiate Partner Network Agreements involving no investment [12], whereby the network in question is usually re-branded with the original operator's name hyphenated to that of Vodafone. By this means, Vodafone has enjoyed brand recognition without needing to lay out huge sums of money, and has been able to introduce its Vodafone live! portal with associated roaming benefits, while the network owner has enjoyed improved subscriber numbers and reduced churn because the Vodafone brand is more attractive than its own. In practice, therefore, Vodafone has a much greater brand recognition factor in accession countries than any other operator and is absent only in Slovakia.

The next in line is Deutsche Telekom's wholly-owned subsidiary T-Mobile. This is not surprising since the geographical position of Germany clearly lends itself to investment in countries close to its borders, many of which are accession countries (with possibly even more to come). It may also be noted that T-Mobile is present in the relatively large accession markets of Hungary, Poland and the Czech Republic. This is an important point because it is immediately noticeable that three of the big five EU incumbent mobile operators, Telefónica Móviles, $O_2$ (its sister company) [13] and Telecom Italia Mobile (TIM – which became a wholly-owned subsidiary of Telecom Italia in June 2005) do not appear in Table 4. For Telefónica Móviles in particular, this is ultimately a question of history, culture and language. Telefónica Móviles (and/or occasionally its parent although there is no longer any difference since it is currently a wholly-owned subsidiary) has long operated overseas, primarily in Latin America with the only exception prior to the end of 2004 being Morocco, its immediate southern neighbour. In other words, apart from a minor reciprocal stake in Portugal Telecom and some toying with 3G licences that had so far resulted in nothing other than fairly substantial write-offs, the company had zero interest in the EU let alone in accession countries. This strategy, it must be said, had served it well up to that point in time.

**Table 4. Operators present in at least two accession countries, 1 January 2007**

|  | Vodafone | T-Mobile | Tele2 | Telia Sonera | TDC | Orange | mobilkom | OTE |
|---|---|---|---|---|---|---|---|---|
| Austria | 1 | ■ | ■[3] |  | ■ | ■ | ■ |  |
| Belgium | 1 |  |  |  |  | ■ |  |  |
| **Bulgaria** | 1 |  |  |  |  |  | ■ | ■ |
| **Cyprus** | 1 |  |  |  |  |  |  |  |
| **Czech R** | ■ | ■ |  |  |  |  |  |  |
| Denmark | 1 |  | ■[3] | ■ | ■ |  |  |  |
| **Estonia** | 1 |  | ■ | ■ |  |  |  |  |
| Finland | 1 |  |  | ■ |  |  |  |  |
| France | ■ |  | ■[3] |  |  | ■ |  |  |
| Germany | ■ | ■ |  |  | ■[3] |  |  |  |
| Greece | ■ |  |  |  |  |  |  | ■ |
| **Hungary** | ■ | ■ |  |  |  |  |  |  |
| Ireland | ■ |  |  |  |  |  |  |  |
| Italy | ■ |  |  | 2 |  |  |  |  |
| **Latvia** | 1 |  | ■ | ■ | ■ |  |  |  |
| **Lithuania** | 1 |  | ■ | ■ | ■ |  |  |  |
| Luxembourg | 1 |  | ■ |  |  | 2 |  |  |
| **Malta** | ■ |  |  |  |  |  |  |  |
| Netherlands | ■ | ■ | ■[3] |  |  | ■ |  |  |
| **Poland** | ■ | ■ |  |  | ■ | ■ |  |  |
| Portugal | ■ |  |  |  |  | ■ |  |  |
| **Romania** | ■ |  |  |  |  | ■ |  | ■ |
| **Slovakia** |  | ■ |  |  |  |  |  |  |
| **Slovenia** | 1 |  |  |  |  |  | ■ |  |
| Spain | ■ |  |  | ■ |  | ■ |  |  |
| Sweden | 1 |  | ■ | ■ |  |  |  |  |
| UK | ■ | ■ |  |  |  | ■ |  |  |
| Total Accession | 5 + 6[1] | 4 | 3 | 3 | 3 | 2 | 2 | 2 |
| Total | 14 + 12[1] | 8 | 9 | 7 | 6 | 10 | 3 | 3 |

[1] Via Partner Network Agreement not involving direct investment.
[2] 3G licence only. No network.
[3] Trading as a MVNO. Sold provisionally in June 2007.
Source: Calculated by authors from data in Table 1.

However, the Brazilian operation, Vivo, half-owned by Portugal Telecom, has struggled to show a profit [14] (and is currently being switched from CDMA to GSM at great expense), so Telefónica has decided to diversify back into Europe. It began with the acquisition of an initial 51.1 per cent stake in EuroTel of the Czech Republic in April 2005 [Roman & Rousek, 2005] – subsequently raised to 69.4 per cent. In March 2006, it completed the purchase of $O_2$ (which also has an independent network in the Isle of Man) [Telefónica, 2006] and thereby brought its proportionate subscriber numbers – that is, total subscribers multiplied by the ownership stake expressed as a percentage – within and outside Latin America into rough equivalence. Subsequently, it won a combined GSM/UMTS licence in Slovakia [15].

As for $O_2$ (changed from $mmO_2$ in March 2005), both prior to and after its divestment from what is now the BT Group it had spent a period of retrenchment involving the shedding of minority interests such that it remained operational in only Germany, the Netherlands and the UK (plus the Isle of Man). Even so, it has to be said that in its earlier life it was never really interested in the accession countries, preferring to get involved in South-East Asia and North America.

Telecom Italia and its then 56.1 per cent subsidiary TIM also had a fairly significant presence in Latin America at the end of 2004 although the great majority of proportionate subscribers were to be found in Brazil. Elsewhere, its presence in a single accession country, the Czech Republic, merely represented a tiny stake in the operator controlled by T-Mobile, and was the least significant of its overseas holdings bar Cuba. The stake was accordingly sold in March 2005 and the sale of its Greek operation shortly thereafter means that Telecom Italia is currently present in Europe only in its home market.

It is also be useful for the purposes of clarification to examine briefly the operations of mobile companies in what used to be termed Eastern Europe since only some of its constituent countries had become accession countries. As of 31 December 2003, four EU incumbents had a significant presence involving investment in Eastern Europe, namely Telenor, OTE, T-Mobile and TeliaSonera. OTE, interestingly, had stakes in Albania, Armenia, Bulgaria, Macedonia, Romania and Serbia, so it had not profited so far from accession nor was it about to do so during 2004 although it proved to be the main beneficiary in 2007. TeliaSonera's accession stakes were in practice entirely in the Baltic countries but its stakes to the east, in Azerbaijan, Georgia, Kazakhstan, Moldova and Russia were also in line to miss the accession boat. For its part, Telenor had twelve overseas interests but, interestingly, it was not focussed upon the Nordic/Baltic area being present in only Denmark, Norway and Sweden (as a MVNO), whereas it had stakes in Albania, Montenegro, Russia and the Ukraine in respect of which it would miss out on accession. T-Mobile accordingly stood out because it had stakes in four prospective accession countries, of which three (Czech Republic, Hungary and Poland) generated more than two million proportionate subscribers during 2003. In addition, it owned stakes in Belarus, Bosnia, Croatia, Macedonia and Russia. Given its limited overseas stakes within the EU at the time (consisting of Austria, the Netherlands and the UK), accession would serve to enlarge significantly its EU coverage even if the USA comfortably generated the third-largest number of proportionate subscribers after Germany and the UK (with Russia in fourth place).

What the above suggests is that there is a useful distinction to be made between the Baltic and Eastern European aspects of the 2004 accession – Cyprus (South) and Malta are of little significance because of their size and lack of potential for the entry of major operators. Taking the three Baltic accession countries as a whole, the eight operators listed in Table 4 generated eight entries but only three operators among them accounted for this – Tele2, TeliaSonera and TDC – although account should also be taken of Vodafone's three Partner

Network Agreements. In contrast, the five broadly Eastern European countries generated ten entries. This was not a significant difference, so it is worth asking whether it resulted from the companies sampled. To answer this, we can return to the data in Whalley & Curwen (2003) which encompassed thirteen major European operators, and these reveal that increasing the sample size makes almost no difference when compared to Table 4. Of the ten accession countries, only two are affected at all by the altered size of the sample, namely Hungary where Telenor had a substantial stake and the Czech Republic where TIM had a very small stake. It is also possible to establish whether any significance can be attributed to the fact that two Nordic countries – Iceland and Norway – were not members of the EU. In practice, Iceland was not significant since the only EU operator there was Vodafone via a Partner Network Agreement, but in Norway we find (predictably) both Telenor and TeliaSonera (trading as NetCom GSM) as incumbents with Tele2 as a MVNO (although it had returned its 3G licence).

In summary, accordingly, the situation was as follows at the time of the 2004 accession: Vodafone had invested in accession countries in the former Eastern Europe (Group A) but had been keen to extend its footprint to the Baltic accession countries (Group B) without investing heavily. T-Mobile had heavily invested in Group A but was wholly disinterested in Group B. Orange was less involved in Group B but equally indifferent to Group A. TDC was slightly interested in both, while both TeliaSonera and Tele2 were heavily invested in Group B while wholly disinterested in Group A. Curiously, Telenor (not in Table 4) was the only Nordic operator acting in a wholly non-Nordic manner where accession was concerned.

The 2007 accession added one large operation in Romania to the EU holdings of both Orange and Vodafone but these operators were already EU-centric in terms of subscribers. OTE, as noted, gained two EU operations but one, CosmOTE in Romania, was very small, while mobilkom gained the largest operator in Bulgaria, MobilTel.

## EXPANSION AND CONSOLIDATION

Given the aforementioned differences in the countries in which the mobile operators identified in Table 4 have chosen to invest prior to the 2004 accession, an inevitable question to ask is whether the accession of twelve new Members States either has resulted so far, or is expected to result, in changes in their strategic priorities. In the first sub-section the focus is on T-Mobile, Orange and Vodafone whose ability to expand further is to a degree interlinked with one another, while the second sub-section concentrates on those other mobile companies with a presence in the accession countries.

### T-Mobile, Orange and Vodafone

If we begin with T-Mobile, then the strategic importance of the Eastern European countries to the company is clear for all to see in Table 5. Indeed, the CEO Kai-Uwe Ricke, basking in predictions of massive cash inflows during 2004, stated in May 2004 that 'Taking into account the EU's enlargement towards the east, we are placing a special focus on this region'. It is possible to calculate the importance of this region to T-Mobile as at 31

December 2003 when it had in total 68.7 million proportionate subscribers. Of these, 26.3 million were in Germany and 43.9 million in total in the pre-accession EU. Accession transferred a further 7.7 million to that total, yielding 51.6 million in total in the post-accession EU. The rest were largely accounted for by the USA (12.8 million) and Russia (3.4 million) with Croatia and Macedonia adding 0.8 million between them.

T-Mobile had a choice between moving into new countries and expanding into existing ones. In both cases, much depended upon the identity of existing shareholders and their willingness to sell. Faced with a cash offer above the market price, many shareholders might have been expected to succumb, but Deutsche Telekom's own shareholders were unlikely to sanction using up cash reserves to support a move into the likes of Moldova. Predictably, T-Mobile was not willing to fight for the 2G licence issued in Bulgaria in May

Hence, the probability was that T-Mobile would prefer to increase its existing stakes. In some cases, the purchase of additional equity would consolidate its existing control over the operator while in other cases the purchase could allow T-Mobile to take control of the operator for the first time. T-Mobile was particularly keen to acquire the 51 per cent of PTC it did not own in Poland, if only to keep one step ahead of Vodafone in a country with a modest penetration rate. Thus, its existing stakes provided T-Mobile with ample incentives and opportunities to continue its Eastern European-focused investment strategy. However, one intriguing prospect lay in the Czech Republic where, despite its majority stake in an incumbent, T-Mobile was alleged to be interested in acquiring EuroTel Praha via a bid for parent Český Telecom. Presumably, if it did so it would be forced to dispose of its existing network which was almost the same size, but this would get around the problem of trying to obtain full ownership of T-Mobile CZ.

**Table 5. T-Mobile, European stakes, 31 December 2006**

| Country | Stake % | Proportionate subscribers |
|---|---|---|
| Austria | 100 | 3,180,000 |
| Bosnia | 29.9 | 93,000 |
| Croatia | 51.0 | 1,101,000 |
| Czech Rep. | 60.8 | 3,070,000 |
| Germany | 100 | 31,398,000 |
| Hungary | 59.4 | 2,632,000 |
| Macedonia | 30.3 | 320,000 |
| Montenegro | 39.9 | 131,000 |
| Netherlands | 100 | 2,552,000 |
| Poland | 97.0 | 11,861,000 |
| Slovakia | 51.2 | 1,127,000 |
| UK | 100 | 16,905,000 |
| Total world | | 99,476,000 |
| *Accession* | | *18,690,000* |

Source: Compiled by authors from operator and other websites.

It is fair to say that, by 2005, many major European operators had decided that non-controlling stakes were often more trouble than they were worth. This would be particularly true of stakes in countries such as Russia where the rule of law could best be described as shaky. Hence, Deutsche Telekom's decision to sell its 10.2 per cent stake in Russia's MTS – and with it its indirect stakes in the likes of Belarus, the Ukraine and Uzbekistan – which was completed in September 2005, was not unexpected. Its travails in Poland, where it embarked upon an immensely long-winded trawl through the law courts of Europe in order to determine whether Vivendi Universal or itself could lay claim to the other half of the shareholding in PTC, were not entirely of its own choosing but did, however, appear to have turned out well in the end since it was able to take control of the shares at well below their market value in November 2006 – although Vivendi has yet to accept this apparent *fait accompli* [16].

2005 saw other activity in the EU and prospective accession countries, but to less good effect. For example, talks about a joint bid for $O_2$ with KPN fell through and Deutsche Telekom declined to make a bid for Turk Telekom. In August, the CEO stated that the operator was not actively seeking new acquisitions, but would continue to evaluate any opportunities arising within its existing footprint. These proved to be fairly plentiful during 2006 when Deutsche Telekom failed to win Serbia's Mobi 63 and expressed an interest in a licence in Romania while its Hungarian subsidiary expressed an interest in acquiring stakes in Bosnia, Romania, Serbia and the Ukraine. Very recently, Deutsche Telekom has indicated that it may bid for a 20 per cent stake in OTE which is being privatized during 2007 although it will not be prepared to over-pay for a minority stake. This is especially the case given that, in late January, Deutsche Telekom announced its second profit warning in six months and embarked upon a massive cost-cutting exercise [Williamson, 2007].

But would any of the other mobile operators identified in Table 4 follow T-Mobile and respond to accession by increasing their geographical coverage? Vodafone had a presence of one kind or another in all but three of the ten accession countries at the beginning of 2004, but given that it had chosen to use Partner Network Agreements as a substitute for direct investments in many cases, the scope for it to invest in more of these markets was actually quite limited. Of the markets where Vodafone was not active as an investor, the most significant was the Czech Republic. Of the three Czech GSM operators, two – Ceský Mobil and EuroTel Praha – were potential acquisitions. The third operator, T-Mobile CZ, was majority-owned by Deutsche Telekom and thus unavailable unless, as noted above, T-Mobile was forced to sell it. In principle, EuroTel Praha could also be dismissed as an acquisition target of Vodafone since it was possibly being targeted by T-Mobile and, in any event, was a subsidiary of the incumbent PTO which was most unlikely to want to be split off from it mobile operations. However, Vodafone had recently indicated that it was willing, and had the financial resources, to acquire the entire operation.

Ceský Mobil was owned by Telesystem International Wireless, which was possibly prepared to sell its 96.4 per cent stake if the price was sufficiently attractive. Nevertheless, it remained to be seen whether Vodafone's shareholders would be prepared to countenance the comparatively expensive acquisition of the smallest of the Czech Republic's GSM operators. Given the existing investment, and the 16 per cent market share of Ceský Mobil, a more attractive course of action looked to be to enter into a Partner Network Agreement. However, Vodafone duly bought 99.9 per cent of Ceský Mobil in May 2005 which it has re-branded as Vodafone CZ.

**Table 6. Orange, European stakes, 31 December 2006**

| Country | Stake % | Proportionate Subscribers |
|---|---|---|
| Austria | 17.5 | 362,000 |
| Belgium | 50.2 | 1,588,000 |
| France | 100 | 24,109,000 |
| Liechtenstein | 100 | negliigble. |
| Netherlands | 100 | 2,115,000 |
| Poland | 43.9 | 5,497,000 |
| Portugal | 19.2 | 2,023,000 |
| Romania | 96.8 | 7,786,000 |
| Slovakia | 100 | 2,690,000 |
| Spain | 77.7 | 8,636,000 |
| Switzerland | 100 | 1,367,000 |
| UK | 100 | 15,333,000 |
| Total world | | 80,580,000 |
| *Accession* | | *15,973,000* |

Source: Compiled by authors from operator and other websites.

As for Latvia and Slovakia, both were comparatively small. As a consequence, it was more likely that Vodafone would enter these markets through the use of Partner Network Agreements rather than an equity investment. However, this assumed that the existing GSM operators would enter into such an arrangement. So far as Latvia was concerned, this was highly unlikely given who owned the two existing operators; Baltkom was owned by Tele2 while LMT was jointly owned by the Latvian state (51 per cent) and TeliaSonera (49 per cent). It seemed inconceivable that either Tele2 or TeliaSonera would sign a Partner Network Agreement with Vodafone as this would expand the brand recognition of their main competitor in the Baltic States. In practice, Vodafone achieved its aim indirectly when Bité GSM, the Lithuanian subsidiary of TDC and a Partner Network of Vodafone, launched a network in Latvia in 2005 [Vodafone, 2006a].

The situation in Slovakia was a little more complicated, not least because Vodafone's ability to enter this market was also dependent on the strategic priorities and intentions of Orange. Orange had only a limited exposure to the mobile markets of the ten accession countries, with a presence across the EU that was increasingly skewed in favour of Western Europe. Orange already possessed two accession country mobile investments; in Poland, where it – or strictly its parent France Télécom – was a majority shareholder in PKT Centertel, and Slovakia where it owned 63.9 per cent of Orange Slovensko, the largest operator. These two investments were, however, somewhat detached from the other investments that Orange had made. Their relative peripherality was further reinforced when subscriber numbers are taken into account; Poland and Slovakia accounted for less than 10 per cent of the wider European subscriber base of Orange. Orange had also invested in Romania, a country that expected (correctly) to be among the next wave of accession countries, though when these subscribers were also included the three countries still only accounted for roughly 17 per cent of the European subscriber base. In contrast, France and the

UK, which were the two largest mobile markets of Orange, accounted for over 75 per cent of its European subscriber base as shown in Table 6.

However, given that Orange was committed to consolidation based upon countries where it held majority stakes and hence control, preferably in conjunction with a top-two ranking, and that parent France Télécom's short-term need for cash had abated somewhat, a wholesale withdrawal from the former Eastern Europe no longer seemed likely. Indeed, Orange moved to mop up the minorities in Romania [Middleton, 2005] and Slovakia [17] during 2005 although the situation in Poland was much harder to disentangle despite a re-branding to Orange in August [France Télécom, 2005; Halaba, 2005; McQuaid, 2005]. Nevertheless, Orange was not successful in building up any further presence in accession/potential accession countries: In March 2005, it was part of a consortium that bid unsuccessfully for 51.1 per cent of EuroTel Praha; it withdrew from the bidding for Turkey's Telsim in December 2005; and it failed to win Mobi 63 of Serbia in July 2006.

Vodafone would potentially have been interested in acquiring in the majority of Orange Slovensko owned by Orange had Orange been forced to sell it, but Vodafone's shareholders were unlikely to be willing to support an acquisition that would add a comparatively small number of subscribers in a market where growth expectations were limited – Slovakia's population was only 5.4 million and already heavily penetrated – but there were attractive investments with more potential elsewhere. One such was Romania where Vodafone was already a minority shareholder in Connex and whose population was four times that of Slovakia. Telesystem International Wireless (TIW), holder of a 63.6 per cent stake, agreed in July 2004 to buy all or part of the 14.4 per cent stake held by Deraso Holdings. In May 2005, Vodafone acquired this entire stake [Krosnar, 2005a], raising its total holding to 99.1 per cent and subsequently to 100 per cent before re-branding as Vodafone Romania in May 2006. Outside of such an acquisition, Vodafone was unlikely to expand into new markets other than through Partner Network Agreements.

The remaining accession country of interest was Poland where Vodafone already held a 19.6 per cent stake in Polkomtel (Plus GSM) as did TDC and various local partners. Vodafone attempted to acquire a further 23 per cent stake in July 2005 [18] but without success, and also expressed interest in the TDC stake when it was put on the market in December 2005 although this has degenerated into yet another legal battle over who has what rights in relation to the sale [19].

In September 2005, Vodafone CEO Arun Sarin stated that he was now broadly happy with the company's European profile but would be bidding to buy Turkey's Telsim at the year-end [Krosnar, 2005b]. However, in October, Vodafone agreed to sell its 100 per cent stake (yielding 1.5 million subscribers) in Sweden to Telenor (replacing it with a Partner Network Agreement) [Vodafone, 2005], and it followed up with the sale of its 25 per cent stake in Belgacom and its 25 per cent stake in Swisscom to their majority owners in November 2006 and December 2006 respectively [Smith & Simonian, 2006; Vodafone, 2006b]. Neither produced a lot of subscribers (Table 7), but it is of interest that the overall effect was to re-balance Vodafone's European holdings towards the accession countries, a factor enhanced by the agreed purchase of Telsim in December [Boland & Edgecliffe-Johnson, 2005] [20] given Turkey's pressure to become an accession country.

**Table 7. Vodafone, European stakes, 31 December 2006**

| Country | Stake % | Proportionate Subscribers |
|---|---|---|
| Albania | 99.9 | 919,000 |
| Czech Rep. | 100 | 2,413,000 |
| France | 43.9 | 7,851,000 |
| Germany | 100 | 30,622,000 |
| Greece | 99.8 | 4,955,000 |
| Hungary | 100 | 2,134,000 |
| Ireland | 100 | 2,178,000 |
| Italy | 76.9 | 20,129,000 |
| Malta | 100 | 188,000 |
| Netherlands | 99.9 | 3,817,000 |
| Poland | 19.6 | 2,324,000 |
| Portugal | 100 | 4,618,000 |
| Romania | 100 | 7,717,000 |
| Spain | 100 | 14,464,000 |
| UK | 100 | 16,939,000 |
| Total world | | 198,584,000 |
| *Accession* | | 14,776,000 |

Source: Compiled by authors from operator and other websites.

## Other Operators

A further company with a presence in the accession countries is Tele2 (see Table 8). Although Tele2 operated in nine EU Member States at the end of 2006 – it subsequently sold its mobile operation in the Netherlands to Versatel in May 2007 and is negotiating the sale of the assets in France – it has made just three investments in the accession countries, namely in Estonia, Latvia and Lithuania, and all prior to the 2004 accession. In other words, as a factor in its basic strategy rather than on account of accession per se, Tele2 has invested in the Baltic States that complement geographically its presence in the nearby Nordic States although none of these Baltic networks yields significant numbers of subscribers. Table 8 also draws attention to a second characteristic of Tele2's investment strategy; namely its tendency to use MVNO arrangements to enter new markets. Of the nine mobile investments in the post-accession EU set out in Table 6, almost half were as a MVNO. Of the five networks owned by Tele2, only one, in Luxembourg, could be found outside of Sweden and the Baltic States. Thus, the geographical preference in terms of ownership is marked as is the preference for control – only in Sweden, where it has an 87.3 per cent stake, does Tele2 not own the entire company although it only has a 51 per cent stake in Croatia, a potential accession country. In addition to the above, it may be noted that Tele2 failed to acquire 3G licences in Hungary (2004) and Bulgaria (2005) and to buy Serbia's Mobi 63 in July 2006.

### Table 8. Tele2, European stakes, 31 December 2006

| Country | Stake % | Proportionate Subscribers |
|---|---|---|
| Austria | 100 | 174,000 |
| Croatia | 51.0 | 158,000 |
| Denmark | 100 | n/a[1] |
| Estonia | 100 | 536,000 |
| France | 100 | 417,000 |
| Latvia | 100 | 1,025,000 |
| Liechtenstein | 100 | 15,000 |
| Lithuania | 100 | 1,708,000 |
| Luxembourg | 100 | 235,000 |
| Netherlands | 100 | 577,000 |
| Norway | 100 | n/a[1] |
| Sweden | 87.3 | 3,062,000 |
| Switzerland | 100 | 27,000 |
| Total world | | 18,679,000 |
| *Accession* | | *3,269,000* |

1. The Nordic region – Denmark, Norway and Sweden – is credited with 4,292,000 subscribers.
Source: Compiled by authors from operator and other websites.

What of the other companies identified in Table 4? For different reasons, neither TDC nor TeliaSonera was likely to expand its geographical footprint as a result of EU expansion. At the end of 2003, TDC had only two remaining investments in accession countries – in Bité in Lithuania and Polkomtel in Poland – having sold in the course of that year its holding in the Czech Republic (as well as in the Ukraine) Hence, it did not appear to see the former Eastern Europe as other than providing opportunities for financial investments. In any event, any additional investments by TDC in accession countries could be ruled out until the uncertainty over its own future was resolved. In mid-2004, SBC Communications Inc. sold 32.1 per cent of the 41.6 per cent of TDC that it owned, but to financial institutions rather than to another operator. Pending the completion of this sale, TDC stated that it would not enter into any negotiations regarding potential partnerships or strategic transactions at Group level. Moreover, these would only resume once the new board had been able to conduct a strategic review of the company.

Its subsequent strategy has been less than clear-cut. For example, in May 2005 Lithuania's Bité acquired a licence in Latvia and launched its network in September, but TDC also failed to win the 3G licence in Hungary in December 2004 and withdrew from the bidding for EuroTel Praha in 2005, so it clearly has no particular interest in the accession countries per se. Indeed, Bité was provisionally sold to Mid Europa Partners in January 2007 in line with the view expressed by the CEO in November 2006 that 'everything outside of Denmark isn't core business…if somebody comes with a price, and the price is right, it will be sold'. [21] Furthermore, the (so far frustrated) desire to sell its stake in Poland has been mentioned previously, so taken with its strong interest in expanding in Finland, Norway and Sweden, it is fair to say that TDC is if anything retreating back to its Nordic roots as shown in Table 9.

**Table 9. TDC, European stakes, 31 December 2006**

| Country | Stake % | Proportionate subscribers |
|---|---|---|
| Austria | 15.0 | 312,000 |
| Denmark | 100 | 1,202,000 |
| Germany | 100 | 3,742,000 |
| Latvia | 100 | 204,000 |
| Lithuania | 100 | 1,839,000 |
| Norway | 80.0 | 11,000 |
| Poland | 19.6 | 2,324,000 |
| Switzerland | 100 | 1,361,000 |
| | | |
| Total world | | 11,074,000 |
| *Accession* | | *7,333,000* |

Source: Compiled by authors from operator and other websites.

For its part, TeliaSonera has for many years been one of the most widespread operators in the mobile world, deriving more proportionate subscribers from Russia and Turkey than from Sweden at the time of the 2004 accession. With the exception of the three Baltic States, none of which understandably produced large numbers of subscribers, TeliaSonera had no other mobile investments in accession countries in 2004, nor did it gain any in 2007, although with its stake in Turkcell and, via jointly-held Fintur Holdings, in countries to the east of the former Eastern Europe, it might end up with further EU holdings during the next stages of accession.

Telia and Sonera, prior to their merger, did take advantage of the 3G licensing process to enter Germany, Italy and Spain, three of the largest Western European markets. However, there followed a period of post-merger repentance involving the writing off of the investments in all three markets and these barely signify in Table 10. Interestingly, despite its widespread empire, TeliaSonera tended to think of itself as the Nordic and Baltic telecommunications leader, but although this might simply have been an appropriate description of its market position in these two regions, and lacking a 'local' partner to offset the risk inherent in investing in the non-Baltic 2004 accession countries, it became possible that TeliaSonera would sell some of its overseas investments, leaving it predominantly as a Nordic/Baltic operator. The April 2004 offer by TeliaSonera to take outright control of Eesti Telekom, although unsuccessful [Brown-Humes, 2004], together with the acquisition of the outstanding 10 per cent of Lithuania's Omnitel in August 2004 [22], the purchase of Orange Denmark in October 2004 and the subsequent sales of stakes in Hong Kong and Namibia, reinforced the feeling that its strategic priorities lay in the Nordic/Baltic Member States and not elsewhere. Nevertheless, TeliaSonera was set to be debt-free by the end of 2004, and had a substantial war chest for acquisitions, so a contraction of its international footprint was not a foregone conclusion [George, 2004].

**Table 10. TeliaSonera, European stakes, 31 December 2006**

| Country | Stake % | Proportionate Subscribers |
|---|---|---|
| Denmark | 100 | 1,123,000 |
| Estonia | 51.6 | 408,000 |
| Finland | 100 | 2,407,000 |
| Latvia | 60.3 | 484,000 |
| Lithuania | 100 | 2,074,000 |
| Norway | 100 | 1,641,000 |
| Spain | 76.6 | 24,000 |
| Sweden | 100 | 4,603,000 |
| | | |
| Total world | | 42,110,000 |
| *Accession* | | *2,966,000* |

Source: Compiled by authors from operator and other websites.

It is worth noting that the Finnish government appeared to have agreed to the effective takeover of Sonera by Telia on the understanding that TeliaSonera would pursue a strategy of growth. Ultimately, because TeliaSonera stated in June 2004 that its ambition was to take majority control of its foreign investments, and given the size of the proportionate subscribers involved, its strategy was dependent primarily upon its relationship with its main partners. For example, the relationship between TeliaSonera and Turkcell's largest shareholder, Çukurova, had at times been fraught [23] – Çukurova's stake was confiscated by the government in 2003 as collateral against debts and was about to be returned in stages commencing in July 2004 – and the situation in Russia was permanently unsettled. Such problems are usually addressed either via a takeover or a withdrawal. It is significant that, in late June 2004, the Finnish deputy CEO of TeliaSonera, with responsibility for pursuing the purchase of majority stakes in Turkcell and MegaFon, was dismissed by the Swedish CEO (George, 2004). At the very least, this indicated that TeliaSonera would not 'overpay' to take control, but to remain a permanent minority investor hardly seemed an attractive proposition as TeliaSonera was prepared to acknowledge.

In the event, TeliaSonera has acted in a somewhat conservative manner of late. It is hopeful of raising its stake in Latvia to 100 per cent [24], but failed to win the third GSM licence in Slovakia in August 2006 and it withdrew from the bidding for Mobi 63 in July 2006. Against all the odds, not to mention analysts' reservations, TeliaSonera finally decided to go ahead with its launch of Xfera, its 3G licensee in Spain, in June 2006 [Roman & Hanson, 2006]. The CEO of TeliaSonera explained that this was really not a dramatic change from before when TeliaSonera was exclusively looking for acquisitions in, or close to, its existing markets in the Nordic, Baltic and Eurasian regions [Hansson, 2006]. Rather, with an extension into continental Europe, the company was 'expanding our thinking a bit'. However, it is unlikely that this will turn out to mean that TeliaSonera intends to buy stakes in the non-Baltic accession countries, although there is currently much discussion surrounding a possible merger with Telenor which has networks in Hungary, Montenegro, Norway, Serbia and Sweden.

**Table 11. mobilkom, European stakes, 31 December 2006**

| Country | Stake % | Proportionate subscribers |
|---|---|---|
| Austria | 100 | 3,630,000 |
| Bulgaria | 100 | 4,270,000 |
| Croatia | 100 | 1,910,000 |
| Liechtenstein | 100 | 5,000 |
| Slovenia | 100 | 421,000 |
| Total world | | 10,234,000 |
| *Accession* | | *4,691,000* |

Source: Compiled by authors from operator and other websites.

**Table 12. OTE, European stakes, 31 December 2006**

| Country | Stake % | Proportionate subscribers |
|---|---|---|
| Albania | 55.0 | 545,000 |
| Bulgaria | 66.7 | 2,182,000 |
| Greece | 66.7 | 3,480,000 |
| Macedonia | 66.7 | 315,000 |
| Romania | 62.9 | 771,000 |
| Serbia | 20.0 | 806,000 |
| Total world | | 8,100,000 |
| *Accession* | | *2,953,000* |

Source: Compiled by authors from operator and other websites.

As shown in Table 11, mobilkom Austria has always tended to look to the east for overseas acquisitions, of which it had three at the end of 2003 among which only Croatia provided more than a modest number of proportionate subscribers. However, since the 2004 accession its record of expansion has been patchy. On a positive note, parent Telekom Austria exercised an option to buy the whole of Bulgaria's MobilTel in July 2005 [25], and won the third GSM licence in Serbia in November 2006 [McDonald, 2006]. In addition, in Slovenia it was able to raise its stake from 75 per cent to 92.2 per cent and then to 100 per cent in May 2006. In contrast, mobilkom made an unsuccessful bid for 51 per cent of Telekom Montenegro (owner of MoNet) in March 2005 and, although mobilkom then provisionally agreed to acquire a 49 per cent stake in Serbia's Mobi 63, the deal fell through, and mobilkom was also unsuccessful when the stake was re-offered in July 2006. Furthermore, mobilkom failed to win a 65 per cent stake in Telekom Srpske in Bosnia in 2006 and to win a licence in Slovakia in July of that year.

Despite these setbacks, mobilkom continues to aspire to become the leading player in south-eastern Europe [Lenningham, 2005; Simonian & Yuk, 2006] and to that end managed to win the third GSM licence in Macedonia in February 2007 [26] – albeit by virtue of being the sole bidder – and is pursuing licences in Kosovo and Romania as well as a stake in OTE.

Overall, therefore, while a minor player during the 2004 accession and a relatively significant one in 2007, it may be that mobilkom will play a more important role in future accessions.

Finally, OTE has always been strategically focused on the Balkans, as shown in Table 12. Hence, although the 2007 accession coincidentally meant that OTE was now present in two accession countries and three Member States, this was not the result of strategic decisions related to accession. Given its profile, OTE may end up with further accession candidates.

## CONCLUSION

The above discussion has focused on the ownership of mobile licences in the enlarged EU. In the course of this a distinction has been made between the original 15 Member States and the twelve accession countries that joined in May 2004 and January 2007. Drawing such a distinction allowed those mobile operators with a presence in the accession countries to be differentiated from those that did not.

The first conclusion that can be drawn is that the largest multiple owners of mobile licences identified by Whalley & Curwen (2003) have, with the exception of Vodafone, only a limited presence in the mobile markets of the twelve accession countries. Both Tele2 and TeliaSonera have focused on the Baltic States while Deutsche Telekom has concentrated its attention on those Eastern European markets that either border, or are close to, its home market. This is not particularly surprising since liberalization offered so many opportunities to expand into the other Member States of the pre-accession EU, and the costs of licence acquisition plus network roll-out were extremely burdensome.

Ten of the 12 accession countries applied to join the EU between 1994 and 1996, with Cyprus and the Czech Republic being the two exceptions that applied much earlier in 1990 [27]. The adoption of EU legislation by the applicant countries liberalized their telecommunications markets so that on the one hand foreign investment was possible while on the other they instituted a regulatory regime that would reassure investors that they would not be discriminated against. As a consequence, the formal accession date did not trigger a wave of foreign investment by mobile operators as they had already responded to inward investment opportunities as they arose. Thus, the second conclusion that can be drawn is that the expansion of mobile operators into accession countries began before they formally joined the EU.

That said, it was reasonable to expect that once the formal accession dates had been announced, the strategic interest of operators could have been refreshed. However, not only were opportunities comparatively limited due to the aforementioned pre-accession inward investment but many operators were struggling with the fall-out from the telecommunications-media-technology (TMT) collapse that became very marked in 2002. Few accordingly had the wherewithal, let alone the will, to expand into the accession countries. One possible expansionary candidate was Vodafone, given its resources and strategy based on its international footprint, while alternative contender Orange was forced to retrench to the point that it became, to all intents and purpose, a Western European-focused operator with a presence in an increasingly scattered set of markets.

While the need to raise capital for its parent has abated, Orange, like TeliaSonera, is no longer interested in playing bit parts and wants to be either a major player or to exit. Exit is nevertheless easier said than done because of the shortage of buyers. Insofar as stake-building is concerned, it does appear to be far more likely that operators will seek to consolidate their positions in existing markets through purchasing additional equity in companies where they already own a stake. Taken together, these points suggest a third conclusion, namely that little change can be expected in accession countries as those operators wishing for whatever reason to exit are unlikely to find buyers while those wishing to consolidate their position in existing markets will be unwilling to pay the premium sought.

The limited future scope for structural change has implications for the competitiveness of mobile telecommunications markets. Most accession markets are effective duopolies, with the largest two operators controlling 70 per cent or more of the market between them. One consequence of this is that the ability of the third or fourth operators in the market to bring about the competition-derived benefits associated with liberalization is limited, while another is that any would-be investor into the market is stuck with the choice either of investing in one of the smaller operators or to become part of the duopoly. Neither is an attractive option, though for different reasons: The revenue and growth potential of the former is likely to be limited while the latter will involve paying a premium.

The unattractiveness of the investment decision may also be compounded by the relatively small size of many accession countries and, by extension, their mobile telecommunication markets. In this respect, it is significant that although Vodafone is present in seven accession countries, it does not own a network in all seven markets. Indeed, Vodafone owns a network in just two markets – Malta and Poland – and is present in the other five through the use of Partner Network Agreements. Those markets where Vodafone has used these Agreements are all characterised by their small size. The Agreements have enabled Vodafone to offer its services in new markets independently of owning a network and to retain a presence in those markets like Sweden that it has exited. As such, they are a key component of Vodafone's pan-EU footprint. When this observation is combined with the propensity of Tele2 to use MVNO arrangements to enter markets, a final conclusion is that multiple licence owners are using a wider variety of entry modes than was previously the case.

## NOTES

[1] See, for instance, Bartle (2005: 68-71) or Thatcher (2001: 563ff) for an overview of the Directives introduced as a consequence of the Green Paper.

[2] Five out of the EU's 15 Member States were granted derogations from this. Luxembourg was given until January 2000 due to the small size of its network while the other four Member States – Greece, Ireland, Portugal and Spain – were given until January 2003 (Curwen, 1997: 57). Having said this, all of the Member States liberalised their markets before the extended deadline with Greece, in January 2001, being the final country to do so (Walden, 2005: 122).

[3] For example, Member States were required to ensure that the terms and conditions of interconnection were published and that number portability was introduced – see, Thatcher (2001: 568-570) for a detailed discussion of the actions taken by the EU.

[4] Eliassen and Sjøvaag (1999) highlight some of the regulatory differences between Member States in the EU.
[5] For details of the six Directives see, for example, Bartle (2005: 72f).
[6] See, Stumpf (2006) or Walden (2005) for a discussion of the new regulatory framework.
[7] The 1993 Copenhagen European Council identified criteria for EU membership. Applicant countries must possess stable institutions, have functioning market economies with the capacity to withstand competitive pressures from elsewhere within the EU and be able to assume the obligations of membership (Blair, 2006: 156). To these three criteria Friis (2003: 182) adds a fourth, namely, that the EU must be able to absorb the Member State without interrupting its continued integration.
[8] Walden (2005: 107) estimates that since 1984 more than 100 different pieces of legislation have been adopted across the telecommunications industry.
[9] See Friis (2003: 183) for a full list of the areas where negotiations occur or Barnes and Barnes (2007) for a description of the access process. Friis implies that the potential for telecommunications-centred disagreements and delays is less than for other areas such as agriculture, energy or regional policy.
[10] This should not, however, be taken as implying that there are no differences between the accession countries and original 15 Member States of the EU. Bak and Szczesniak (2007) identify two areas where differences remain even though many barriers to foreign investment were removed when they joined the EU.
[11] GSM is a generic term encompassing an uplink at 890-915 MHz combined with a downlink at 935-960 MHz and an uplink at 1710-1785 combined with a downlink at 1805-1880 MHz, where the latter are also known as PCNs. PCNs were typically licensed after 1997 either to ameliorate spectrum shortages in GSM900 or to permit the entry of new operators. Fortunately, what was then Eastern Europe chose almost universally to adopt the same spectrum bands and technologies as were being enforced across the EU. Eastern European countries also chose UMTS for 3G, which initially required the use of 1885-2025 MHz combined with 2110-2200 MHz, although other spectrum bands have subsequently been added.
[12] See, for example, the newsletter released by Vodafone on 22 February 2006 relating to its new Agreement with Bulgaria's Mobiltel.
[13] BT was certainly the mobile incumbent in respect of analogue telephony but, somewhat surprisingly, Vodafone was the first operator to launch a GSM service (in July 2002) and hence has a claim to be known as the GSM incumbent. mmO$_2$, as it was known after being hived off from BT, launched only in January 2004 and the close proximity of the other three launches after that of Vodafone explains why, uniquely, the UK has four mobile operators of roughly the same size.
[14] See, for example, 'Vivo shares fall 39% since merger', available at www.cellular-news.com of 10 August 2006.
[15] See 'Telefónica O$_2$ Slovakia wins mobile licence with lowest bid', available at www.telegeography.com of 3 August 2006.
[16] See, for example, 'Elektrim says Deutsche Telekom now controls PTC', available at www.cellular-news.com of 19 June 2006; 'Vivendi offers €2.5 billion for Deutsche Tel PTC stake', available at www.cellular-news.com of 26 September 2006; and 'Polish court backs Vivendi appeal in PTC row', available at www.cellular-news.com of 19 January 2007.

[17] See 'France Telecom buys Orange Slovakia minority holders', available at www.cellular-news.com of 20 October 2006.
[18] See 'Vodafone, TDC currently hold 19.61% in Polkomtel', available at www.totaltele.com of 25 July 2005.
[19] See, for example, 'Vodafone files to block TDC sale of Polkomtel shares', available at www.telegeography.com of 28 February 2006.
[20] The purchase was completed in May 2006.
[21] See 'TDC CEO said telco may sell operations outside Denmark', available at www.totaltele.com of 1 November 2006.
[22] See 'TeliaSonera pays $63.5 million for rest of Lithuanian Omnitel', available at www.totaltele.com of 17 August.
[23] And remained so – see, for example, 'TeliaSonera launches new suit against Çukurova', available at www.telegeography.com of 22 August 2005, 'TeliaSonera initiates legal action against Russian Alfa Group', available at www.telegeography.com of 28 November 2005, and Ostrovsky & Bergstrom (2005).
[24] See 'TeliaSonera given permission to buy LMT, but not Lattelekom', available at www.telegeography.com of 11 May 2006, and Kaza & Hansson (2006).
[25] See 'Telekom Austria closes purchase of Mobiltel', available at www.telecomdirectnews.com of 14 July 2006.
[26] See 'Macedonia selects Austrian Mobilkom as third mobile operator', available at www.telegeography.com of 5 February 2007.
[27] For details of the application and subsequent progress of the twelve accession Member States see, for example, ec.europa.eu/enlargement.

## REFERENCES

Bak, M., Kulawczuk, P., & Szczesniak, A. (2007). Institutional and regulatory barriers to Foreign M&A in Central Europe. In, K. Meyer. & S. Estrin (Eds.), *Acquisition Strategies in European Emerging Markets.* Basingstoke: Palgrave.

Barnes, I., & Barnes, P. (2007). Enlargement. In M. Cini (Ed.), *European Union Politics*, Second Edition. Oxford: OUP.

Bartle, I. (2005). *Globalisation and EU Policy-making. The neo-liberal transformation of telecommunications and electricity.* Manchester: Manchester University Press.

Blair, A. (2006) *Companion to the European Union,* Routledge, England.

Boland, V., & Edgecliffe-Johnson, A. (2005). Win in Telsim auction hurts Vodafone stock. Available at http://news.ft.com of 13 December 2005.

Brown-Humes, C. (2004). TeliaSonera in cash offer for Eesti Telekom. Available at http://news.ft.com of 14 April.

Cottrell, R. (2003). A survey of European enlargement. When east meets west. *The Economist*, 22 November.

Curwen, P (1997). *Restructuring Telecommunications*, Basingstoke: Palgrave.

Curwen, P. (2002). *The Future of Mobile Communications: Awaiting the third generation.* Basingstoke: Palgrave.

Curwen, P., & Whalley, J. (2007). Tele2 and the strategic role of virtual operations, *info*, 9, 54-69.

Dinan, D. (2006). Governance and institutional developments: In the shadow of the constitutional treaty. *Journal of Common Market Studies*, 44, 63-80.

Eliassen, K., & Solvang, M. (1999). *European Telecommunications Liberalisation.* London: Routledge.

France Télécom (2005). France Telecom to take full ownership of 47.5% of TPSA. Press release of July 27.

Friis, L. (2003). EU enlargement...and then there were 28". In E. Bomberg & A. Stubb (Eds.), *The European Union: How does it work?* Oxford: OUP.

George, N. (2004). Shadows over an unhappy Nordic marriage. Available at http://news.ft.com of 29 June.

Grabbe, H., & Hughes, K. (1998). *Enlarging the EU Eastwards.* London: The Royal Institute of International Affairs.

Halaba, M. (2005). Poland's TPSA launching Orange brand on 19 September. Available at www.totaltele.com of 15 September.

Hansson, M. (2006). TeliaSonera extends acquisition strategy to Europe. Available at www.totaltele.com of 25 April.

Hitran, A. (2006). Romania – a vibrant market on the doorstep to the European Union. In J. Kittl, E. Lichtenberger, E-O. Ruhle & C. Schleps (Eds.), *Telecommunications Markets in Central and South Eastern Europe. Market Developments and Regulatory Frameworks*, Germany: EUL Verlag, Köln.

Kaza, J., & Hansson, M. (2006). TeliaSonera aims to end impasse on Latvian ownership. Available at www.totaltele.com of 8 March.

Krosnar, K. (2005a). TIW to complete Oskar, MobiFon sales by end June. Available at www.totaltele.com of 20 May.

Krosnar, K. (2005b). Vodafone looking to expand further in Europe, Asia – CEO. Available at www.totaltele.com of 30 September.

Lenningham, M. (2005). Mobilkom plans €2bn trip down Danube. Available at www.totaltele.com of 16 February.

McDonald, N. (2006). Telekom Austria wins Serbian licence. Available at www.ft.com of 8 November.

McQuaid, D. (2005). TPSA to start rebranding mobile unit to Orange 1 September. Available at www.totaltele.com of 30 August.

Middleton, J. (2005). France Telecom increases Romanian holding for US$523 million. Available at www.telecoms.com of 14 April.

Ostrovsky, A., & Bergstrom, R. (2005). Alfa jumps in with Turkcell bid. Available at http://news.ft.com of 31 March.

Peet, J. (2007). Fit at 50? A special report on the European Union. *The Economist.* 17 March.

Rachman, G. (2001). A survey of European enlargement. Europe's magnetic attraction. *The Economist.* 19 May.

Roman, D., & Hansom, M (2006). TeliaSonera move to Spain rattles local operators. Available at www.totaltele.com of 16 June.

Roman, D., & Rousek, L. (2005). Telefonica outbids rivals with Cesky Telecom offer. Available at www.totaltele.com of 31 March.

Sedelmeier, U., & Young, A. (2006). Crisis, what crisis? Continuity and normality in the European Union in 2005. *Journal of Common Market Studies*, *44*, 1-5.

Shin, D., & Bartolacci, M. (2007). A study of MVNO diffusion and market structure in the EU, US, Hong Kong and Singapore. *Telematics and Infomatics*, *24*, 86-100.

Simonian, H., and Yuk, P. (2006). Telekom Austria vows to continue Balkan search. Available at www.ft.com of 24 August.

Smith, G., & Ionian, H. (2006). Vodafone sells Swisscom Mobile stake. Available at www.ft.com of 19 December.

Stumpf, U. (2006). Markets susceptible to ex ante regulation: methodology and Commission recommendations. *Communications & Strategies*, *64*, 41-60.

TDC (2004). SBC intends to sell TDC shares. Available at http://tdc.com/about/press/releases.

Tele2 (2003). *Annual Report 2003*. Stockholm: Tele2.

Telefónica (2006). Nota de Prensa: Telefónica successfully completes the acquisition of $O_2$. 23 January.

TeliaSonera (2003). *Annual Report 2002*. Stockholm: TeliaSonera..

Thatcher, M. (2001). The Commission and national governments as partners: EC regulatory expansion in telecommunications 1979-2000. *Journal of European Public Policy*, *8*, 558-584.

Vodafone (2005). Sale of Vodafone Sweden. Available at www.vodafone.com on 31 October 2005.

Vodafone (2006a). Vodafone welcomes Latvia to its global community. Available at www.vodafone.com of 11 April.

Vodafone (2006b). Sale of 25% interest in Proximus. Available at www.vodafone.com of 25 August.

Walden, I. (2005). European Union Communications Law. In I. Walden & J. Angel (Eds.), *Telecommunications Law and Regulation*. Oxford: OUP.

Whalley, J., & Curwen, P. (2003). Licence acquisition strategy in the European mobile communications industry. *info*, *5*, 45-57.

Whalley, J., & Curwen, P. (2006). Third generation new entrants in the European mobile telecommunications industry. *Telecommunications Policy*, *30*, 622-632.

Williamson, H. (2007). Deutsche Telekom hit by profits warning. Available at www.ft.com of 28 January.

In: Telecommunications Research Trends
Editors: H. F. Ulrich, E. P. Lehrmann, pp. 45-74

ISBN: 978-1-60456-158-6
© 2008 Nova Science Publishers, Inc.

*Chapter 3*

# THE PRIVATIZATION OF INTELSAT: THE TRANSITION FROM AN INTEGOVERNMENTAL ORGANIZATION TO PRIVATE EQUITY OWNERSHIP

*Patricia McCormick*
Department of Radio, Television, and Film,
Howard University, Washington, D.C. USA

## ABSTRACT

This chapter examines the privatization of the world's most prominent international satellite organization, namely the International Telecommunications Satellite Organization (INTELSAT), which underwent a profound restructuring in November 2000 when it was separated into two entities. INTELSAT's intergovernmental treaty arrangement was fundamentally altered in March 2000 with the enactment of U.S. legislation, the Open-market Reorganization for the Betterment of International Telecommunications Act or ORBIT Act, which set forth specific criteria relating to the privatization of INTELSAT, which transpired on 18 July 2001. Simultaneously, a continuing intergovernmental organization, retaining the name the International Telecommunications Satellite Organization with the acronym 'ITSO', was designated to serve for a minimum period of twelve years in a supervisory and monitoring capacity, ensuring that the private company meet its public service and lifetime connectivity obligations to those developing countries largely dependent on the services and pricing mechanisms of Intelsat, Ltd.

This research examines the transformation of INTELSAT from an intergovernmental organization, owned by governments and by extension, citizens, to its current ownership by a private equity firm, within the larger context of the privatization of our global commons. This research analyzes the complicated consequences of the notable structural and ownership alterations of INTELSAT as well as the influence exerted by Lockheed Martin Corporation in the restructuring process. Methodologically, this research employs institutional ethnography in conjunction with the case study. In the context of international regime change and increased cooperation between national governments and the private sector, this work critically examines the process of privatizing INTELSAT and post-privatization developments as well the ability of ITSO to effectively oversee its charge.

## INTRODUCTION

The reform and privatization of the telecommunications sector, a policy process which diffused rapidly globally throughout the 1990s, has been accompanied by the privatization and restructuring of the world's most prominent intergovernmental, treaty-based satellite organization, the International Telecommunications Satellite Organization (INTELSAT). In November 2000, the 144 Member States of INTELSAT made the historic decision to restructure the organization into two separate entities. The actual satellite system or INTELSAT was privatized by transferring substantially all of its assets and liabilities, including satellites and orbital fillings, to a new company established for this purpose, Intelsat, Ltd., incorporated in Bermuda, and its subsidiaries.[1] The other entity, which retains the name International Telecommunications Satellite Organization (ITSO), is a highly modified revision of the intergovernmental organization. With a stipulated lifetime of at least twelve years, ITSO is designed to supervise and monitor the private company to ensure that it meets its public service and lifetime connectivity obligations to those developing countries largely dependent on the services and pricing mechanisms of INTELSAT. The privatization of this instrumental international satellite organization has fundamentally changed its mission to one of an international regulatory authority overseeing international public service telecommunications services.

This research examines the ramifications of the structural alterations of this organization, especially for developing countries, since more than 90 developing countries remain dependent on INTELSAT for their telecommunications needs. This research thus assesses the effectiveness of ITSO in addressing issues related to universal access, since the existing inequities in the global distribution of power and wealth will grow increasingly entrenched unless effectual international interconnection mechanisms are instituted. The mechanisms employed by INTELSAT essentially used the profits from the lucrative, high-density international traffic routes to subsidize service in rural and economically deprived or low-density route areas. While the efficacy of this subsidization process has been questioned, most developing countries, with such notable exceptions as Brazil, China, India, and Indonesia, do not have the indigenous capabilities for developing satellite technology and have difficulties in paying transponder fees and acquiring other commercial satellite services (Thussu, 2001). The obstacles facing developing countries remain unresolved as commercial imperatives prevail over development concerns in a market driven environment. This chapter traces the historical development of INTELSAT and critically analyzes legislation of the United States, notably, the Open-market Reorganization for the Betterment of International

---

[1] Please note that henceforth in this document, INTELSAT shall refer to the satellite organization prior to privatization and Intelsat, Ltd. or simply Intelsat will refer to the organization and its subsidiaries after its privatization.

Telecommunications Act (ORBIT Act), which in accord with the expanded embrace of neo-liberalism and agreements of the World Trade Organization (WTO), have affected significant changes within the satellite industry, including the privatization of INTELSAT.

## METHODOLOGIES

The work of international institutions, such as the World Trade Organization, the International Monetary Fund, and the World Bank as well as the International Telecommunication Union, is increasingly one of the most influential factors governing our world, with decisions generally made by representatives of ruling groups, thus its endeavors are biased in relation to those in power (Escobar, 1995). As Escobar (1995) writes, our knowledge is ideological in the sense that the international organization's conceptions and means of description represent the world as it is for those who rule it, rather than for those who are ruled. The relationship between the two is socially constructed by bureaucratic and textual mechanisms that are anterior to their interaction, though the interaction is presented as "facts", which can be categorized by professional discourse and presented in standardized ways (Escobar, 1995). Ethno-methodologists contend that institutional texts cannot be taken as objective records of external reality, but must be understood in relation to institutional uses and goals and in the context of their production and interpretation (Escobar, 1995). Institutional ethnography thus attempts to discern the ways in which institutional procedures shape or produce socioeconomic and socio-cultural practices.

Since this research is interdisciplinary in nature with several contemporary, interrelated issues being examined, methodologically, this work attempts to employ institutional ethnography in conjunction with the case study, which draws upon multiple data sources to investigate a specific phenomenon. Multiple sources of evidence are the primary strength of case studies since they facilitate the development of converging lines of inquiry, a process of triangulation. This study, which employs economic, political, and communication concepts and theories, is based on three sources of evidence: documentation, archival records, and elite interviews. Case studies are not generalizable to populations or universes, but to theoretical propositions, thus the goal of the research is to generate analytic generalizations regarding the reform and restructuring of intergovernmental organizations or international regimes.

## REGIME ANALYSIS

A regime is traditionally defined as a form of governance which comprises a set of norms, principles, rules and decision-making procedures that govern a particular issue area, such as the use of the global commons (Vogler, 1995). Space and the electromagnetic spectrum are part of our global commons, like seas, watercourses, and the atmosphere. The military and commercial utilization of space does not, however, generate widespread public interest or extensive involvement of non-governmental organizations as do other environmental issues. Radio frequencies or satellite orbital slots cannot be depleted as can fish stocks or minerals, but they are a relatively scarce and inordinately valuable resource that engenders competition for the best slots and specific frequencies. All wireless services use and thus require radio frequency spectrum. Demands for spectrum are escalating as a result of expanded applications in new technologies and as a result of the pro-competitive, liberalized market mandated by agreements of the World Trade Organization (Haller and Sakazaki, 2000). The allocation, assignment and coordination of the radio frequency spectrum and orbital locations has grown increasingly complex, since having access to and authority to use the spectrum and specific orbital slots has implications for national security and foreign policy as well as economic, technological and societal implications for virtually every country and corporation in the world (Haller and Sakazaki, 2000). In this context of space and spectrum as part of our global commons, INTELSAT and its successor can thus be defined as a common property resource regime.

While this work does not intend to delve into the theories of international regimes, it is nonetheless necessary to differentiate meritorious, but contrasting points of view. Neoliberals contend that interdependencies and mutual interests conspire to promote increased international cooperation. Certainly establishing a common property status for international space and airspace and creating international technical standards reduces the transaction costs of conducting commerce. This congruence of interest does not, however, devalue the validity of neorealist contentions that mutual interests are not sufficient for regime development. Neorealist proponents contend that the presence of a dominant state or group of states which has the power to impose acceptance of the regime and compliance on other states is essential, because the hegemonic state or grouping will only support an international regime if it acquires greater relative gains (Zacher, 1996).

Cowhey (1990) contends that domestic politics and policies are the primary source of regime change as hegemonic players influence the change of institutional structures, affect the centralization of power in a regime, and alter the jurisdiction of the regime vis-à-vis that of an individual state. Liberalization of telecommunications as a domestic policy has led, at the international level, to the introduction of trade in services into the General Agreement for Tariff and Trade (GATT) Uruguay Round 1986-1994. The series of accords that were approved in April 1994 included a General Agreement in Trade in Services (GATS) and a Telecommunications Annex. Adherents to these GATT Agreements went on to establish within the institutional body of the World Trade Organization, which effectively succeeded GATT, an Agreement on Basic Telecommunications and a Reference Paper, which placed national regulations in the multilateral arena for the first time. Arguably, more significant telecommunications deliberations now occur in the liberalization-oriented World Trade Organization than in the International Telecommunication Union (ITU).

The privatization of INTELSAT occurred in accord with WTO policies as well as the domestic policies of its dominant shareholding Signatories, most markedly the U.S. and its legislation, the Open-market Reorganization for the Betterment of International Telecommunications Act or ORBIT Act, thus providing additional evidence corroborating Cowhey's (1990) assertions regarding institutional change. The ORBIT Act and subsequent privatization of INTELSAT also supports Lyall's (1989) views regarding the role of the U.S. when he wrote nearly two decades ago, "the law of England as applied to maritime matters through the Admiralty Court had significant impact and became for most purposes international maritime law ... History, therefore, leads me to suspect that U.S. Law may become international space law ... By reason of its technical skills, its domestic market and its entrepreneurial attitudes, the USA is a major leader in space matters. That 'lead' may result in much U.S. Law becoming the language in which problems are discussed and solved."

The existence of a hegemonic leader is not, however, as Volger (1995) states, the distinguishing characteristic of an effective regime, but the perception of high levels of mutual vulnerability interdependence. In this regard, it was not considered viable for the U.S., for example, to have left INTELSAT, as it had UNESCO, due to economic dependence on the vital, global satellite system. Cognizance of this position may lead to compromise, however, equally arguable is that it portends for hegemonic states, notably the U.S., to strive for greater influence over other regime members. The interests and activities of states, particularly of dominant states, and corporations are often indistinguishable, and Cox (1987) views the senior officials of such institutions as forming a transnational managerial class who have created a globally hegemonic business culture whose interests transcend the state-market dichotomy.

Issues of stability and change have long been of concern to regime theorists, as Volger (1995) writes, and it should be noted that the re-positioning of ITSO as overseer of Intelsat, Ltd.'s public service and lifetime connectivity obligations, arguably, signifies the adaptive process of the regime; but one can, however, contend that the privatization of the satellite system, INTELSAT, represents actual regime change as it entails an alteration of norms and principles, especially with the pending expiration of ITSO. While reform of ITSO did not include changing the achievement of consensus, the privatization of the satellite system entailed an utter alteration of its structure and purpose as it seeks to provide end-to-end services through its recently established international network of teleports, leased fiber and points of presence (PoPs) for corporate business communications as well as for the broadcasting and entertainment industries. In such lucrative and competitive markets, it becomes necessary for Intelsat, Ltd. to give priority to its key customers, such as major Internet service providers, cable networks such as CNN, and international record carriers, rather than rural networks in developing countries where the return on investment is not commercially viable (Thussu, 2001).

This transformation of values is in accord with the larger, subtle, but fundamental shift of power from nation states to transnational corporations. This is in accord with what Geri (2001) refers to as the New Public Management, in which the state, devoted to securing conditions that enable the domination of multinational firms, engages in administrative reforms which include divestment of state owned enterprises, deregulation, competition in services, and facilitation of the role of the private sector in international organizations, if not, their actual privatization. In the span of a few years, not only INTELSAT, as an intergovernmental, treaty-based organization was disbanded as it was simultaneously transformed into a private corporate enterprise, but also other intergovernmental satellite organizations, notably the International Maritime Satellite Organization (INMARSAT) and the European Telecommunication Satellite Organization (EUTELSAT).

## GEOSYNCHRONOUS SATELLITES

The fundamental principles by which INTELSAT and other geosynchronous satellites and systems were formed are based in accord with the concepts outlined by renown author Arthur C. Clarke in the paper titled "Extra-Terrestrial Relays – Can Rocket Stations Give Worldwide Radio Coverage?" published in *Wireless World* in October 1945. After some 25 years Clarke's proposition that three satellites positioned over the Atlantic, Pacific and Indian Ocean regions could provide global communications was realized. Communications satellites placed in geosynchronous or geostationary orbit, approximately 22,300 miles above the equator, travel at the same speed as the earth rotates so that the satellite seems to retain its position. An antenna pointed in the direction of the satellite can thus remain stationary. Each communications satellite contains transponders or repeaters, which receive signals from the earth, translates and amplifies them, and then retransmits the signals to an earth station or ground station, which are equipment on the earth that can transmit or receive satellite signals.

Signaling in geostationary fleets generally takes place over two frequency bands, the C-band and the Ku-band. Since signals in the Ku-bank are more adversely affected by rain, C-band signals are better suited for wide-area coverage and, as the principal frequency for switched telephony and for transmission to cable headends, have traditionally been the prime frequency for INTELSAT (Einhorn, 1998). The C-band, however, readily encounters interference from adjacent satellites and terrestrial wireless signals that employ the same frequencies. Satellite providers, thus, must not only limit the power of their beams, but must locate earth stations away from urban areas, which then require miles of terrestrial feeders to reach an urban populace. Due to less interference in conjunction with more powerful signals and less expensive and smaller ground equipment, highly focused Ku-band signals can be successfully employed with very small aperture terminals (VSATs), direct-to-home (DTH) TV broadcasting reception antennae, an area which has significantly expanded satellite markets, mobile applications and in countries where physical space is scarce, as in Europe and Japan (Einhorn, 1998).

Despite the ability to deliver more tailored and smaller spot beams to serve different markets, there is, however, a slight delay inherent in communications transmission, especially notable with voice communications, with the use of geosynchronous satellites. Although video and audio broadcasting more readily tolerate the echo, delay and signal error of satellite communications, fiber optic cables, which preclude these drawbacks, are the preferred medium for switched voice/data, especially along high-density routes. The growth in switched demand for INTELSAT capacity has thus slowed dramatically since the late 1980s, particularly along transoceanic routes between North America and Western Europe, and North America and Asia (Einhorn, 1998). In spite of the declining cost of fiber, it is not yet economically feasible in thin switched access routes. Although the telecommunications network in most developing countries does not sufficiently extend to rural areas and links that presently exist are used to capacity, resulting in traffic congestion, the states can ill afford the capital investment required to extend these lines. Satellite systems, however, make it possible to provide point-to-multipoint or point-to-point communication services to any site with cost independent of distance. Satellites are completely cost-insensitive to distance anywhere within the coverage area of a specific satellite, whether the two ends of a link are separated by ten miles or 3000 miles, the cost is the same. For virtually every other form of communications, distance is one of the key elements in providing a telecommunications link, and an important consideration for rural communications, thus many developing countries rely upon satellites for switched interconnections.

Furthermore, installation and maintenance of the ground segment are required only at the terminal site rather than at multiple points in between. Since any other point in the network is reached in a single "hop" through the satellite, reliable interconnection depends solely on the two stations involved. This contrasts with terrestrial systems which depend on a series of sequential links, the failure of any one of which can disrupt the connection. Failure of any one earth station or dish affects only that location and has no negative impact on the rest of the network; it remains intact, thus reliability and access to remote areas are facilitated. This makes the satellite system a most robust system and one independent of trunking hierarchies. Facilities can be placed in those areas most needing communication services, regardless of distance from the urban center, which reduces the bias of conventional terrestrial systems. Satellites can also enable immediate connectivity in short-term, emergency situations and can be employed with minimally developed infrastructure, such as poor roads and scarcity of prime power, and are not affected by rugged or inhospitable terrain, such as mountains, jungles, deserts, ice and snow.

Satellites are thus a critical communications component for many developing countries seeking to strengthen and diversify their economies, since the telecommunications sector is decisive in enabling countries to achieve social and economic development goals as well as compete in the international economy, as effective use of electronic communication permits improved coordination in the distribution and production of goods and services. Both case studies and statistical analyses have indicated a strong correlation between telephone penetration rates and GNP per capita. Since markets depend on the timely flow of information, the telecommunications network is arguably the most basic form of infrastructure, with a pervasive effect on the efficiency of an economy. The telecommunications infrastructure serves as a platform and a catalyst for other industries, thus the sector facilitates forward and backward linkages within the economy. It is an essential factor, a requisite for most businesses, especially foreign firms considering direct investment in a developing state. Since satellites can establish immediate multipoint connectivity without investments in physical conduit, they well serve governments, corporations, non-governmental organizations and others operating in those developing countries that lack an established domestic terrestrial system (Einhorn, 1998).

## INTELSAT – THE EARLY YEARS

### The U.S. Communications Satellite Act of 1962

The United Nations General Assembly, in recognition of the beneficial applications of satellites and in response to the 1960 U.S. launch of the first telecommunications satellite, Echo I, for switched transmission, which was followed by Telstar and Relay in 1962 for wideband and broadcasting transmission, passed a resolution in 1962 that called for a system of 'communications by satellite' which would be 'available to the nations of the world ... on a global and nondiscriminatory basis' (Einhorn, 1998). President Kennedy thus envisaged a global satellite network that would aid developing countries and rely on considerable involvement of government agencies, instead of being simply an organ of private enterprise. Although the final version of the Communications Satellite Act of 1962 was largely a comprised version of the legislation originally envisioned by Kennedy, it did represent a major pillar in U.S. Cold War foreign policy objectives, as it positioned the U.S., rather than the Soviet Union, as the principal benefactor in bringing the benefits of this nascent technology to the rest of the world. The inception of the satellite industry in the United States, however, begot public advocates and private businesses to struggle over its control throughout a series of congressional hearings in 1961-1962. The resultant Communications Satellite Act of 1962, as Pelton and Edelson (1997) note, represented hard-fought concessions on behalf of all legislators as it was filibustered for weeks before cloture was invoked. The Act, however, essentially represented a victory for private interests as it culminated in significant participation of the private sector in the emerging satellite industry, since it authorized the formation of the Communications Satellite Corporation or COMSAT to administer satellite communications for the United States.

Senators Wayne Morse (D-Oregon) and Carey Kefauver (D-Tennessee) and Congressman Emanuel Celler (D-New York) formed an alliance against the private ownership of COMSAT in conjunction with the American Communication Association, a union of telecommunications workers, as well as Assistant Attorney General Lee Loevinger and renown academic scholars, Herbert Schiller and Dallas Smythe (Museum of Broadcast Communications, 2007). This alliance voiced concerns that a privately owned COMSAT would strengthen the private sector's control over public airwaves. They called for increased public participation in the congressional hearings as well as government ownership of the satellite company. Senator Robert Kerr (D-Oklahoma), in staging an opposing alliance with major communications corporations, including RCA and AT&T, the latter who ostensibly sought to extend their telephony monopoly to satellite communications, proposed a bill that called for the privatization of satellite communications, since space communications, as Kerr contended, offered beneficial new business opportunities for the private sector (Museum of Broadcast Communications, 2007).

Amidst this controversy of competing pressures, COMSAT emerged as a compromise. COMSAT was incorporated as a publicly traded company in 1963, though the Act prohibited greater than 50 percent aggregate common carrier ownership. Federal agencies, including the State Department, the Federal Communications Commission (FCC), the National Aeronautics and Space Administration (NASA), and the Office of Telecommunications Policy (now the National Telecommunications and Information Administration (NTIA)), and the President were each charged with monitoring duties and the President was granted the authority to appoint three directors to the COMSAT Board. The organization, though designed to operate as a private business while being regulated to serve the public interest, has historically favored the corporate interests of its mandate as it straddles the often conflicting objectives of private enterprise and the public good.

Since building a global telecommunications system was thought to be beyond the capacity of any single country or private entity, COMSAT was charged with raising private financing and securing foreign partners to develop a global satellite communications system. Although U.S. negotiators initially considered creating a new global satellite system through a series of bilateral agreements, the Committee on European Post and Telecommunications would only accept a multilateral agreement (Pelton and Edelson, 1997). Two years of international negotiations thus ensued which resulted in August 1964 in the formation of the International Telecommunications Satellite Cooperative (INTELSAT) as the world's first commercial satellite operator, which thereafter established the first global satellite communications system. At the time INTELSAT was considered a major innovation in the domain of international organizations since, as Pelton and Edelson (1997) write, its structural organization, international procurement procedures, contractual services, and operations were modeled on a commercial corporate structure, not the United Nations system. Ownership and voting rights, specifically in the Board of Governors, were related to usage of the system, with the United States guaranteed a dominant role since COMSAT was the largest investor.

COMSAT was also responsible for the technical management of the system. In 1965, COMSAT and an ad hoc partnership of 44 nations successfully launched into geostationary orbit the first commercial communications satellite, "Early Bird." Several more geosynchronous satellites were launched in subsequent years, until, in 1971, the cooperative, originally comprised of 11 countries, was converted into an international agreement entered into by 86 nations that established the International Telecommunications Satellite Organization, INTELSAT, as an intergovernmental treaty-based organization effective in 1973 (Intelsat, 2003). The special legal status of INTELSAT granted it immunity from paying taxes to any national government as well as immunity from jurisdiction, which prevents courts from considering lawsuits against INTELSAT; archival and testimonial immunity, which protects INTELSAT from being forced to submit documents or testimony of its employees; and immunity of assets, which precludes courts from enforcing monetary judgments against INTELSAT (Katkin, 2002). Its treaty status also assisted in ensuring its access to the domestic markets of member states, a valuable privilege or asset not readily conferred upon the entry of its competitors in forthcoming years.

## Governance

Operating as a user-owned cooperative, INTELSAT had a multi-tiered governance structure. This configuration, reflective of INTELSAT's dual roles as an intergovernmental treaty organization and a commercial provider of telecommunications services, consisted of the Assembly of Parties, the Meeting of Signatories, the Board of Governors, and the unitary Executive Director General. The Assembly of Parties represented the principal organ of INTELSAT, the pinnacle of its structure, as it comprised each country that was party to the INTELSAT Articles of Agreement, which, at the time of its privatization, constituted 144 member states with eight Western countries accounting for half the controlling shares and the U.S. holding the largest investment of 20.4 percent (Thussu, 2001). While each country had one vote at the Assembly of Parties, which met biennially to establish general policy and long-term objectives and was needed to ratify major decisions, this body, according to Hahn and Kroszner (1990), generally followed the counsel of the Board of Governors, which consisted of Signatories to the Operating Agreement.

Each member state designated a single telecommunications entity to be a Signatory to INTELSAT. Each Signatory was responsible for the requisite capital contributions to finance its state's share of INTELSAT, and to conduct the obligatory commercial and technical operations to provide the transmission capacity and communications services of the satellite system to carriers and users in their respective nation (Katkin, 2002). Generally the government-owned and operated Post, Telephone and Telegraphy (PTTs) monopolies were so designated, though in the case of the U.S., COMSAT served as its Signatory to INTELSAT. COMSAT, like the other Signatories, then had the sole right of access to INTELSAT. The long distance telephone service providers, news organizations, private network operators, and radio and television broadcasting networks could not lease services directly from INTELSAT, but were required to go through its state's Signatory if it sought to employ international satellite services (Hahn and Kroszner, 1990). In reselling capacity to customers, these

Signatory monopolies could thus overprice international service and restrict competitive entry (Einhorn, 1998).

The structural organization of INTELSAT, arguably, further secured the interests of these monopolies, since designated representatives from each Signatory body comprised the Meeting of Signatories, which, like the Assembly of Parties, adhered to the one-country-one-vote rule. The Meeting of Signatories met annually to, among other duties, establish rules governing the rates and adjustments of INTELSAT satellite transmission capacity, adjust capital contribution ceilings, authorize new earth stations, consider the financial implications of proposed programs, and resolve disputes among member states (Katkin, 2002).

The general, day-to-day operation and administration of INTELSAT, however, as Hahn and Kroszner note (1990), resided with those who most used the system, since representation on the INTELSAT Board of Governors was apportioned in accord with the Signatory's share ownership, itself based on each Signatory's utilization of INTELSAT satellite transmission capacity. Since the principal responsibility "for the design, development, construction, establishment, operation and maintenance of the INTELSAT space segment", as Katkin (2002) notes, was borne by the Board of Governors, the stipulations regarding representation on the Board consequentially meant that those industrialized, hegemonic countries, specifically the U.S. and the U.K., extended the greatest influence over the direction of the organization. Akin to the conflicting goals and agendas of developed and underdeveloped states in most international forums, conflicts arose in INTELSAT between the few wealthy investors that financed most of INTELSAT's investments and the majority of smaller Signatories that owned very little capacity. The hegemonic position of the major Signatories, notably the U.S., in the organization is in accord with the work of Cowhey (1990) who contends that the domestic policies of the dominant, prevailing players are the primary source of international regime change.

Directly accountable to the Board of Governors for the performance of all management functions was the fourth level of governance, the Director General, who exercised executive responsibility over daily operations (Katkin, 2002). This post was elected at the Assembly of Parties. Several candidates were nominated from various geographic regions, and then there were successive rounds of balloting, in which the candidate with the lowest vote total each round was eliminated, until one candidate received a two-thirds majority of the vote or until there were only two candidates remaining, in which case the candidate with the higher vote total would then become the Director General-elect, to serve as the chief executive and legal representative of INTELSAT.

The quadripartite structure of INTELSAT, arguably, hindered its development and ability to compete in latter years. Einhorn (1998) contends that the administrative process for resolving the conflicting needs and agendas of its numerous Signatories regarding competitive rulemaking, tariff-setting, capacity additions, and technological updates, among other issues, was slow, cumbersome, highly political and costly to administer. With the advent and subsequent growth of competing separate satellite systems, INTELSAT was thwarted in its ability to rapidly respond to competitive offerings that bundled products and services and featured price discounts. Since, as Pfeifenberger and Houthakker (1998) note, INTELSAT only provided space-segment services and neither owned nor operated earth stations and interconnecting terrestrial facilities, it could not market itself as an all encompassing service provider.

## The Emergence of Separate Satellite Systems

At the time of INTELSAT's creation, the prevailing economic rationale held that natural monopolies in the provision of public utilities were the most efficient. In virtually every country in the world, providers of telephony and telegraph services were either private monopolies, crown corporations, or, most commonly, government Ministries of Post and Telecommunications (PTTs). This historical structure, in accord with the natural monopoly argument that contended it was more economically efficient to construct a single satellite network connecting different countries rather than building several competing satellite systems, gave rise to the formation of INTELSAT as a single monopoly provider of satellite services. The large initial costs or high sunk costs in building and operating satellites coupled with the relatively low cost of adding additional users in conjunction with the arguments about economies of scale and unnecessary duplication of costly facilities, contributed to the general assumption that competition in satellite communications was neither feasible nor desirable (Hahn and Kroszner, 1990). Maintaining INTELSAT's role as the premier, if not sole global satellite operator, also dovetailed with U.S. foreign policy objectives and, thus, the U.S. itself was an early proponent of safeguarding INTELSAT's special and unique role from possible competing systems.

At the same time, though, potential competition from other systems was never intended to be totally foreclosed. Section 102 [47 U.S.C. 701.] of the U.S. Communications Satellite Act of 1962 expressly provided for the eventual creation of additional communications satellite systems as it did not "preclude the creation of additional communications satellite systems, if required to meet unique governmental needs or if otherwise required in the national interest." Although Katkin (2002) contends that the accession of the U.S. to the INTELSAT Agreement may have in effect delayed the deployment of separate satellite systems serving the U.S., on 28 November 1984, President Ronald Reagan, in accord with his administration's policies promoting deregulation, increased market competition, and less government involvement in commercial enterprises, which by then had overtaken foreign policy considerations, made a determination that separate international satellite systems were required in the country's national interest. As Orion and the Pan-American Satellite Corporation (PanAmSat), in addition to four other companies, made their initial applications to the FCC, INTELSAT responded by encouraging a boycott, as it adopted a resolution on 31 January 1985, urging its members to refrain from participating in the construction of any separate international satellite systems not owned or operated by the PTTs (Hahn and Kroszner, 1990). Despite such resistance, the FCC proceeded to make orbital slots available and, with the launch of PAS-1 in 1988, PanAmSat (PAS) became the first private satellite operator in direct competition with INTELSAT and it achieved full global coverage in 1994 (Pfeifenberger and Houthakker, 1998).

Contrary to its strong opposition to private satellite systems, INTELSAT did not present such a defiant position against the European PTTs, which in conjunction with others, had initiated planning in the late 1970s for the creation of a European satellite system largely modeled on INTELSAT. The European Telecommunication Satellite Organization, EUTELSAT, headquartered in Paris, was created in 1977 as an intergovernmental organization to develop and operate a satellite-based telecommunications infrastructure for Europe. Similarly, after studying the operational requirements for a satellite system devoted to maritime purposes, the Convention for the International Maritime Satellite Organization

(INMARSAT) was adopted in September 1976. Hence, on 16 July 1979, INMARSAT, headquartered in London, effectively came into being as an intergovernmental organization to improve maritime communications, and, thus, alleviate distress and emergencies and generally enhance safety of life at sea, increase the efficiency and management of ships, and provide maritime public correspondence services and other communications for the maritime community (http://www.imo.org/Conventions/contents.asp?doc_id=674&topic_id=257). Initially, these systems were operated by INTELSAT Signatories, and thus presented little direct competition. INTELSAT, however, sought to contain the growth and development of competing private satellite systems. Although the INTELSAT Articles of Agreement recognized that separate satellite systems might be of value in the future and Article XIV(d) specified procedures for the coordination and approval of separate satellite systems, this aspect of its charter proved to be an effective anticompetitive instrument as it enabled INTELSAT to assume a quasi-regulatory role in regard to its competition.

Article XIV(d) required potential competitors to coordinate and acquire approval for their activities from INTELSAT. The lengthy review process, which included assessing their competitors' business plans before beginning operation, enabled INTELSAT to preempt its competition or, at the least, delay its entry, a claim made, albeit unsuccessfully, in PanAmSat's antitrust suit against COMSAT, and, by extension, INTELSAT (Hahn and Kroszner, 1990). Further evidence of INTELSAT's ability to thwart its competition resides in its Charter's stipulation that separate satellite systems, in addition to meeting technical compatibility and noninterference criteria, essentially demonstrate that their system would not cause INTELSAT 'significant economic harm,' a relative and ambiguous concept since the calculation of such harm was not denoted. As Hahn and Kroszner (1990) write, INTELSAT employed the ambiguity of 'economic harm' to its advantage as it discriminated in its calculation, impeding the entry of private satellite systems by subjecting them to an arduous approval process with harm calculated as the maximum traffic diversion, but more readily granting approval, due to a calculation of negligible economic harm, to those regional systems owned and operated by the PTT Signatories, including EUTELSAT, the Arab Satellite Organization (ARABSAT) and Indonesia's Palapa system.

The quandary, in essence, lay in linking separate satellite systems with the existing domestic networks, specifically, the public-switched network, from which INTELSAT derived the majority of its revenue. The Reagan administration devised a policy, the separate systems restrictions, that licensed new competitive systems, but limited their traffic to new video and business systems, while granting INTELSAT up to seven years to prepare for competition in basic telecommunications services (Pelton and Edelson, 1997). The separate systems restrictions, which initially prohibited private satellite systems from providing services that accessed the public switched network, were designed to protect INTELSAT from competition in its "core" market of long-distance telephone traffic and thus ensure subsidized service to developing countries. This provision was not only never applied to competing terrestrial networks, such as transoceanic fiber optic cables, but Hahn and Kroszner (1990) challenge the validity of the cross-subsidy in satellite circuits as they contend that INTELSAT's pricing policy was employed to impede competition rather that to subsidize the less developed nations.

## Pricing: Concerns and Criticism

INTELSAT's mandate to charge uniform prices for similar services was considered a subsidy to the thin-route users in developing countries, because prices were not related to the actual costs of service provision. Hahn and Kroszner (1990), however, contend that INTELSAT's average costs were predatory because the system was overbuilt. Since the cost of overbuilding the system could be allocated among all the users, high-density countries had no incentive to accurately estimate capacity needs and, as Hahn and Kroszner (1990) note, the manufacturers and launchers of satellites, principally located in industrialized states, pressured their governments to support high estimates to increase the demand for their products, with the end result being that users on low-density routes paid higher average prices. Einhorn (1998) concurs with the view that the uniform pricing policy encouraged excess capacity, as he writes that the practice of geographically averaging tariffs, presumably to shield countries with low volume, generated uneconomic bypass and strand investment. Excess capacity also enabled INTELSAT, as a protected monopolist, to undercut its competitors' prices and introduce new classes of services at a minimal additional cost.

While INTELSAT's prices were geographically nondiscriminatory, as Einhorn (1998) states, tapered tariffs provided discounts for both duration and volume. Furthermore, Hahn and Kroszner (1990) assert that INTELSAT's technological choices were more appropriate and cost-effective for high-density use and that developing countries thus bore a very high ground segment cost per circuit for their low-density routes. Despite the considerable capital investment of constructing large earth stations borne by developing countries, INTELSAT assumed no liability for any damages to the ground arrangements that contracting countries could suffer (Gruenwald, 1998). Einhorn (1996) further charges that, in the absence of competition, a nonprofit consortium arrangement should design tariffs to protect against exorbitant prices, but like other monopolies, INTELSAT enjoyed a substantial, if not excessive, return on its investments, averaging 14 to 18 percent per year, protection of which was not warranted. Pelton and Edelson (1997) contend that this unnecessarily high rate of return, which only directly benefited members with net ownership over usage, served to establish the 'wholesale' rate for international communications levied by INTELSAT at a much higher rate that, in turn, created a basis for inflating national retail rates through a said Signatory. While the rate of return was high, the arrangement, one can contend, did serve one laudatory purpose as it provided a means whereby smaller countries could maintain their status as member countries without being forced to assume increased investment obligations as their relative utilization levels increased over time. In any event, the success of PanAmSat in negotiating with INTELSAT Signatory countries in South America and the Caribbean can be viewed as providing additional evidence against the efficacy of INTELSAT's service subsidization, though INTELSAT did provide personnel training and technical assistance to its Signatories, especially in developing countries, and thereby acquired an understanding of their needs through these interactions.

Although INTELSAT thus served as an institutional framework that facilitated information and technology transfer (Einhorn, 1998), its marketing and pricing arrangements allegedly grew outdated and the organization was unable to contend with its new competitors. As companies from countries throughout the world were able to build, launch, and operate satellites, what became of critical importance was a firm's sales and marketing ability, not one of INTELSAT's strengths. Furthermore in concurrence with the emergence of competition from separate satellite systems, INTELSAT began to face substantial competition in the market for international communication transmission capacity from the proliferation of transoceanic fiber-optic cables. Capable of delivering many of the same services as satellites, but at lower costs, the first transoceanic cable, extending across the Atlantic Ocean from the U.S. to the U.K., was completed in 1988 and the first trans-Pacific cable became operational in 1991 (Katkin, 2001). These technological developments, in conjunction with INTELSAT's wieldy politicized bureaucratic governance structure and questionable uniform pricing practice as an effectual subsidization mechanism, coupled with broader changes in the global environment in accord with the dictates of the World Trade Organization, since market changes do not occur in a vacuum but generally reflect policy changes, together provided impetus for the reform of the organization.

## INTELSAT – THE ERA OF RESTRUCTURING

### Resolution to Reform

The goals of the reform were, as Einhorn (1996) writes, two-fold – to enable INTELSAT to more effectively market its services while simultaneously forfeiting uneconomic advantages and immunities its had henceforth enjoyed, since the privileges INTELSAT derived from its special legal status proved to be a source of strife among its private competitors. Several options that would enable INTELSAT to successfully adapt to the new competitive environment were proposed and some were integrated into propositions to reform the organization. These various alternatives, as outlined by Pelton and Edelson (1997), included separating basic services from value-added and high-growth services; restructuring the tariff system to enable greater pricing flexibility, which could entail setting limits on charges or a price cap for thin routes and developing countries while other service rates could be commercially determined by the market; streamlining operations for maximum efficiency; acquiring other international sources of support for development activities; exploring strategic partnerships to develop new services and products; and restructuring the organization's capital financing to allow new sources of investment.

In regard to allowing non-Signatory users and investors, in 1992, INTELSAT introduced four new types or levels of direct access to INTELSAT satellites by non-Signatory carriers and users. Although the first two levels were confined to the exchange of technical and operational information, levels 3 and 4 pertained to access to communication services. Level 3 direct access allowed customers to enter into a contractual agreement with INTELSAT for ordering, receiving and paying for INTELSAT space segment capacity at the same rates charged to INTELSAT signatories. Level 4 direct access enabled customers, in INTELSAT member states only, to make a capital investment in INTELSAT in proportion to its utilization of the INTELSAT system, as well as obtain INTELSAT space segment capacity at INTELSAT tariff rates, though rights to participate in INTELSAT's governance process were not generally accorded to a level 4 customer unless special arrangements were made by the Party and the official Signatory representing the country. Indeed, INTELSAT only allowed direct access in countries where it was authorized by the Signatory representative. As of 1999, level 3 direct access was available in 65 countries and level 4 in 29 countries for a total of 94

Signatory states, though, at that time, neither level was available in the U.S. as COMSAT retained its monopoly position with the sole right of access to INTELSAT (FCC, 1999).

Despite INTELSAT's effort to increase its customer base through direct access, there remained the lack of true competitive access to and ownership of INTELSAT at the national level as well the lack of a competitive charging structure that was responsive to the market and correlated to geography, volume and diverse telecommunications sectors (Pelton and Edelson, 1997). In lieu of these enduring quandaries, British Telecom had suggested in the early 1990s that INTELSAT be fully privatized, which was followed by a similar proposal by COMSAT in 1993. In the face of substantial resistance to these proposals, the INTELSAT Executive recommended to the Assembly of Parties that a Working Party be formed to advise and implement competitive reform (Einhorn, 1998). In August 1995 the Assembly of Parties convened to consider the Working Party's recommendation to create a commercial subsidiary, which led, among other things, to the U.S. Clinton Administration and COMSAT proposal in March 1996 to restructure INTELSAT into two groups, one being a continuance of the intergovernmental satellite operator committed to maintaining INTELSAT's core services under the existing INTELSAT agreement; and the other, a separately chartered private affiliate to offer broadcast and private network services, in which the current Signatories' ownership would be rapidly and substantially diluted, by up to 80 percent, by the offering of shares to the public, and for which treaty-granted protections would be abolished, and each entity would be assigned assets aimed at providing switched and non-switched business respectively, but with no restrictions prohibiting competitive entry into any service (Einhorn, 1996).

Although this proposal with its draconian ownership changes was overwhelmingly rebutted, it ultimately did lead to the creation of New Skies Satellites, N.V. as a spin-off company and other endeavors to reform and restructure its organization. In an attempt to increase its efficiency, in 1995 INTELSAT initiated a review and subsequent revision of the internal mechanisms of two of its major service lines – lease and channel/carrier - to make the organization more responsive to its customers. Since staff handling channel/carrier orders, known as booking, had been assigned to specific satellites, a customer with multiple services had to work with several staff members, thus increasing the likelihood of errors, whose resolution in turn required customers to negotiate back and forth between several staff (Congram, Slye, and Glidden, 1999). In response to these problems, a team was formed, led by non-booking staff, which redesigned the booking process and simplified the bureaucratic maze for customers. Appreciating the difference in values held by its multicultural staff, INTELSAT made new customer/region assignments based on staff business culture and/or language when possible. These regional assignments, in conjunction with booking staff providing customer consulting services, significantly reduced throughput time, with 98 percent of new service transactions processed in one to three days, as compared with the previous 15 to 20 days for most such transactions, and billing errors became virtually nonexistent (Congram, Slye, and Glidden, 1999). In regards to lease services, a group was formed in May 1996 to redress the lease process, since a lease is customized to customer specifications, and to distinguish its role from sales. As a consequence of changes made to lease procedures, including incorporating the lease staff as members of their respective regional sales team, the productivity of both the sales staff and lease staff increased (Congram, Slye, and Glidden, 1999). These efforts at reform demonstrate a commitment to the conviction that management, as opposed to ownership, is the key to efficiency.

The principle economic issue in the privatization debate concerns efficiency. The property rights school suggests that privatization in the form of a change in ownership, an alteration of the structures of property rights, will improve the incentives for productive efficient performance (Cook and Kirkpatrick, 1988). It recommends the privatization of public enterprises operating in competitive markets, for competition compels companies to improve their performance, that is, increase allocative and productive efficiency in order to profit. It would thus seem that an improvement in the economic performance of the public enterprise sector is more likely to result from an increase in market competition than from a change in ownership (Cook and Kirkpatrick, 1988). In accord with this view, in April 1997 INTELSAT's Assembly of Parties affirmed a Board of Governors decision to eliminate, except for technical interference, the anticompetitive Article 14(d) of its Charter (Einhorn, 1998).

INTELSAT further promoted market competition in the next year, 1998, when it spun off five satellites to a separate entity, New Skies Satellites, N.V., based in the Netherlands. Although the creation of New Skies failed to remove INTELSAT's benefits, such as immunity from antitrust laws and exemption from taxation, some viewed it as an initial step in the privatization process. Analysis of privatization cannot, however, be confined to debates concerning economic efficiency, but must address the larger political context for such decisions and the ascendance of what Cox (1987) terms the transnational managerial class of dominant states and their corporations who have created a globally hegemonic presence promoting neoliberal reform policies in such multilateral institutions as the WTO. Shiva (2005) contends that 1995 marks the year when corporate globalization became the legal constitution of the world for it was the year in which the WTO was established.

Indeed, the principle historic step toward creating a more competitive commercial global satellite market was the 1997 World Trade Organization Protocol to the General Agreement on Trade in Services (GATS). Sixty-nine countries, accounting for more than ninety percent of the world's basic telecommunications revenue, initially signed the WTO Agreement (Haller and Sakazaki, 2000). Collectively representing the majority of INTELSAT's ownership and traffic, these countries agreed, among other commitments, to provide unconditional market access to all international telecommunications services and facilities. In this marked shift from a state-centric to a market-oriented view of communications among major global powers, PTT political and market authority subsequently eroded as countries agreed to secure competition in telecommunications services, including non-discriminatory market access to satellite markets. Effective on 1 January 1998, by WTO decree, the world's skies opened to satellite competition (Spiegel, 1997).

In accord with the WTO Agreement, in 1999, the FCC granted U.S. earth station operators limited three-year authorizations to operate with New Skies in the U.S. market, conditional on the company taking steps to become independent of INTELSAT, including an Initial Public Offering (IPO) to dilute the ownership of INTELSAT signatories, which was successfully concluded in October 2000 (Abelson, 2005). Also in 1999, the U.S. Congress chartered the Commission to Assess U.S. National Security Space Management and Organization to assess U.S. national security space and ways in which it could be strengthened (Haller and Sakazaki, 2000). According to the White House (1999) as stated in *A National Security Strategy of Engagement and Enlargement*, "We are committed to maintaining U.S. leadership in space. Unimpeded access to and use of space is a vital national interest – essential for protecting U.S. national security, promoting our prosperity and ensuring our well-being. . . . We will maintain our technological superiority in space systems, and sustain a robust U.S. space industry and a strong, forward-looking research base." Such views were, arguably, incorporated in the Open-market Reorganization for the Betterment of International Telecommunications (ORBIT) Act, which was agreed to or passed by both the U.S. House and Senate in the 106th Congress on the 24 of January 2000, though the legislation was initially proposed in 1997 in concurrence with the WTO General Agreement in Trade in Services.

## The ORBIT Act and the Role of Lockheed Martin Corporation

In June 1997 House Commerce Committee Chairman Thomas J. Bliliey (R-VA) introduced in the 105th Congress Bill H.R. 1872 to overhaul the Communications Satellite Act of 1962 to promote competition and privatization in satellite communications by requiring the privatization of INTELSAT and INMARSAT. In support of Bliliey's Bill, on 24 March 1998, at a hearing on International Satellite Reform before the Subcommittee on Communication of the U.S. Senate, John Sponyoe, the Chief Executive Officer of Lockheed Martin Global Telecommunications, presented a statement in which he described a strategic partnership with COMSAT, which, it should be noted, was at a time when the telecommunications market was thriving and military sales were flat. According to Sponyoe (1998), Lockheed Martin, one of the largest defense contractors in the U.S., had created a wholly owned subsidiary, Lockheed Martin Global Telecommunications (LMGT), in an endeavor to focus its business efforts in the commercial telecommunications services market and become a leading provider of global telecommunications services, and, upon the creation of LMGT, had announced its intention to acquire COMSAT. "The rationale for this acquisition is very straightforward – to marry our own technological and entrepreneurial assets with Comsat's 37 years of experience as a provider of satellite services" (Sponyoe, 1998). Lockheed Martin, in essence, proposed a business plan that required an enabling Act of Congress, since the Satellite Act of 1962 prohibited greater than 50 percent aggregate common carrier ownership in COMSAT.

In May of 1998, Bliliey's Bill H.R. 1872 passed as amended the House of Representatives and was then referred to the Senate committee, though it was not addressed by the Senate during the 105th Congress and thus failed to be enacted as law, which was a setback to Lockheed Martin's stated object of acquiring COMSAT. According to Sponyoe (1998), Lockheed Martin was "poised to make a 2.7 billion investment in COMSAT. . . . [and was] not pursuing this investment for the purpose of preserving the status quo at COMSAT – far from it. We want to buy COMSAT, transform it into a normal US commercial business operation, and integrate it into LMGT to form a strong new competitive entrant in the global telecommunications services marketplace." In addressing the fact that Lockheed Martin's acquisition of COMSAT would give it a significant interest in INTELSAT, Sponyoe (1998) stated that Lockheed Martin had "no intention of buying COMSAT to acquire an 18% share either in a 'mini-United Nations' or of a diminishing asset. . . . [W]hatever perceived advantages INTELSAT may or may not have in its current incarnation, these advantages are certainly not reflected in its steadily decreasing market share. . . .Indeed, INTELSAT's current position in the US-international market vis-à-vis other satellite and terrestrial competitors is so far from anything that could be accurately termed 'dominant' that I have to wonder whether its current structure might not pose a greater threat to itself than to its competitors. The INTELSAT Lockheed Martin wants to be part of is one that can soon be a viable commercial system that operates in a manner indistinguishable from any other commercial system. . . . This is why we see our combination with COMSAT as a means for achieving not only own business objectives, but major US policy objectives as well."

Diane Hinson, Vice President and General Counsel for INTELSAT, said in an interview on 4 May 1998, that INTELSAT Signatories were "a little baffled . . . at how the United States Congress purports to pass legislation that governs them" (Gruenwald, 1998a). Some INTELSAT officials, who supported the Bill's goals, opposed the unilateral dictation of which services INTELSAT could provide to U.S. customers, since H.R. 1872 proposed to restrict the ability of INTELSAT and INMARSAT to provide lucrative new satellite services, such as high speed Internet access or direct to home TV broadcasting in the U.S., if they failed to privatize (Gruenwald, 1998a). Although privatisation and reform of the telecommunications sector had diffused rapidly internationally in the 1990s, many states had yet to fully engage in this policy diffusion process, and PTT Signatories to INTELSAT were loath to allow competition to reduce the prices they charged consumers (Lynch, 1998). Sponyoe (1998), however, contended that "as a result of market incentives and the persistent market opening efforts of the U.S. government over many years, there is an *irreversible* (author's emphasis) trend toward market liberalization and privatization around the world." Sponyoe (1998) further charged that "the marketplace imperatives that compel this transformation are well understood by INTELSAT management and its *leading* (author's emphasis) Signatories." Indeed, due to pressure from both within and without the organization, the 24$^{th}$ INTELSAT Assembly of Parties in 1999 resolved to transform INTELSAT from a public intergovernmental treaty organization into a private corporation. In light of this resolution, the U.S. Department of State and some legislators warned that unilateral U.S. legislation mandating INTELSAT privatization would now be unnecessary and potentially counterproductive (Katkin, 2002).

Such views may account in part for that the fact that it took eighteen months, until September of 1999, for Lockheed Martin, frustrated by months of delay in securing the required regulatory approvals to complete the first phase of the merger, to finally gain FCC authorization to acquire, through its wholly owned subsidiary, Regulus, LLC, COMSAT Government Services, a COMSAT subsidiary, and to purchase up to 49 percent of COMSAT stock. Following these initial FCC regulatory approvals and COMSAT shareholder approval, the first phase of the transaction was completed, which entailed a tender offer for 49 percent of the outstanding shares of COMSAT common stock that was valued at approximately US$1.2 billion (http://www.spaceandtech.com/digest/sd2000-21/sd2000-21-005.shtml, 2000). President and CEO of COMSAT, Betty Alewine, hailed the deal as a 'strong pro-competitive merger' that would increase global telecommunications competition, since Lockheed Martin's space assets, including rocket and satellite building capabilities, in combination with COMSAT's satellite services network, would strengthen Lockheed Martin's position as an industry power and thereby increase its ability to compete with The Boeing Company (http://findarticles.com/p/articles/mi_m3457/is_20_11/ai_55938901, 1999).

The goal of S. 1328/H.R. 1872, the ORBIT Act, to promote competition in the international satellite communications market by encouraging the timely and pro-competitive privatization of INTELSAT and INMARSAT was strongly supported by the Clinton Administration. Thus, on 17 March 2000, the ORBIT Act, Pub. L. No. 106-180, which set forth specific criteria relating to the privatization of INTELSAT and INMARSAT, was enacted by Congress and signed by President Clinton. Then, following FCC authorization on 31 July 2000, COMSAT assigned its licenses and authorizations to Regulus, LLC, thereby enabling Lockheed Martin's successful purchase of the remaining shares of the COMSAT Corporation in a stock swap valued at US$790 million, as the second phase of the transaction, the one-for-one tax-free exchange of Lockheed Martin common stock for COMSAT common stock, was completed, and COMSAT stock ceased to be trade on August 3 (http://www.spaceandtech.com/digest/sd2000-21/sd2000-21-005.shtml, 2000). Owing to the lengthy approval process, however, COMSAT shareholders lost an estimated US$300 million in purchase value because of a drop in Lockheed Martin stock, allegedly due to the company's two-year toil to overcome complex congressional and regulatory barriers but also attributable to other problems within Lockheed Martin (http://www.spaceandtech.com/digest/sd2000-21/sd2000-21-005.shtml, 2000). A series of launch failures in 1999 destroyed approximately US$4 billion in rockets and payloads, which led to an inquiry that found poor management practices and quality-control problems that prompted the company to restructure and sell many non-core operations, since its profits were two-thirds lower in 1999 than the prior year (http://homepages.wmich.edu/~a2olexse/webresearch.htm, 2002). In 2000, the Pentagon reportedly bailed out Lockheed Martin by agreeing to buy 24 C-130J transports and awarding it other Pentagon contracts (http://homepages.wmich.edu/~a2olexse/webresearch.htm, 2002). Despite its internal difficulties, Lockheed Martin's perseverance enabled it to become the U.S. owner and the largest shareholder in INTELSAT and INMARSAT, which was fully privatized on 15 April 1999, as well as New Skies.

Lockheed Martin then proceeded to use its shareholding stature to vigorously advance the rapid and complete privatization of INTELSAT, which transpired on 18 July 2001, with the simultaneous creation of ITSO, by transferring substantially all of its assets and liabilities to a new company created for this purpose, Intelsat, Ltd., incorporated in Bermuda. Intelsat Global Sales & Marketing Ltd., headquartered and organized under the laws of England and Wales, promotes the company's varied service offerings, while the subsidiary, Intelsat L.L.C., a Delaware corporation, provides technical, marketing and business support services to Intelsat, Ltd. and its subsidiaries as well as files applications with the FCC for licenses to operate its Fixed Service Satellites (FSS) in the C- and Ku-bands, while the United Kingdom is responsible for licensing Broadcasting Satellite Services and FSS in Ka-bands. As an intergovernmental organization, INTELSAT was not subject to any given national licensing authority, but in its privatization, the unlicensed, international treaty organization became a U.S. telecommunications carrier, thus strengthening the U.S. position in the global marketplace as INTELSAT's 22 satellites and accompanying optimal orbital locations were transferred to the U.S. Registry under the procedures of the International Telecommunication Union. This complicated legal structure was principally designed as an optimum structure for tax purposes, but one must question its efficacy as well as the integrity of Lockheed Martin to keep its stated commitments in acquiring COMSAT to build a strong new competitive entrant in the global telecommunications services marketplace, for within five months of the privatization of INTELSAT, in December 2001, the company announced its decision to exit the telecommunication services business.

In the aftermath of the attack on the Twin Towers of the World Trade Center in New York and in light of continued overcapacity in the telecommunications industry and deteriorating market conditions in Central and South America, Lockheed Martin recommitted itself to its core business of U.S. defense, government information technology and homeland security, and thus eliminated the Lockheed Martin Global Telecommunications (LMGT) administrative structure and immediately reassigned certain of the former LMGT businesses and investments to other units of Lockheed Martin while initiating sales of the remaining operations, including its stake in Intelsat, Inmarsat, New Skies and other satellite ventures (http://www.spaceandtech.com/digest/flash2001/flash2001-106.shtml). At the time, Lockheed Martin took a $1.7 billion charge to write down the COMSAT purchase and, as the company began to dismantle its, arguably, expensive foray into commercial satellite services, it sought to delay Intelsat, Ltd.'s initial public equity offering, for otherwise it would have to sharply write down its Intelsat investment as it was carried on its books at a much higher value than that implicit in the IPO (de Selding, 2004).

## INTELSAT, LTD. – POST-PRIVATIZATION

### IPO Delays

One provision of the ORBIT Act required that the privatized company conduct an initial public offering of equity securities that would "substantially dilute" the aggregate ownership of the former Signatories of INTELSAT (ORBIT Act, S.376, 106 Cong., 2d Sess., 2000). These entities were either private telecommunications companies or governmental agencies of the countries party to the intergovernmental agreement that formally established INTELSAT, and these entities became shareholders of Intelsat, Ltd. upon its privatization. On 18 December 2003, Intelsat, Ltd. obtained an extension from the Federal Communications Commission to extend the date by which the company must complete an initial public equity offering from 31 December 2003 to 30 June 2004.

In February 2004, the company announced its intention to conduct an initial public offering of its ordinary shares in an amount of up to US$500 million, but then withdrew its planned IPO when an amendment to the ORBIT Act was signed into law on 18 May 2004, changing the date by which Intelsat, Ltd. needed to conduct an IPO to as late as 31 December 2005, thus giving the company the opportunity to explore other options, for which it retained Merrill Lynch and Company and Morgan Stanley (http://www.intelsat.com/aboutus/ press/release_details.aspx?year=2004&art=20040521_01_EN.xml&lang=en&footer=61, 2004). Senator Conrad R. Burns (R-Montana) introduced the legislation that permitted Intelsat, Ltd. to delay its IPO by up to 18 months because several U.S. investors in Intelsat, notably Lockheed Martin as the largest single shareholder with 24 percent of shares, stood to lose millions of dollars since the market conditions were not conducive for conducting a successful IPO (de Selding, 2004). The quandary Intelsat faced was that its near billion dollar purchase in March of 2004 of six North American satellites and related customer service contracts of Loral Space and Communications Corporation included services, such as direct-to-home TV broadcasting and high speed Internet access, that Intelsat was forbidden to offer in the U.S. before its IPO was completed under the terms of the ORBIT Act. Although the FCC granted Intelsat a six-month grace period to continue offering such services so as not to interfere with the Loral purchase, the acquisition placed pressure on Intelsat to complete an IPO as its projected revenues from the acquisition would be jeopardized.

## Zeus – Private Equity Acquisition of Intelsat

Another amendment to the ORBIT Act was thus signed into law on 25 October 2004, providing an alternative method by which Intelsat could comply with the privatization requirements of the Act, that is, the dilution of share ownership, by adding new subsections (F) and (G) to Section 621(5). Then, on 22 December 2004, the FCC issued an Order granting applications filed by Intelsat and Zeus Holdings Limited (Zeus), a company formed by a consortium of private equity funds advised by Apax Partners, Apollo Management, Madison Dearborn Partners and Permira, to transfer control of certain FCC authorizations from Intelsat to Zeus, as the Commission concluded that approval of the applications would serve the "public interest, convenience, and necessity", the nebulous terms employed by the Communications Act of 1934 and retained in subsequent legislation (Abelson, 2005). On the following day, 23 December 2004, in conjunction with this Order, Intelsat filed a Petition for Declaratory Ruling and attached certification, pursuant to Section 621(5)(F) of the ORBIT Act, that upon consummation of the transaction with Zeus, Intelsat would be in compliance with the said requirement as it would achieve substantial dilution of the aggregate amount of former signatory financial interest in Intelsat, which the FCC granted on 8 April 2005 (FCC, 2005). On 28 January 2005, Intelsat announced the successful closing of the amalgamation under Bermuda law of Intelsat and a subsidiary of Zeus, with the result that those individuals and organizations that held shares in Intelsat immediately prior to the closing of the transaction were entitled to receive US$18.75 in exchange for each share (Intelsat, 2005). Hence, the former intergovernmental treaty based satellite organization became a private equity acquisition, arguably to ensure the financial well-being of its largest shareholder, Lockheed Martin.

According to Pearlstein (2006), Zeus purchased Intelsat for about US$3 billion, putting up US$515 million and borrowing the rest, and then during 2005, the group paid itself two dividends, one being about US$340 million, which the company had to borrow, and the other being about US$198 million, which was financed from the company's free cash flow. The new private equity owners also paid themselves about US$70 million in management fees, which resulted in the four firms comprising Zeus essentially recouping all the cash they had invested in Intelsat while retaining virtually all of the stock by the end of 2005. Within a period of two years, the private equity owners, as Pearlstein (2006) writes, drained much of Intelsat's equity value as they engaged in a series of acquisitions which increased the company's debt to more than US$11 billion.

## Mergers and Acquisitions

With John Sponyoe, former CEO of LMGT, serving as the Chairman of Intelsat's Board of Directors upon its privatization, Intelsat, Ltd. boarded the merry-go-round of mergers and began to acquire components from, ironically, Lockheed Martin. In an effort to be more market-oriented and assemble a new ground-based infrastructure to complement its global satellite system, in March 2002, Intelsat signed an agreement to acquire the World Systems business unit of Lockheed Martin Corporation, including earth stations located in Clarksburg, Maryland and Paumalu, Hawaii, and also purchased the teleport facilities and related assets of Comsat Digital Teleport, Inc. (Mark, 2002). In building a Global Connectivity Solutions portfolio, Intelsat, while continuing to be a leading provider of satellite communication services offering video, data and voice connectivity, made other acquisitions as well, so that in May 2002, within a year of its privatization, it began providing end-to-end solutions through its newly established network of teleports, leased fiber and points of presence (POPs) around the globe. This global infrastructure enables Intelsat to earn revenue through the delivery of hybrid services that combine space segment, teleport facilities, cable and other land facilities for end-to-end telecommunications, though the company continues to earn revenue principally by selling satellite transponder capacity and seeks to sustain its leadership position in the fixed satellite services sector of the satellite industry.

To further expand its capabilities, in September 2002, Intelsat made an unsuccessful bid to acquire the recently privatized European operator, EUTELSAT, then, in May 2004, Intelsat acquired Lockheed Martin's COMSAT General business, which provides satellite-centric telecommunications services and equipment, concentrating on international fixed and mobile satellite systems for clients with quick response and high availability needs, for approximately US$90 million (Lockheed Martin, 2004). INTELSAT, as noted, also acquired the U.S. domestic satellite fleet operated by Loral Skynet in 2004. Although INTELSAT's privatization as a means of promoting competition was the stated purpose of the ORBIT Act, in actuality, the industry has since witnessed increased consolidation, with, of course, the approval of the FCC, as exemplified by Intelsat's acquisition of PanAmSat in July 2006 for US$6.4 billion, including US$3.2 assumed debt, creating in the U.S. the largest global satellite company with a fleet of 51 satellites.

## ITSO – THE RESIDUAL INTERGOVERNMENTAL ORGANIZATION

The state of indebtedness resulting from this merger was of serious concern to the International Telecommunications Satellite Organization (ITSO), the residual intergovernmental organization that monitors Intelsat's compliance in providing international public telecommunications services in conjunction with adherence to three core principles, namely, as stated in the ITSO Agreement, "to maintain global connectivity and global coverage; serve its lifeline connectivity customers, and provide non-discriminatory access to the Intelsat system" (ITSO, 2001a). ITSO must thus ensure that Intelsat maintain the technical interconnection capability to carry communications to and from any country or territory within and between the five regions, as defined by the ITU, of Africa, the Americas, Asia and the Pacific, Europe, and the Middle East. ITSO must also assure that all users and prospective users enjoy fair and equal opportunity to access the Intelsat system. Finally, ITSO must ensure Intelsat's adherence to its Lifeline Connectivity Obligations (LCO) customers, which includes those countries that are either low income, as classified by the World Bank, or have a low teledensity, as defined by the ITU. At the Intelsat $25^{th}$ Assembly of Parties in November 2000, 69 countries qualified for LCO protection under the income or teledensity eligibility criteria, and 41 additional countries or territories were identified as qualifying for "Petition Eligible" LCO Protection for either all or part of their international traffic. These countries then had until 1 August 2000, to petition the Assembly of Parties for LCO protection on the basis that either there was no cost effective alternative provider of a service equivalent to Intelsat's services or, among other criteria, were facing emergency conditions, such as an earthquake or war (Katkin, 2002). Also, new countries, created after 18 July 2001, were deemed eligible to join ITSO and apply for LCO protection.

According to ITSO (2000), LCO protection, which runs through July 2013 at which point it terminates regardless of the continuance of ITSO, consists of:

- Protection against price increases above a 2000 price level reference;
- Protection of the satellite capacity used for the LCO program; and
- Reduction of prices when the 15% trigger on a Pricing index is reached. The Pricing Index reflects, on an annual basis, the overall price increases or decreases of Intelsat customers without LCO protection.

ITSO has questioned Intelsat's commitment to the ITSO Agreement as the company has pared down its public service obligations and decreased satellite coverage in several sensitive regions, including Asia and Africa; and in an April 2006 filing to the U.S. Securities and Exchange Commission, Intelsat reported that nine percent of its US$3.8 billion in backlog consisted of Lifeline Connectivity contracts (de Selding, 2006a). Intelsat's rivals have no public service obligations, which, as de Selding (2006a) notes, represent a "ball and chain" that reduces Intelsat's competitiveness, especially for a company now owned by private equity investors and required to meet debt covenants.

ITSO contended that the increase in Intelsat's debt level due to the purchase of PanAmSat, to which it objected, but had no authority to prevent, made the company unstable and could thus trigger a bankruptcy filing or a creditor-forced sale of assets, and, as there were no legal covenants preventing Intelsat from divesting any of its satellites, the new owner would not be bound to adhere to the core principles established by the ITSO Agreement. In seeking conditions on the PanAmSat merger, ITSO thus petitioned the FCC to modify the space station licenses held by Intelsat for the use of certain orbital locations and associated radio frequency assignments to ensure that Intelsat and any successor satellite operator using the orbital locations and associated frequencies would be bound by the ITSO Agreement, essentially forcing Intelsat to assure that some of its assets would continue to be available for public-service use, even if owned by another company, in the event of a bankruptcy. ITSO asked that Intelsat place a lien, a letter of credit or some other binding guarantee on at least five satellites and orbital slots that could together continue to offer global coverage, but the FCC refused to consider the issue during the regulatory review of the proposed merger with PanAmSat, and suggested that ITSO's concerns could be more properly addressed in a separate proceeding initiated under Section 316 of the Communications Act (de Selding, 2006). ITSO has subsequently initiated this proceeding and it is still pending. Nonetheless, the ruling by the FCC in the PanAmSat merger context demonstrates the powerlessness rendered onto the international treaty organization in the wake of continued U.S. hegemony.

ITSO, which is organized along the lines of the former INTELSAT, albeit without retention of the Meeting of Signatories and Board of Governors, is comprised of an Assembly of Parties, that today consists of 148 member states which meet on a biannual basis; an Executive Organ, headed by the Director General, responsible to the Assembly of Parties; an ITSO Advisory Committee (IAC), which includes representatives of 19 Parties selected by the Assembly of Parties and other stakeholders; and a Panel of Legal Experts, 11 experts and 11 alternates elected by the Assembly of Parties, that resolves disputes in connection with the

Treaty Agreement (ITSO, 2000). The Assembly of Parties has the option to terminate the Treaty Agreement effective upon ITSO's 12th Anniversary in July 2013, which would require formal approval of two-thirds, effectively 100 countries, of the member states. ITSO's continuance beyond its projected lifetime of 12 years withstanding, it is crucial that this intergovernmental organisation, with its laudable public services mission, actually achieve tangible results and is not a mere paper tiger. The value of the organization, which views promoting the development of telecommunications services and competition as a vital means of securing international public telecommunications services to all countries, at the individual State level is questionable, as are the benefits developing countries derive from the role ITSO plays at the multilateral level in ITU and WTO fora, where it promotes equitable access to orbital/spectrum resources, *notably for commercial satellite systems* [author's emphasis] committing to international public service (ITSO, 2000a).

While ITSO's compromised role in such fora, straddling as it is between the needs of developing states and the private sector, is not surprising in this era of unprecedented corporate enclosure of the commons, from water resources and forests to biodiversity and knowledge, it is nonetheless the case that a privatized commons, in this case space, is no longer a commons, it is private property. The Outer Space Treaty, formally the Treaty on Principles Governing the Activities of States in the Exploration and Use of Outer Space, Including the Moon and Other Celestial Bodies, which represents the basic legal framework of international space law, states in Article I that outer space shall be used for the benefit and in the interest of all countries and shall be in the province of all mankind (Treaty, 1967). The root of the present conflict rests in policies emanating from hegemonic nation states and the essentially undemocratic global governing bodies of the World Bank, the International Monetary Fund, and the WTO, that reflect the shift from social values, represented in the Outer Space Treaty, to commercial values, as evident in the private equity ownership of Intelsat. The increasing levels of interdependence and vulnerability enhanced by technological change necessitates innovative, democratic forms of global political authority and governance, especially in regard to the commons, not private corporate control and influence in the multilateral institutions forming the global trade architecture whose strategies are facilitated by individual ascendant states, such as the U.S. Indeed, U.S. foreign policies have the potential to portend a greater threat to the ITSO agreement than Intelsat's debt.

## U.S. Foreign Policy

As Katkin (2002) notes, the former treaty-based INTELSAT was immune from U.S. trade policies, but the private, U.S.-licensed Intelsat, Ltd. is subject to such policies, thus, in the event that the provision of service to a lifeline customer, that is, a country dependent on Intelsat for carriage of their intercontinental telecommunications traffic, as in accord with the ITSO Agreement, should conflict with a subsequently enacted U.S. policy of law, the U.S. law will supersede Intelsat's prior obligations. In essence, lifeline countries, which include countries not allied with the U.S., such as Afghanistan, Sudan, North Korea, Somalia, and Cuba, could be cut off from the global telecommunications network by the imposition of any future U.S. economic trade sanctions. Although one might reason that the U.S. would forbear from such actions to minimize negative diplomatic ramifications, in 2002 the U.S. blocked deployment of a planned Ku-band Intelsat satellite (Intelsat APR-3), capable of providing strategic landmass coverage of China, Russia, India, and the Middle East, and to which SINOSAT, a commercial telecommunications agency of the Chinese government, had purchased the right to use six transponders. The Intelsat APR-3 satellite was to have been launched by the China Great Wall Industry Corporation, a Chinese government agency, but the U.S. State Department denied Intelsat's application for an export license that would have permitted the launch, though it did not violate any U.S. law or policy, and Intelsat, thus, terminated its deployment (Katkin, 2002), providing evidence that privatization does not eliminate politicization. Privatization has also not incurred stability in terms of ownership.

## BP Partners/Serafina – From Private Equity to Private Equity

The credit-rating service Moody's said on 8 January 2007 that Intelsat's refinancing transactions would not change Intelsat's continued high level of debt, thus leaving Intelsat's debt rating unchanged, as Moody's contended that, given Intelsat's private equity investor shareholder mix, the company was more likely to pay dividends than use its cash flow for debt reduction (de Selding, 2007). Indeed, Intelsat has decided to increase its capital spending plan for 2007, seeing more reasons to focus on investment than for reducing its debt, which was arguably a good decision for the company's investors as the debt did not deter the bidding for Intelsat when the company placed itself up for action. Several firms bid on the company, including EchoStar Communications and Liberty Media, which placed a joint bid, Loral Space & Communications Inc., the Carlyle Group, Providence Equity Partners, Australia's Macquarie Bank, Ltd. and the Blackstone Group LP, which reportedly offered US$6 billion to buy Intelsat in April of 2007, thus triggering the heated bidding war, which resulted in the sale of Intelsat to yet another private equity firm (de Selding, 2007a).

Although the practice of a private equity firm buying a company from another private equity firm, referred to as a secondary buyout, is still relatively rare, on June 20 2007, Intelsat announced the signing of a definitive agreement for the purchase of 76 percent of Intelsat, Ltd. by BP Partners, an international private equity firm based in the United Kingdom but organized under the laws of Guernsey, in a transaction valuing the company's equity at approximately US$5.03 billion, for which BP Partners received US$5.11 billion in financing commitments from Credit Suisse, which advised Intelsat during the transaction, Banc of America Securities, and Morgan Stanley (Intelsat, 2007). BP Partners is to assume Intelsat's debt, which, as of 31 March 2007, stood at approximately US$11.4 billion, and an additional US$3.85 billion in debt as a result of the anticipated financings at closing, thus the enterprise valuation implied by the transaction is approximately US$16.4 billion. Upon closing, the current shareholders of Intelsat, Zeus Holdings Limited (Zeus), comprised of Apax Partners, Apollo Management, Madison Dearborn Partners and Permira, are expected to receive approximately US$4.6 billion in cash while continuing to be a minority partner in Intelsat with 24 percent of the shares. Since Intelsat received early notification that the Department of Justice approved its sale to BP Partners, closing of the transaction is expected in the fourth quarter of 2007 or the first quarter of 2008. The deal is the largest in the 21 year history of BC Partners and its first acquisition of a company with U.S. headquarters (Goldfarb and Davies,

2007). BP Partners, which was formed in 1986, operates through integrated teams based in Geneva, Hamburg, London, Milan, Paris, and a recently opened office in New York.

To facilitate the transfer of control of Intelsat, Serafina Holdings Limited was recently formed in Bermuda and is effectively controlled by BP Partners, as it holds approximately 71 percent of the equity and voting interests in Serafina through 41 subsidiary investment funds (the "BCP Funds"). Two funds controlled by Silver Lake Group, L.L.C., a U.S.-based investment firm, will collectively hold approximately 16.84 percent of the equity interests in Serafina. Other investors in Serafina include Banc of America Capital Investors, V, L.P., CSFB Strategic Partners III, L.P., which is indirectly controlled by Credit Suisse, and 13 members of Intelsat's management team. Since virtually all of the foreign investment in Serafina would come from entities or individuals whose home markets are WTO Member states, the FCC, in accord with U.S. WTO commitments, must presume that such investment would serve the U.S. "public interest, convenience, and necessity," and thus approve Serafina's proposal to acquire Intelsat through its wholly-owned subsidiary, Serafina Acquisition Limited, also a Bermuda company. Upon summation of Serafina's acquisition of all of the equity and voting interests in Intelsat, Intelsat Bermuda, an indirect, wholly-owned subsidiary of Intelsat Holdings, Ltd., will transfer substantially all of its assets and liabilities to Intelsat Jackson Holdings Ltd., a new wholly-owned subsidiary of Intelsat Bermuda, including all of the existing indebtedness of Intelsat Bermuda, and the debt that ensues from the acquisition of Intelsat by Serafina will also be assigned (by contract, merger or otherwise) to Intelsat Bermuda (Intelsat and Serafina, 2007).

Although private equity firms often acquire a company to break it up into smaller units and thereby, not only recouping their initial investment, but earning an increased amount of capital, Intelsat is expected to remain intact, as BP Partners intends to retain Intelsat's existing management team and operational staff, including the current Chief Executive Officer, Dave McGlade, who joined Intelsat in April 2005. Its debt withstanding, Intelsat, as the world's leading provider of fixed satellite services worldwide, serving the media, network services and government customer sectors, is, according to its own accounts, considered the gold standard in satellite operations, with the hallmark of the system being its transponder reliability, with a 99.997 transponder availability rate since 1985 (Intelsat, 2003a). The company undoubtedly seeks to retain this record, while continuing to upgrade its fleet, thus ensuring its reputation in the field of satellite services.

## CONCLUSION

Although the ITU recognizes "the right to communicate" as fundamental, countless number of people around the world in developing countries remain without access to communications services. The private Intelsat, Ltd. seems unlikely to reduce that disparity. ITSO is hampered in its ability to ensure that commercial imperatives don't trump global egalitarianism in a market-driven environment, where, in spite of competition from fiber, there is growing demand for satellite services, particularly for broadband and Internet services. Geostationary communications satellites also continue to be the primary provider of telecommunications services in areas with low-density populations, particularly rural areas in developing countries, which persist in having few wire line networks because the return on investment is not commercially viable, but consolidation in the satellite industry has resulted in fewer players providing services. The admirable goals of the U.S. Communications Satellite Act of 1962 and the commendable aims of the Outer Space Treat have been lost in the neoliberal environment, and the world still requires a competitively neutral international transfer mechanism for directly subsidizing economically disadvantaged users in developing states.

The entire process of privatizing INTELSAT, including the sale of Intelsat to Zeus Holdings Limited, as well as the ensuing series of acquisitions by Intelsat, was largely undertaken to benefit U.S. corporations, notably Lockheed Martin, not to create a system better able to meet the telecommunications needs of the economically underprivileged countries of the South. If the wealth and socio-economic health of a state is defined by its ability to participate in the networked economy, these states can be considered impoverished, but ITSO wasn't endowed with the power to ensure that developing countries' technological dependency and marginalization in this era of accelerated technological change is in fact reduced. Rather, the privatized Intelsat, whose Lifeline Connectivity Obligations end in a few short years, insufficient time for developing countries to overcome patterns of dependency, represents the triumph of neoliberalism and its market-based mechanisms as a solution to socio-economic problems. Alternatives to the demise of intergovernmental organizations and the ascent of corporate power are limited, as Soederberg (2006) notes, by the authority structures underpinning the United Nations system, whose global governance, as evident in the WTO, has been complicit with the dominance of neoliberalism as the rules and laws of its legal regime, formulated by the U.S. and other hegemonic industrial countries, facilitate the reproduction of capitalist social relations and thus the continued privatization of the commons.

## REFERENCES

Abelson, D. (2005). Written Statement of Donald Abelson, Chief, International Bureau, Federal Communications Commission on the ORBIT Act: An Examination of Progress Made in Privatizing the Satellite Communications Marketplace Before the Subcommittee on Telecommunications and the Internet, Committee on Energy and Commerce, United States House of Representatives. Washington, D.C.: FCC.

Congram, C., Slye, P. and Glidden, P. (1999). Transformation at INTELSAT: sometimes the tortoise beats the hare. *Team Performance Management: An International Journal*, Vol. 5, No. 6, 194-203.

Cook, P. and Kirkpatrick, C. (1988). Privatisation in Less Developed Countries: An Overview. In *Privatisation in Less Developed Countries*. Eds. P. Cook and C. Kirkpatrick. Brighton: Wheatsheaf Books Ltd.

Cowhey, P. (1990). The International Telecommunications Regime: The Political Roots of Regimes for High Technology. *International Organization*, Vol. 44, No. 2, 169-199.

Cox, R. (1987). *Production, Power and World Order: Social Forces in the Making of History*. New York: Columbia University Press.

de Selding, P. (2004). IPO Delay Could Gut Revenues From Intelsat's New Fleet. *Space News*. Available: http://www.space.com/spacenews/archive04/ipoarch_051004.html. [7 April 2006].

de Selding, P. (2006). ITSO Again Seeks Measures To Hedge Against Intelsat Bankruptcy. *Space News*, Vol. 17, Issue 30, 8.

de Selding, P. (2006a). ITSO Questions Intelsat's Commitment to Public Service. *Space News*, Vol. 17, Issue 16, 16.

de Selding, P. (2007). Intelsat Juggling Satellites in Face of Launch Delay. *Space News*, Vol. 18, Issue 2, 6.

de Selding, P. (2007a). Intelsat Boosts Capital Spending Plan for 2007. *Space News*, Vol. 18, Issue 13, 5.

Einhorn, M. (1996). Intelsat: A Reform Proposal. *Space Communications*, Vol. 14, Issue 2, 137-145.

Einhorn, M. (1998). Restructuring and competition in international telecommunications: the case of INTELSAT. *Information Economics and Policy*, Vol. 10, Issue 2, 197-218.

Escobar, A. (1995). *Encountering Development The Making and Unmaking of the Third World*. Princeton, NJ: Princeton University Press.

Federal Communications Commission. (1999). In the Matter of Direct Access to the INTELSAT System. IB Docket No. 98-192. File No. 60-SAT-ISP-97. Report and Order. Adopted September 15, 1999.

Federal Communications Commission. (2005). Intelsat, Ltd. Files Petition for Declaratory Ruling and Certification Pursuant to Section 621(5)(F) of the Open-Market Reorganization for the Betterment of International Telecommunications Act, as amended (the "ORBIT Act"). Pleading Cycle Established IB Docket No. 05-18. DA 05-88.

Geri, L. (2001). New Public Management and the Reform of International Organizations. *International Review of Administrative Sciences*, Vol. 67, No. 3, 445-460.

Goldfarb, J. and Davies, M. (2007). Update 2-BC Partners buys Intelsat $4.6 billion majority stake. Reuters. Available: http://today.reuters.com/news/articleinvesting.aspx?type=bondsNews&storyID=2007-06-1. [23 June 2007].

Gruenwald, J. (1998). Panel Moves to Privatize Satellite Organizations. *Congressional Quarterly Weekly*, Vol. 56, Issue 12, 744.

Gruenwald, J. (1998a). Satellite Organization Pins Hopes on Senate to Stop Privatization Bill. *Congressional Quarterly Weekly*, Vol. 56, Issue 19, 1236.

Hahn, R. and Kroszner, R. (1990). Lost in Space U.S. International Satellite Communications Policy. *Regulation The CATO Review of Business and Government*, Vol. 13, Issue 2, 57-66.

Haller, L. and Sakazaki, M. (2000). *Commercial Space and United States National Security*. Prepared for the Commission to Assess United States National Security Space Management and Organization. Washington, D.C.: Commission to Assess United States National Security Space Management and Organization.

Intelsat. http://www.intelsat.com.

Intelsat, Ltd. (2003). Our History 1970s – A Decade of Expansion. Available: http://www.intelsat.com/aboutus/ourhistory/yr1970s.aspx. [19 January 2004].

Intelsat, Ltd. (2003a). Resources. Available: http://www.intelsat.com/resources/satellites.aspx. [19 January 2004].

Intelsat, Ltd. (2004). Intelsat Announces Decision to Withdraw Planned Initial Public Offering of Shares and Intention to Explore Strategic Alternatives. Available: http://www.iintelsat.com/aboutus/press/release_details.aspx?year=2004&art=20040521_01_EN.xml&lang=en&footer=61. [20 July 2004].

Intelsat, Ltd. (2005). Intelsat Announces Completion of Acquisition by Zeus Holdings Limited. Available: http://www.intelsat.com/aboutus/press/release_details.aspx?year=2005&art=20050128_01. [24 May 2005].

Intelsat, Ltd. (2007). Press Release – BC Partners to Acquire Majority of Intelsat. Available: http://www.spacered.com/newa/viewpr.rss.spacewire.html?pid=22894. [23 June 2007].

International Maritime Organization. (2006). Convention on the International Maritime Satellite Organization, 1976. Available: http://www.imo.org/Conventions/contents.asp?doc_id=674&topic_id=257. [23 January 2006].

Intelsat Holdings, Ltd. and Serafina Holdings, Ltd. (2007). Consolidated Application for Consent to Transfer Control of Holders of Title II and Title III Authorizations. 10 August.

ITSO. (2001). About Us. Available: http://www.itso.int/php_docs/tp11_itso.php?dc=aboutus. [19 January 2004].

ITSO. (2000). ITSO Treaty Agreement. Available: http://www.itso.int/php_docs/tp11_itso.php?dc=agreement. [19 January 2004].

ITSO. (2000a). Mission & Role. Available: http://www.itso.int/php_docs/tp11_itso.php?dc=mission. [19 January 2004].

ITSO. (2000b). Services. Available: http://216.119.123.56/dyn4000/dyn/docs/itso/tp11_itso.cfm?location=&id=345&link_src=HPL&lang=english. [15 August 2007].

Katkin, K. (2002). *Universal Global Interconnection After INTELSAT*. Presented at the 30th Research Conference on Communication, Information and Internet Policy, Washington, D.C.

Lyall, F. (1989). *Law and Space Telecommunication*. Dartmouth, Aldershot, Gower.

Lockheed Martin Corporation. (2004). Lockheed's COMSAT General to be Acquired by Intelsat. Available: http://www.defense-aerospace.com/cgi-bin/client/modele.pl?prod=38658&session=dae.21650007.1152057590.RKsA9sOa9dUA. [4 July 2006].

Lynch, M. (1998). Space Balls. *Reason*, Vol. 30, Issue 1, 14-15.

Mark, R. (2002). Intelsat to Acquire Lockheed Martin Unit. Available: http://www.internetnews.com/bus-news/article.php/992961. [4 July 2006].

Museum of Broadcast Communications. (2007). Communications Satellite Corporation. Available: http://www.museum.tv/archives/etv/C/htmlC/communication/communication.htm. [22 January 2007].

n.a. (1999). Congress Keeps Lockheed Martin-Comsat Merger On Hold Until Legislation Passes – Brief Article. Available: http://findarticles.com/p/articles/mi_m3457/is_20_11/ai_55938901. [7 April 2006].

n.a. (2002). Hoover's Company Profile Database – American Public Companies. Available: (http://homepages.wmich.edu/~a2olexse/webresearch.htm. [7 April 2006].

n.a. (2000). Lockheed Martin Completes Acquisition of Comsat Corporation. *SPACEandTECH Digest*. Andrews Space & Technology. Available: http://www.spaceandtech.com/digest/sd2000-21/sd2000-21-005.shtml. [7 April 2006].

n.a. (2001). Lockheed Martin Quits Telecom Business. *SPACEandTECH Digest*. Andrews Space & Technology. Available: http://www.spaceandtech.com/digest/flash2001/flash2001-106.shtml. [7 April 2006].

ORBIT Act. S.376. 106 Cong., 2d Sess. (2000).

Pearlstein, S. (2006). Sweet Deals Buried Intelsat in Debt. *The Washington Post*, August 18, 2006, D01. Available: http://www.washingtonpost.com/wp-dyn/content/article/2006/08/17/AR2006081701578_pf.html. [14 August 2007].

Pelton, J. and Edelson, B. (1997). Reinventing INTELSAT and INMARSAT. *Aerospace America*, Vol. 35, No. 1, 28-35.

Pfeufenberger, J. and Houthakker, H. (1998). Competition to International Satellite Communications Services. *Information Economics and Policy*, Vol. 10, NEED ISSUE NUMBER 403-430.

Shiva, V. (2005). *Earth Democracy*. Cambridge, MA: South End Press.

Spiegel, P. (1997). Lost in Space. *Forbes*, Vol. 160, Issue 6, 123.

Sponyoe, J. (1998). Statement of John Sponyoe, Chief Executive Officer, Lockheed Martin Global Telecommunications, Hearing on International Satellite Reform, Before the Subcommittee on Communications, Committee of Commerce, Science, and Transportation, United States Senate.

Soederberg, S. (2006). *Global Governance in Question*. London and Ann Arbor: Pluto Press.

Thussu, D.K. (2001). Lost in Space. *Foreign Policy*, Issue 124. Available: http://cassell.founders.howard.edu:2294/citation.asp?tb=1&_ug=dbs+aph+sid+8C51B260%2DD91F%2D44CD%2D9836%2D8D. [31 January 2004].

Treaty on Principles Governing the Activities of Sates in the Exploration and Use of Outer Space, Including the Moon and Other Celestial Bodies. (1967). Available: http://www.state.gov/t/ac/trt/5181.htm. [1 August 2007].

Vogler, J. (1995). *The Global Commons A Regime Analysis*. NY: John Wiley & Sons.

White House. (1999). *A National Security Strategy of Engagement and Enlargement*. Washington, D.C.: White House.

Zacher, M. and Sutton, B. (1996). *Governing Global Networks*. Cambridge, UK: Cambridge University Press.

In: Telecommunications Research Trends
Editors: H. F. Ulrich, E. P. Lehrmann, pp. 75-90

ISBN: 978-1-60456-158-6
© 2008 Nova Science Publishers, Inc.

*Chapter 4*

# CONTRACTING FOR MUNICIPAL BROADBAND SERVICES: A COMPARATIVE STUDY

*Carol Ting*
Ohio University, Athens, OH, USA

## ABSTRACT

In the past few years, a number of municipal governments have launched citywide broadband networks to promote economic development, digital inclusion and economization of city services. Such projects often face the challenges of building a sustainable business model and dealing with long-term transaction cost issues arising from information problems, market power and lock-in.

Building on the literature of transaction cost theory and history of public utility networks, this chapter examines franchise agreements of major municipal broadband projects and analyzes local governments' approaches to implementing their goals and addressing transaction cost issues with deployment of broadband networks. The study identifies significant variation in most of these major aspects and suggests subjects for long-term research.

## I. INTRODUCTION

As broadband networks become an increasingly important infrastructure for modern economies, national and local governments have been confronted with the problem of digital divide between different geographic markets and income groups: market forces and competition serve metropolitan and affluent areas well; however, broadband deployment in smaller markets tends to lag behind and statistics consistently show an income gap in broadband adoption rate.[1] In the past two years, many local governments have taken up the

---

[1] One recent example can be found at: http://www.pewinternet.org/pdfs/PIP_ICT_Typology.pdf. Retrieved on June 9, 2007.

task of expediting broadband deployment and providing community-wide broadband access to citizens through franchise agreements or public ownership.

Local governments undertaking community-wide broadband franchise projects face significant challenges. First, in most major markets there is plenty of competition; franchisees not only have to compete with existing providers, but they also have the unique burden of providing blanket area coverage. Creating a sustainable business model under such conditions is difficult. Moreover, as studies on history of traditional utility industries show, the complexity and stakes involved in these long-term franchises inevitably give rise to friction that results in transaction costs not explicitly included in the price tag of a project (e.g., cost of conflict resolution and costs caused by changes in market condition or technology).

Through examining the contract terms and conditions of a number of municipal franchises, this study explores how cities and municipalities[2] are addressing the goals of their broadband initiatives and the difficulties facing these public-private broadband initiatives. The paper is organized as follows. Section II describes goals of municipal broadband initiatives and related issues, particularly transaction cost issues arising under public-private cooperation. Section III analyzes cities' and municipalities' approaches to achieving their goals and addressing these issues through contracting. Section IV discusses the implications and subjects for further study. Section V concludes.

## II. MUNICIPAL BROADBAND FRANCHISING: GOALS AND ISSUES

### 2.1. Goals: Economic Development, Digital Inclusion and Economization of City Services

Economic development probably is the universal rationale for municipal broadband initiatives. Many cities consider a city-wide network an infrastructure that powers economic growth, and they view ubiquitous network access as a way to attract businesses and investment. The statement of mayor of Provo, Utah, one of the few US cities with a city-owned FTTH network, reflects this view:

"Bandwidth, access to the big pipes, broadband networks are as critical to this new century as roads, canals and railroads were to the 19th century, telephones and airports were in 20th. Fiber infrastructure is the economic development infrastructure investments that will be critical to the community success in the 21st century."[3]

To a great extent, big cities' broadband initiatives are driven by digital inclusion—providing Internet access to economically disadvantaged groups. As the Internet gradually becomes the de facto platform for consumer services and an essential tool for enhancing human capital, inequality in access to the Internet between income groups has raised concerns that this digital divide will reinforce inequality in opportunities for social mobility (DiMaggio et al., 2004). Many cities undertake their broadband initiatives partly as a response to such concerns—as will be discussed in Section III, franchises in major metropolitan areas tend to have a strong element of digital inclusion.

---

[2] In the following text, the terms cities and municipalities may be used interchangeably for succinctness.
[3] See http://www.lafayetteprofiber.com/FactCheck/ProvoMayor.html. Retrieved on June 9, 2007.

Lastly, many cities are already involved in operation of city-wide distributed networks (either through public ownership or leasing) for their daily operations such as internal data communication and public safety. Expanding their existing networks to public access may reduce the total cost of these services through exploitation of economies of scope (Gillett et al., 2004).

## 2.2. In Search of a Sustainable Business Model

At present, the first major challenge to municipal broadband franchising is establishing a sustainable business model—the recent unraveling of some high-profile projects due to franchisees' financial woes shows this harsh reality.[4]

The demand and market for a service in its early stage can be quite unpredictable, which poses high financial risk to franchisees. What makes broadband franchising especially challenging is the fact that mostbroadband franchises are overbuilds. Most cities (at least in the most profitable areas) are already served by DSL and cable[5] Internet access providers. A franchisee building the third pipe (be it FTTx[6] or wireless) will have to compete with these established companies, and the obligation to provide blanket coverage and low-cost service tends to add to the cost pressure on the franchisee.

The financial risk for a franchisee can be reduced if the franchise includes anchor tenancy requirements; that is, the franchisor must purchase other services (e.g. network services for public safety, traffic control, etc.) from the franchisee, which provides a stable revenue source. However, a persisting problem is that many cities, while talking about the potential of economization of city service, do not have a plan of integrating the public access and city service networks.

## 2.3. Transaction Cost Issues

Broadband and traditional utility networks share some important characteristics—long-term, high-level capital investment and strong scale economies—which often lead to tendencies towards monopoly and transaction cost issues. Drawing on transaction cost theory, historic studies on public ownership have shed much light on the comparative merits of franchising and public ownership in utility industries (water, electricity, and gas). First examined by Coase (1937), transaction cost is a general term covering implicit costs that are

---

[4] For example, Earthlink's recent down-sizing and re-evaluation of its municipal networking strategy led to the unraveling of San Francisco's WiFi plan and also cast doubt on its commitment to signed WiFi projects in some major cities. See "Municipal Wi-Fi: Reality bites." (Aug. 30, 2007). *TCMNet.com*. Retrieved on September 1st, 2007 from: http://www.tmcnet.com/usubmit/2007/08/30/2900204.htm. "S.F. citywide Wi-Fi plan fizzles as provider backs off." (Aug. 30, 2007). *San Francisco Chronicle*. Retrieved on September 1st, 2007 from: http://www.sfgate.com/cgi-bin/article.cgi?f=/c/a/2007/08/30/MNEJRRO70.DTL. "Earthlink's Wi-Fi delays put project in question." (Aug. 18, 2007). *Houston Chronicle*. Retrieved on September 1st, 2007 from: http://www.chron.com/disp/story.mpl/chronicle/5064518.html. These are not isolated cases; Milwaukee's contract with Midwest Fiber Network and MetroFi's contracts in Anchorage, Alaska and Toledo, Ohio are also on the list of recently aborted projects.

[5] Hybrid Fiber-Coaxial (HFC), this is the broadband access technology provided by the cable service operators.

[6] Fiber-to-the-Home/Premise. These include a number of fiber optic technologies that can deliver data rates higher than that of DSL and HFC.

not included in the price tag of a product/service but nonetheless must be incurred during the transaction process; such costs can range from the time and energy spent on searching for buyer/seller to the subtle but significant cost associated with conflict resolution. Williasm's elaboration (1986) on the effects of transaction costs on contracting and market organization has great influence on recent studies on ownership and regulation in the utility industry. For example, Priest (1993) investigates history of utility regulation in the U.S. and contends that regulation was introduced to address the many transaction cost issues related to long-term contracting and tendency towards consolidation. Jacobson (2000), on the other hand, argues that these same issues also gave rise to public ownership in utility industries.

Hansmann's framework (1988) of ownership analysis provides a systematic approach to analysis of transaction cost issues, and this paper adopts his framework to analyze the potential transaction cost issues with broadband networks. In Hansmann's framework, market contracting (or franchising in this context) are susceptible to issues stemming from information problems, market power and lock-in, and public ownership comes with the cost of monitoring government agents and collective decision making. The choice between franchising and public ownership should depend on which arrangement minimizes the total cost.

## *Information Problems*

Information problems that can result in transaction cost usually fall in three categories: demand/technology uncertainty, asymmetric information and monitoring service quality.

Demand/technology uncertainty often introduces friction to utility franchising when they result in the need to significantly modify contract terms or upgrade facility. In the early stage of a new technology/service, it is often difficult to predict how the technology and demand for it will evolve. As demand increases or technology advances, the original service may become inadequate or obsolete, but the franchisor and franchisee may have very different preferences over possible solutions. Governments tend to prefer large-scale expansion capable of accommodating long-term growth, but the franchisees, who are under the pressure of short term financial performance and are risk-averse, understandably prefer incremental expansion.[7] Such difference in interests is a potential source of conflicts that are costly to resolve. The long amortization horizon necessitated by the high fixed cost of utility networks makes it more difficult to address future contingencies with contracting, and disputes are more likely to result from such incomplete contracting. Such problem can be further complicated by perception of opportunistic behavior by either side or suspicion over abuse of market power.

Meanwhile, information asymmetry is also a major cause of dispute. Due to their specialization and experience, private companies usually have more accurate information about the true cost of a project than government officials do; admittedly, these companies also have incentive to exploit this information advantage. Such information asymmetry and incentive to abuse it often generate conflicts in a long-term contractual relationship. For instance, a franchisee's plead for renegotiation for more generous terms (such as extension of construction deadlines or price increases) is often met with suspicion, and head-on clashes

---

[7] *Infra* note 8.

between governments and franchisees due to mutual distrust were not uncommon in the history of utility networks.[8]

Difficulty in monitoring the service quality provided by the franchisee can potentially be another source of dispute. For example, assessing whether a city's waterworks can function adequately in a major conflagration is a complex task requiring high level of expertise and extensive testing, in such cases, disputes can easily arise unless assessments from impartial experts are available. The fire insurance industry has played exactly this role. In setting insurance rates for different markets, the fire insurance underwriters provide assessments of the fire-fighting capacity of waterworks (Smith, 1927), which reduces the monitoring cost for governments and potentially have prevented disputes that might otherwise have emerged.

All these information problems may be relevant to broadband franchises. With moderate growth in demand, network expansion or upgrade can be carried out incrementally, but if the "exa-flood" demand explosion[9] (Swanson, 2007) really occurs, extensive renegotiation of contract terms might be necessary. There is also considerable technology uncertainty over broadband as technologies are constantly improving and costs going down; technology leapfrogging can bring significant benefits as well as difficulty in franchise management.[10] This is especially true with wireless broadband technologies because WiFi, WiMax and newly proposed 3rd Generation and 4th Generation mobile access technologies are competing in multiple dimensions including cost, mobility and capacity (Cisco, 2007), which creates high technology uncertainty.

To some extent, the proliferation of broadband providers across markets may provide benchmarks that local governments can reference in their negotiation with private providers. However, information asymmetry (or the perception thereof) will always be present in a franchising relationship unless a franchisor can obtain good cost estimates from a project with similar size, technology and network architecture.

Lastly, monitoring broadband service quality is not always straight-forward. This Internet is made up of many networks managed by different entities, and typical Internet traffic usually travels across multiple networks. Therefore, the performance perceived by an end-user is determined by the accumulated delay/jitter introduced by all the networks that her traffic passes through, which means low connection speed is not necessarily due to problems within the local broadband access networks. It is possible to monitor the speed of individual connections, and carriers do sell connection services to enterprises with Service Level

---

[8] Jacobson's account (2000) of the clash between the government of San Francisco and its waterworks franchisee Spring Valley Water Company during 1897-1903 provides a good example for both demand uncertainty and asymmetric information. In response to growing population and demand for water, the City, aiming at long term growth, pushed for acquisition of the farther but abundant water source of the Hetch Hetchy Valley in the Sierra Nevada. On the other hand, Spring Valley was much more conservative and preferred to make incremental expansion from a nearby but more limited source. Meanwhile, suspecting overcharges by Spring Valley, the City also repeatedly ordered price cuts and took action to acquire its own waterworks facility. These moves were seen by Spring Valley as an exploitation of the investment it already made, and it fought back with threats to cut future investments for system improvement. The conflicts escalated and lasted for many years until the establishment of public ownership in the 1920s.

[9] This term refers to an explosion of demand for bandwidth (driven mostly by high definition video applications) that necessitates major upgrade and expansion of our current infrastructure.

[10] It is not hard to imagine a scenario where a government responds to a new technology by issuing a new franchise. Such a new franchise threatens the sunk investment made by existing franchisees and can be perceived as unjust taking by the government. Non-exclusivity clause can ease such concerns but it inevitably reduces incentive for private investment.

Agreements (SLAs) that guarantee connection speed, latency, down-time, etc. However, the cost of such monitoring is high and currently not practical for residential services.

## *Market Power*

Utility industries have a tendency towards monopoly because the high cost of passing lines/pipes through customer premises generates strong scale economies, which favor incumbents with larger customer base.[11] To mitigate market power and ensure adequate level of service, governments have long resorted to franchising, with which a government grants access to right-of-way in exchange for service commitments such as price limits, service quality and coverage in underserved areas. However, negotiation and enforcement of such contract terms are challenging and costly, and there are plenty of historic examples where disputes were fueled by accusations of abuse of market power (Jacobson, 2000;[12] Ting, forthcoming). On the other hand, while the utility industries are prone to monopoly, historic statistics of electricity prices suggest that (inter- and intra-product) competition may limit market power (Stigler & Friedland, 1962).

In the broadband context, the level of market power depends both on the market size and technology employed. In most cities, broadband customers usually have a number of choices from private providers, including DSL, HFC and wireless; major cities may also have FTTx and BPL[13] services. The competition between these technologies has a disciplining effect on the private providers. On the other hand, the potential for inter-product competition in smaller markets is limited by the lower demand. Therefore, potential abuse of market power is a more pressing concern in smaller markets.

Meanwhile, the choice between wireline and wireless networks also can have important implications in terms of the market power of a franchisee. Just like utility industries, for wireline broadband networks the need to run cables to each customer premise translates into high cost and strong scale economies.[14] Wireless broadband technologies such as WiFi also require sizable capital investment for backhaul networks and setup of access points, but relatively speaking the overall cost is lower[15] and scale economies weaker because there is no need to run wires to each household. The stronger scale economies associated with wireline networks generate a stronger tendency towards consolidation and greater market power concerns.[16]

---

[11] Market power in utility industries also arises from lock-in; here we follow Hansmann's framework and discuss lock-in in the next subsection.

[12] Jacobson, 2000. pp. 56, 88, 102, 113-114.

[13] Broadband-over-Power-Lines. This is a relatively new broadband access technology that carries data through electricity distribution networks.

[14] For example, Verizon estimates the cost of deploying its FiOS network to 18 millions US households by 2010 will be around $18 billion (net capital expense). See Verizon Communications Inc., FiOS Briefing Session, Sep. 27, 2006. Retrieved on Sep. 1, 2007 from: http://investor.verizon.com/news/20060927/20060927.pdf

[15] For example, the cost for Houston's wireless project is estimated to be around $50 million. See "Earthlink's Wi-Fi delays put project in question." (Aug. 18, 2007). *Houston Chronicle*. Retrieved on September 1st, 2007 from: http://www.chron.com/disp/story.mpl/chronicle/5064518.html. Cost comparison of such project and FiOS is not straight-forward because we do not know how Verizon calculate depreciation and other embedded assumptions, but these numbers still illustrate the different capital requirements for these technologies—The number of households in Houston in 2000 was 717,945 (http://www.hellohouston.com/Census.Cfm), and it will take extremely unrealistic assumptions of average number of people per household and depreciation parameters to make the average costs for these two networks comparable.

[16] The current telco-cableco duopoly exists because they started as totally different infrastructure networks serving different needs. The consolidation tendency may be more obvious when all types of application migrate to FTTx-based networks.

## Lock-in

Utility networks not only exhibit strong scale economies, but they are also characterized by high asset specificity (Williamson, 1986). The pipelines and cables are not only expensive to build, they are also highly specific assets that cannot be economically repurposed (network design is often optimized for the terrain and population density of the specific deployment locations). Such investment, once made, fundamentally transforms the relationship between the franchisor and franchisee. Before such investment is made, all those competing for the franchise are on equal-footing, but once investment is made and specific assets are acquired, both the franchisor and franchisee will incur high financial or political cost if they back out of the deal. The franchisor and franchisee are therefore locked in a long-term relationship; the relationship may even last beyond the amortization horizon of a network.[17] When disputes occur in such a relationship, they often become intertwined with perception of abuse of market power and/or information asymmetry and result in recurring conflicts that compromise the performance of the service.[18]

Again, the technology choice between wireline and wireless broadband affects the potential of lock-in. The higher capital requirements for wireline networks not only necessitates longer contract duration, it also makes the parties involved more reluctant to terminate a contract, whether by default or expiration. In contrast, wireless networks are less costly and they tend to have shorter amortization horizon, which reduces the impact and potential of lock-in. Compared to wireless networks, wireline networks' stronger tendency towards consolidation and lock-in translates into higher potential transaction cost; this might partially explain why the existing community FTTx networks tend to be publicly owned while franchising is the most common arrangement under which community wireless networks are built.[19]

The above issues associated with franchising can add to the total cost of a franchise project, and such costs must be weighed against the costs associated with public ownership in consideration of a local government's broadband initiatives. Public ownership entails the cost for monitoring government agents (to ensure performance), as well as the cost of seeking consensus and taking collective action (so that constituents with different interests can agree on establishing public ownership). These costs are not further discussed since our focus is on franchising.

---

[17] Theoretically the government can enter into a franchise agreement with a new franchisee once the old franchise expires, but duplication of such expensive networks may lead to price increase, which is not politically popular.

[18] The previous example of the long dispute between the waterworks franchisee in San Francisco and the City government can also be seen in this light. *Supra note 8.*

[19] Of course, another factor is whether private providers can be tapped to build such networks. Existing FTTx networks are concentrated in small-medium-sized markets where incentive for private investment is not particularly strong. For example, Provo, UT and Lafayette, LA both have population of around 100,000, and both claim that they resort to public ownership because private providers refused to deploy fiber optic networks in their community.

**Table 1. Basic information of selected cases**

|  | Anaheim | Farmers Branch | Minneapolis | Philadelphia | Portland | Riverside |
|---|---|---|---|---|---|---|
| Population (000) | 328[1] | 28 | 383 | 1,518[1] | 537 | 286 |
| Area[2] (square miles) | 44.3 | 12.5 | 58.7 | 135 | 134 | 85.6 |
| Business model | Subscription[3] | Subscription[3] Wholesale | Subscription[3] Wholesale City anchor tenancy | Subscription[3] Wholesale | Subscription[3] Wholesale City anchor tenancy | Subscription[3] Wholesale City anchor tenancy |
| Provider | Earthlink | NeoReach | US Internet | Earthlink | MetroFi | AT&T |

[1] Based on Census 2000 data.
[2] The actual coverage usually depends on which areas are designated as developed areas that must be covered.
[3] The franchisee can also sell advertising.
Source: RFPs and company press releases.

## III. COMPARISON OF MUNICIPAL BROADBAND CONTRACTS

This section examines a selection of municipal broadband franchise agreements and compares how municipalities address their major goals and issues with contracting. As most municipal broadband networks are still in a very early stage, this exploratory study focuses on identifying early adopters whose long-term performance may inform us of the effects of contractual terms on the public-private relationship. The franchises studied in this section are all beyond the pilot stage[20] and they must provide community-wide access to the public (as opposed to just for municipal and public safety uses). The franchises examined include Anaheim, Farmers Branch, TX, Minneapolis, Philadelphia, Portland, and Riverside, CA.[21] These cases vary significantly in terms of size of the community, business model and provider (Table 1). On the other hand, cost seems to be a major driver for early franchises as all these early adopters base their access network on WiFi technology.

It is worth noting upfront that since some of the goals and issues are closely related, the corresponding contractual measures taken by cities also tend to serve multiple purposes simultaneously.

---

[20] All these projects are in the process of expanding to provide city-wide coverage.
[21] The actual number of operational public-access networks is larger than six, but as a first step we start from projects whose franchise agreements are available online.

## 3.1. Economic Development, Digital Inclusion and Economization of City Services

Driven by the goals of promoting digital inclusion and economic development, most franchise agreements contain some requirements for low-cost service and city-wide coverage. On the other hand, through anchor tenancy arrangements some cities are integrating their city service functions and public access on the same network, which has the potential of realizing long-term cost savings.

In terms of coverage, we observe an array of different approaches. Cities with a strong focus on digital inclusion may take a more targeted approach and specify designated digital inclusion areas where the deployment is required to be at least at the same rate as other areas (e.g. Philadelphia). A less stringent alternative is the no-red-lining rule (e.g. Minneapolis). On the other hand, within this small sample, more cities leave out such specific requirement and demand only city-wide coverage without requirements for geographic breakdown or individual deployment schedule. This approach might give the franchisee more flexibility in planning construction so that they do not need to spread out the construction team across the city, which might realize some cost savings. On the other hand, franchises often allow for up to 2 years to complete construction in the coverage area, and it is possible that some low-income neighborhoods will have to wait just that long to have service.

Cities also have very different approaches on digital inclusion. Anaheim takes a laissez faire approach and does not include any price specification; Farmers Branch does not have services specifically for low-income users, either. Minneapolis' approach is quite unique in this sample: it requires the franchisee to contribute $500,000 and provide training service to its Digital Inclusion program personnel; free services will be available in designated community centers, and also citizens will have free access in some public locations on a time-limited basis. But the contract leaves out details about the uses of the Digital Inclusion Fund and whether there will be city-wide access service for low-income users. The other cities require either free service (supported by advertising) or low price service for qualified low-income users (e.g. Philadelphia), in addition to regular subscription service (Table 2). Listed data rates of such free services usually do not exceed 1Mbps/1Mbps (download/upload). While some contracts explicitly state that such free services are offered on a best-effort basis (which practically means individual connections slow down as more people access the network), none of the rest provides guarantee on connection speed, as discussed in 2.3.

**Table 2. Cities' free or low-cost subscription services**

|         | Anaheim | Farmers Branch     | Minneapolis | Philadelphia | Portland         | Riverside |
|---------|---------|--------------------|-------------|--------------|------------------|-----------|
| Service | n.a.    | 512Kbps/ 512Kbps   | n.a.[2]     | 1Mbps/1Mbps  | 1Mbps/ 256Kbps   | 512Kbps   |
| Price   | n.a.    | Free[1]            | n.a.[2]     | $9.95/month  | Free             | Free      |

[1] For government users only.
[2] Free services are available in designated community technology centers.
Source: RFPs and company press releases.

Both Minneapolis' and Riverside's contracts include anchor tenancy provisions, which contract the cities' public safety and internal communication operations to the franchisees. The revenues[22] from these city services support the build-out of the wireless infrastructure; this not only provides some financial certainty to the franchisee,[23] in the long term sharing the network capacity with public access services also has the potential to realize scope economies. Portland's approach is somewhat different; its contract does not stipulate the services the city will purchase in spite of the strong anchor tenancy component in it. On the other hand, the other contracts do not include any concrete plans of economization through network sharing. This lack of commitment on the part of municipalities and their reluctance to enter into anchor tenancy contracts[24] raise the question "why while cities are touting these wireless networks' benefit of lowering costs for city operations, they are reluctant to act on it?" The fact that some municipalities have public safety networks funded by agencies such as the Homeland Security Department[25] suggests one possible answer: there is a fragmentation over cities' network operations and coordination might be an obstacle to network integration. Lack of serious long-term planning and concerns over the quality issues with WiFi as an access technology (discussed in Section IV) might also be part of the puzzle.

## 3.2. Business Model

A few cities employ other business models such as relying on grants or their major employers to deploy wireless networks,[26] but these models tend to be location-specific and might work better for smaller communities. For most cities, franchising is still likely to be the main vehicle for deploying city-wide networks.

Under the franchise model, the city can choose to establish a non-profit organization to manage the operation of the franchise network. Compared to direct government involvement, the main benefit of this approach is to free city officials from the burden of implementing the contract. This is the model chosen by Philadelphia, which established Wireless Philadelphia to implement the contract with Earthlink and its Digital Inclusion program. So far, the effect of the choice of non-profit manager versus direct government involvement is not obvious, and the major difficulty regarding business model is mainly revenue sources, regardless of who is in charge of implementing the contract.

As mentioned in the previous sub-section, whether the contract includes anchor tenancy arrangements is the major difference in terms of revenue source. So far this business model seems to work more smoothly as it provides some financial assurance to franchisees, and the city service components also give more room for the two sides to experiment and work out

---

[22] In the case of Riverside, the city waives permit and attachment fees for using city assets as well as costs for power as in-kind contribution. Farmers Branch also waived such charges in exchange for free service to city employees and associates.

[23] For example, Minneapolis' anchor tenant payment is $1.25 million per year and Riverside will pay $4 million during the initial five-year contract term.

[24] So far, contracts with anchor tenancy provisions are still the exceptions rather than the norm. A few agreements recently fell apart because the cities refused to become anchor tenants of the networks; e.g. Chicago and Anchorage. See Is the bloom off municipal WiFi? Chicago scraps public wireless network. (Aug. 28, 2007) *NetworkWorld blog.* Retrieved on Sep. 10 from: http://www.networkworld.com/community/node/18848

[25] Anaheim's public safety network funded by the Homeland Security Department is an example.

[26] Ellison, C. (June, 2007.) Finding The Best Business Model. *MuniWireless Magazine.* Retrieved on Sep. 10 from: http://www.muniwireless.com/article/articleview/6135/1/3/

issues at this time of uncertainty. For example, Minneapolis' WiFi network is reported to have played a critical role in the relief efforts for the I-35 bridge collapse;[27] one can view such news as self-promotion by city officials and the franchisee; nevertheless such perception indeed helps to create a favorable condition for the business model to work.

## 3.3. Transaction Cost Issues

### *Information Problems*

In order to address issues that may arise from demand/technology uncertainty, most franchise agreements contain provisions that allow the parties to make changes and adjustments to network and product specifications. However, mechanisms and the level of details (in terms of procedures and scope of issues covered under such provisions) vary significantly. On the one end of the spectrum, there is the institutional approach of joint governance through committees made up of equal number of representatives from both sides (e.g. Philadelphia and Anaheim, both of which are contracting with Earthlink). On the other end, some contracts contain little information about how to reach consensus about adjustments and changes (e.g. Farmers Branch). Theoretically, and also historically, disagreement over changes (such as price increase or upgrades) in the long run tend to lead to more formal procedures that resemble regulation (Priest, 1993), and it will be interesting to see whether a similar pattern will emerge here.

Although most of the contracts we studied do not address specific types of uncertainties, some do try to anticipate demand growth and technology changes more specifically. For instance, Wireless Philadelphia stipulates that "if a default […] is caused in whole or in part by rapid growth of the number of ISPs and/or subscribing users then all of the applicable cure periods shall be twice as long […]" Portland also specifically states that the franchisee has obligation to expand service as the city annexes and populates new developments. Although no contract specifically addresses the possibility of an explosion in bandwidth consumption, most contracts contain general language based on which the franchisor can initiate a request for adjustment. For example, Anaheim's contract states that "From time to time, the City and EarthLink may amend the Service Level Agreement in response to technology or market factors as the City and EarthLink mutually agree in writing, based upon recommendations by the Steering Committee."

In anticipation of technological progress, Minneapolis requires that the franchisee "…to maintain the highest standard of service reasonably available and economically feasible such that the entire Network will have been totally refreshed within 5 years of its inception if such a refresh is technically available and economically sound. USIW [franchisee] will upgrade the network periodically before this and will continue to provide upgrades after the 5 year period…" Riverside also requires renegotiation for network upgrade upon the expiration of the initial 5-year term. Most other cities, on the other hand, only use more general terms such as "updated with industry standard equipment", leaving the upgrade timeframe to the discretion of the managers of the program. Although all these conditions are subject to

---

[27] Reardon, M. (Aug. 8, 2007). Citywide Wi-Fi network put to test in Minneapolis. *CNet News.com*. Retrieved on Sep. 10 from: http://www.news.com/Citywide+Wi-Fi+network+put+to+test+in+Minneapolis/2100-7351_3-6201561.html?tag=item

interpretation, Minneapolis' approach seems to provide more certainty and how such different approach affect future investment is also worth watching.

The perennial information asymmetry problem is usually addressed by reserving the right for franchisors to inspect franchisees' books with prior notice, and some also specifically stipulate the right to audit franchisees' accounts at franchisors' own expense.

In dealing with the problem of monitoring service quality, franchisees usually are required to provide records on network operation and service quality. In general, most contracts require such reports on a periodic basis (e.g., Philadelphia requires monthly reports regarding system performance as specified in Service Level Agreement, subscriber satisfaction, market penetration and low income subscriber penetration rates).

### *Market Power*

In general, cities often seek to address market power concerns by specifying requirements on the following items: (1) price, (2) service quality (connection speed, system reliability, etc.), (3) open access requirements and (4) most favored customer clause.

(1) *Price.* Cities with anchor tenant agreements often commit in advance the annual payment over a period of time with price escalation limitations (e.g. Minneapolis and Riverside). Portland, on the other hand, does not pre-specify any annual payment but instead using pricing restriction on individual service orders such as guaranteed maximum price, fixed price and cost overrun warning. Pricing structure for retail subscription services in these contracts tend to be simple. In addition to free, ad-supported service, all of them include subscription services, prices of which are pre-specified based on commercial services at comparable speeds (Table 3).

(2) *Service quality.* Although speed is also a major concern for end users, guarantee of connection speed is not economically feasible at the moment. As mentioned in Section II, monitoring quality of network access is not a trivial task, and guaranteeing connection speed is more difficult—even commercial wireline broadband service providers do not guarantee data rate for retail service. Given the difficulty in monitoring and guaranteeing quality of wireless connection, Service Level Agreements usually focus on coverage, system reliability, customer support and network maintenance instead of connection speed, which basically is provided on a best-effort basis whether explicitly stated or not. Since quality requirement is not location-specific, these requirements seem to be far more uniform than any other dimension of franchising contracts.[28]

(3) *Open access.* With Anaheim's exclusive contract with Earthlink as an exception, all the cities studied seek to promote competition through open access provisions requiring franchisees to provide wholesale services to unaffiliated ISPs. There is substantial variation in their wholesale pricing policy, though: for example, Riverside stipulates wholesale price at $12/user/month (subject to annual review). Other cities in general leave wholesale pricing to the franchisee's discretion. Same can be said about interconnection policies: most do not spell out arrangements required for

---

[28] For example, here are the typical Service Level Agreement items: 95% outdoor and 90% indoor coverage, with 99.9% reliability in the access networks and 99.99% in the backhaul networks, and 24x7 technical support. Some also include requirements of systems availability and system down time.

interconnection between the franchisee and retail ISPs, but some do attach strings to franchisee's wholesale service: Portland requires the franchisee to provide Layer 2 handoff and Virtual ISP options[29] to wholesale customers; Minneapolis, on the other hand, also requires open access to hot spot vendors, as well as net neutrality rules prohibiting the franchisee from limiting bandwidth, content and practices aiming at hampering competition.

(4) *Most favored customer.* Some cities seek most favored customer status so that their terms and rates will automatically match the best terms the franchisee currently offers in similar franchises. For example, Portland's contract stipulates that "[the] City shall be offered the Licensee's then-current, most-competitive rates which may represent reduced cost to the City." Farmers Branch requires that the franchisee must provide the same level of benefits or rates if it should enter into a franchise agreement that provides more favorable terms to similar municipalities in the same county (Dallas County, TX). This is an interesting approach, although its overall effect is not clear. Obviously, cities that adopt these provisions first will not be worse off, but such provisions might reduce the incentive for franchisees to offer better terms to other franchisors. If such provisions are widely adopted, they may have the potential negative effect of dampening competition.

## *Lock-in*

With more competition and lower capital investment, the lock-in problem is a less pressing concern for wireless networks. As expected, duration of wireless franchises usually is shorter than that of traditional utility networks (where 20-50 years is common) and cable franchises (commonly 15 years). Among the franchises studied, the initial term is typically 5 or 10 years, with renewal expectancy (most commonly two additional 5-year terms). Most contracts also set forth conditions for termination, which usually kicks in after a period of substantial failure by one party to fulfill its obligations.

A related issue is conversion to public ownership. Some franchise agreements (e.g. Riverside, Minneapolis) also include provisions that give the franchisor first right to purchase the network in the event of an ownership transfer, as long as the franchisor's offer is at least equal to other bids in price and terms. Conversion to public ownership is always an option whether it is specified in a franchise agreement or not, but Riverside's approach may provide greater assurance to the city: "Should AT&T thereafter receive a bona fide offer from another entity for the Network, AT&T will allow City submit an additional bid for such Network on terms and conditions that exceed, taken in the aggregate, that of the other entity and, if City does submit such a bid, then AT&T shall accept such bid." Such assurance can be especially important if the network is also used for mission-critical services such as public safety.

---

[29] The difference is that with the Layer 2 handoff option, the wholesale customer (an ISP) needs to have its own network and equipment, which may allow the ISP to provide its own value added service. The Virtual ISP option, on the other hand, does not require the ISP to involve in network and traffic management at all; the ISP in this case is just a marketer of the same service provided by the franchisee.

### Table 3. Retail services and prices

| Anaheim | Farmers Branch | Minneapolis | Philadelphia | Portland | Riverside |
|---|---|---|---|---|---|
| 1Mbps/1Mbps @ 6.95/month | 512Kbps/128Kbps @ $29.95/month or $3.95/hour, $5.95/day and $9.95/week | 1Mbps/1Mbps @ $19.95/month 3Mbps/1Mbps @ $29.95/month 6Mbps/1Mbps @ $35.95/month | 1Mbps/1Mbps @ $19.95/month or 3Mbps/1Mbps @ $21.95/month | 1Mbps/256 Kbps @ $19.95/month | 1Mbps for $7.99/day and $15.99/week |

Source: Contracts and company websites.

## IV. Discussion and Future Studies

As municipal broadband networks are still in a very early stage, identifying the approaches taken by municipalities in addressing their goals and dealing with difficulties provides merely a glimpse of the current state of affairs; nevertheless, such study lays the groundwork for long-term observation, which can further our understanding of how local governments' involvement can shape the development of capital-intensive infrastructure networks.

Some interesting questions can be asked at this point. First, to what extent is the recent setback in municipal networks a result of using the "wrong" technology? A lot of the complaints about municipal wireless services focus on spotty coverage and unstable signal quality,[30] but these issues to a great extent was a result of using a technology that is probably stretched beyond its limits. A major driver of these wireless franchises is the low cost of WiFi technology, which unfortunately can be offset by issues associated with performance. WiFi technology was designed to provide short haul, low-power, and line-of-sight [31] communication. It is cheap because it uses the 2.4 GHz unlicensed band and because the large user-base drives down equipment cost, but the short haul, low-power, and line-of-sight nature severely limits its capability as a last mile access technology. While mobile cells can be up to a few miles in radius, WiFi signals are typically confined within a few hundred feet to an access point. Hilly terrain and high density of high-rise buildings are additional problems as they tend to negatively impact the reception, connection speed and signal quality. These issues can severely degrade service quality, and manufacturers are constantly tweaking the technology; some are turning to WiMax, a more powerful version of WiFi that has longer reach and better ability to penetrate objects and structures. If WiMax does eliminate the performance issues, the mobility it offers may differentiate itself and gives it a fair chance to compete with DSL and cable services. How cities and franchisees use these new technologies to solve their problems, and what are the accompanying institutional issues (e.g. contract re-negotiation) is therefore an important issue to watch.

Second, what is the merit of the franchise overbuild approach? Historically, most franchise overbuilds are either short-lived (as in electricity) or did not even materialize (as in

---

[30] For example, see Municipal Wi-Fi: a failure to communicate. (May 21, 2007) *Businessweek.com*. Retrieved on Sep. 10, 2007 from: http://www.businessweek.com/print/magazine/content/07_21/b4035084.htm?chan=gl

[31] Many technologies in the microwave range require line-of-sight, which means that there must be an unobstructed path between a sender and a receiver.

cable TV). Given current market condition and technology costs, to the extent that a franchisee is providing a close substitute to existing broadband services, the cost pressure from its obligation to provide blanket coverage and low-cost service is likely to plague the venture because the franchisee cannot charge its retail customers more to subsidize the low-cost services. On the other hand, the equation may change with an anchor tenancy arrangement, which may allow some subsidization to take place. But in the case of a pure subscription/advertising model, the current market condition seems to suggest that the overbuild model may not serve the stated goals of these broadband projects very well.

The recent financial woes at Earthlink (and others to a lesser extent) have cast serious doubt on the future of municipal broadband initiatives. Such development actually provides an opportunity to study the institutional issues of infrastructure franchising. Cases similar to the Earthlink situation was not uncommon in the history of utility industries, although detailed historic records are hard to come by, we do know that transaction cost issues often arise in these situations. Regardless of the impact on the momentum for municipal networks, studies on how events unfold in a franchising relationship where the franchisee is under heavy financial pressure will likely provide additional insight into the effect of transactional cost issues in infrastructure industries. Similarly, the long-term development of projects that are already operational also provides valuable information—they offer a natural experiment with which we can observe the effect of those very different approaches taken by cites to implement the goals and address transaction cost issues.

## V. CONCLUSION

Economic development, digital inclusion and economization of municipal services are the three major driving forces behind the municipal broadband initiatives launched in recent years. Due to regulatory restrictions and opposition from the private sector, most of these initiatives are implemented through franchising agreements between local governments and private providers, and they often face the challenges of creating a sustainable business model and dealing with transaction cost issues stemming from information problems, market power and lock-in. This study examines how municipalities address the goals of municipal broadband franchises and difficulties through contracting. It is found that, even within a small sample, cities have taken quite different approaches to addressing these issues, and the differences raise interesting questions for future studies.

First is the effect of technology on the success of broadband franchising: current problems with broadband franchises seem to be closely related to the limits of WiFi technology. How will new technologies change this picture and does such changes create any institutional issues? Second, an examination of franchise agreements shows an inconsistency between the rhetoric of economization of city services and the actions actually taken. If cities do see a cost advantage in integrating their own network operations with public broadband access service, what is preventing them from doing so? Lastly, the franchise agreements differ in terms of what goals they serve, measures taken to serve these goals, as well as how they address potential transaction cost issues. Studies on the long-term effect of these differences can provide important insights to further our understanding of the role of government in the expansion of capital-intensive infrastructure networks.

# REFERENCES

Cisco Systems (2007). Rise of the 4G network enabling the Internet everywhere experience. White paper. Available at: http://www.cisco.com/en/US/netsol/ns704/networking_solutions_white_paper0900aecd805c247c.shtml

Coase, R. (1937). The nature of the firm. *Economica*, 4(16), 386-405.

DiMaggio, P., Hargittai, E., Celeste, C. & Shafer, S. (2004). Digital inequality: from unequal access to differentiated use. *Social Inequality*. Edited by Kathryn Neckerman. New York: Russell Sage Foundation. pp. 355-400.

Gillett, S., Lehr, W. & Osorio, C. (2004). Local government broadband initiatives. *Telecommunications Policy*, 28(7-8), 537-558.

Hansmann, H. (1988). Ownership of the firm. *Journal of Law, Economics, and Organization*, 4, 267-305.

Jacobson, C. (2000). *Ties that bind: Economic and political dilemmas of urban utility networks, 1800–1990*. Pittsburgh, Pa.: University of Pittsburgh Press.

Priest, G. (1993). The origins of utility regulation and the 'theories of regulation debate.' *Journal Of Law & Economics*, 36(2), 289-24.

Smith, H. (1927). The fire prevention work of stock fire insurance companies. Annals of the American Academy of Political and Social Science. 130, 103-107.

Stigler, G. & Friedland, C. (1962). What can regulators regulate? The case of electricity. Journal of Law and Economics, 5(1), 1-16.

Swanson, B. (Jan. 20, 2007). The coming Exaflood. *Wall Street Journal*.

Ting, C. (forthcoming). Municipal broadband initiatives: lessons from history. In *The Economics of Digital Markets*, edited by Gary Madden. Springer-Verlag.

Williamson, O. (1986). The economic institutions of capitalism. New York: The Free Press.

In: Telecommunications Research Trends
Editors: H. F. Ulrich, E. P. Lehrmann, pp. 91-106
ISBN: 978-1-60456-158-6
© 2008 Nova Science Publishers, Inc.

*Chapter 5*

# INFLUENCING INFRASTRUCTURE PERFORMANCE THROUGH CROSS-BORDER NETWORKS OF REGULATORY AGENCIES

### *Jacqueline Horrall*[*]
University of Pittsburg at Greensburg,
Greensburg, PA, USA

#### ABSTRACT

Increasingly, networks of sectoral regulatory agencies in telecommunications, energy, transportation, and water have been providing regional public goods (RPGs). Developing countries especially stand to benefit from shared resources permitted by the provision of RPGs. RPGs are typically facilitated by regional networks and include data sharing for benchmarking, best practice techniques, capacity building and training, development studies , and the facilitation of events and meetings. The development of transnational networks and the RPGs they provide appears to be a strong answer that would encourage regional infrastructure development. The effects would be most noticeable in developing countries that face capacity, resource and financial and other constraints to national infrastructure development. The success of cross-border provision of public goods, however, may prove to be challenging, because of issues that arise with the provision of public goods. These problems include adverse selection, moral hazard, the prisoners' dilemma, and free-rider problems. Generally, any kind of problems having to do with collective action agreement, such as agreements on data collection standards, cross-country conflict of regulatory policies and legislation form noticeable obstacles to successfully providing cross-border public goods. Improving the provision of cross-border public goods may only be achieved by tackling provision problems that are not only financially related, but from the production end as well.

---

[*] Assistant Professor of Economics (University of Pittsburg at Greensburg). Contact corresponding author at jhorrall@pitt.edu

**Keywords:** Regional public goods, regulatory networks, collective action, provision problems

## 1. INTRODUCTION

Infrastructure development in the form of telecommunications, energy and water are essential to encourage and maintain economic development. Over the last decade and extending into the current one, developing countries have focused increasing attention and resources on improving and reforming these sectors as a means of developing sustainable growth within each nation. Much emphasis has been placed on improving the physical aspect of networks, but as much effort has also been placed on providing autonomous, transparent, and effective regulation of these traditionally public utilities once they were privatized. According to Brown, Stern, Tenenbaum and Gencer (2006, p. xi), between 1990 and 2006, more than 200 regulatory commissions were created worldwide. Traditionally, these utility regulators were national regulators whose jurisdiction did not extend beyond the nation's geographical limits. In addition advances in sectoral utility provision make it possible to develop effective and strong network infrastructure in developing countries. For instance, advances in technology have created the opportunity for developing nations to create effective information infrastructure, and advances in wireless telecommunications technologies are offering opportunities to build interconnections with parts of the world that could not formerly afford to develop or build access to basic telecommunications. The creation of an effective cross-boarder infrastructure, however, requires the resolution of certain, cultural, political, sociological and economic matters. (Riggs, 1996) Included among these is the cross-border regulation of RPGs. Developing countries especially stand to benefit from shared resources permitted by the provision of RPGs. Regional institutions are perceived to be better able to overcome capacity problems, informational and financial shortcomings by pooling and sharing resources. Even in the absence of significant resource limitations, countries could benefit from regional cooperation, because globalization has resulted in greater interdependence among people across borders. (Kaul, Conçeeicao, Le Goulden, and Mendoza, 2003) For instance cellular phones could not be used internationally without the harmonization of standards and digital formats that permit network compatibility. In 1994 the Southern Africa Transport and Communications – Technical Unit (SATCC-TU) in recognition of the importance of enhanced regional cooperation began an initiative to encourage greater harmonization among member countries, with a view to developing common standards, developing and maintaining common facilities, sharing expertise, and generally moving towards an environment that would enable regional integration of markets to progress. (Goulden and Msimang 2005). RPGs are typically facilitated by regional networks and include data sharing for benchmarking, best practice techniques, capacity building and training, development studies, and the facilitation of events and meetings. The development of transnational regulatory networks and the RPGs they provide appears to be a strong answer that would encourage regional infrastructure development. Sandler (2006) provides a detailed description of the supply of transnational and regional public goods, but the role of regional regulatory networks and the goods they provide are not as well documented.[1] The effects would be most noticeable in developing countries that face capacity, resource and financial and other constraints to national infrastructure development. The success of cross-border provision of public goods, however, may prove to be challenging, because of issues that arise with the provision of public goods. These problems include adverse selection, moral hazard, the prisoners' dilemma, and free-rider problems. Generally, any kind of problem having to do with collective action agreement, such as agreements on data collection standards, cross-country conflict of regulatory policies and legislation form noticeable obstacles to successfully providing cross-border public goods. Improving the provision of cross-border public goods may only be achieved by tackling provision problems are not only financially related, but technical (production related) in nature as well.

While there is a lot of research on the issues associated with cross-border public goods, more focus has been given to global public goods, and where RPGs are discussed, they are not usually in relation to infrastructure related products. The use of collective action in the provision services, and in the use of resources are becoming very attractive particularly to developing countries with less than ideal telecommunications, water, transportation and energy sectors. As a case in point, in 2005 a workshop called Knowledge Sharing for Development: Africa Regional Program was held in Cairo, Egypt. The purpose of the workshop was to explore tools for research communication and knowledge sharing challenges, and to build relationships among professionals with similar interests in research communication and knowledge sharing. Included in the workshop, was a session on how regional and international networks function. (Global Development Network, 2005). Improving our knowledge of RPGs and the challenges they face is essential to improve production and supply within regions that could benefit most from cross-border collaboration of this nature.[2] A discussion of these infrastructure related RPGs is necessary to broaden the awareness of some of the issues that are specific to these types of goods and to encourage further attention and discussion.

The rest of the paper is as follows: Section 2 discusses the need for regional regulatory networks and Section 3 clarifies the definition of RPGs. Section 4 and 5 discuss the financial and technical problems associated with the provision of infrastructure related RPGs and presents the conclusion.

---

[1] According to Berg and Horrall (2007), since 1990 17 associations that provide RPGs have been formed including SATRC (South Asian Telecommunications Regulators' Council), IRG (Independent Regulators Group), ARIAE (Asociación Iberoamericana de Entidades Reguladoras de la Energía, Latin-American Association of Regulatory Agencies for Energy),TRASA (Telecommunications Regulators Association of Southern Africa), Regulatel (Foro Latinoamericano de Entes Reguladores de Telecomunicaciones) SAFIR (South Asia Forum for Infrastructure Regulation), AFUR (African Forum for Utility Regulators) CEER (Council of European Energy Regulators), ERRA(Energy Regulators Regional Association), ADERASA (Association of Water and Sanitation Regulatory Entities of the Americas), OOCUR (Organisation of Caribbean Utility Regulators), ERG (European Regulators Group for Electronic Communications Networks and Services), **ARICEA (Association of Regulators for Information and Communication Services of Eastern and Southern Africa) with COMESA**, EAPIRF(East Asia and Pacific Infrastructure Regulatory Forum), and RERA (Regional Electricity Regulators Association), Southern Africa.

[2] As Goulden and Msimang (2005) explain, the ability of regulatory agencies to effectively regulate newly liberalized (telecom) markets depends on the capacity of each country. Given the newness of regulation in many Southern Africa Development Community (SADC) countries, the challenge of developing the human resource capacity to effectively regulate the is acute, making the adoption of a regional approach to dealing with the problem essential.

## 2. The Need for Regional Regulatory Networks (RRNs)

The need for regional cooperation and collaboration among neighboring countries is based on the belief that the right combination of country-based cross-boarder measures can lead to outcomes that are superior to those that could be achieved on the basis of national measures alone. (Ferroni, 2002). The collective provision of a good with the intention that the benefits will be shared among participants put the good in the public domain, and thus explanations for the existence of cross-national regulatory networks are embedded explanations for why public goods exist. Primary among explanations for the existence of RRNs is the failure of national markets to provide the good in question effectively. From a regional standpoint this also explains the need for cross-border regulatory networks in situations where there is a multinational market failure. With increased globalization, boundaries of markets are becoming increasingly blurry and there are recognizable benefits that can be accrued from cross border interactions and cooperation that could not be autonomously obtained. (Inter-American Development Bank, 2005). For example The South American Regional Infrastructure Plan addresses transportation, telecommunications and energy needs along corridors that would link the continent's countries. There is a strong economic and financial case for the coordination of transport infrastructure among neighboring countries so that remote regions and landlocked countries can have access to urban centers. (Ferroni, 2002). Countries could benefit from significant cost savings and increased reliability if power grids could be integrated so that electricity could be jointly produced and distributed across borders. The international cooperation in the provision of goods is however generally problematic, because countries preferences for cooperation and RPGs may be heterogeneous and thus differing incentives may influence varying levels of participation that has inefficient outcomes. Producing a good that requires collective action also introduces incentives issues such as free-riding. When there are market failures such as this, another source of provision is necessary. Traditionally, the other source of production is believed to be the government. In the case of cross-border production however, governments may not be the obvious solution. A formal facility (RRNs) that can develop and enforce regional policies and actions to ensure adequate input into the provision of cross-border goods becomes essential. RRN are suited for this role, because they are structured to monitor compliance with normatively and coercively sanctioned expectations.

Even in cases where infrastructure related goods are not shared physically by countries, regional regulatory networks are important to facilitate the sharing of knowledge, information and experience as a means of improving national infrastructure. In other words, cross-border regulatory networks could be viewed as a regional public good itself, in the sense that it provides services across borders that are at least partially non-rival or only partially exclusionary. For instance, the Asian Regional Consultation on Work of the International Task Force on Global Public Goods met at the Asian Development Bank (ADB) headquarters in 2005.

As a result of this meeting, members of the task force had the opportunity to gain a new understanding and new perspectives on the challenges and opportunities of providing RPGs. (van der Linden, 2005) The public good provided by RRNs crosses national boundaries, but their benefits may be national in the form of the modernization of a country's infrastructure (including energy, water, and telecommunications) that is necessary for sustainable economic development. Networks of regional regulators are therefore important for complementing or encouraging physical network development, by facilitating the sharing of knowledge and experience among regions that share similar experiences and challenges. RRNs can influence decision making in infrastructures within regions and create opportunities for countries to explore common issues more effectively. (Berg and Horrall, 2007).

RRNs are essential for standardizing policy within regions that share network (such as an information highway, pipelines used for transportation of oil), or that are joined economically. For instance, the emergence of the European Union (EU) necessitated the creation of formal agencies with the power to develop appropriate rules that would encourage compliance within member states. RRNs, in this case were essential to coordinate the regulatory activities of EU member nations, thereby facilitating uniform and standard agreements on broad policies.

## 3. Understanding Regional Public Goods (RPGs)

The definition of a regional public good can be derived from the standard textbook definition of a public good. A good is considered to be public if it is non-rival in its consumption and non-exclusionary in its benefits. (Kaul et al, 1999). As an extension, a regional public good is a service or resource that is consumed across national borders and whose benefits are therefore shared by neighboring countries and are non-rival in consumption and non-exclusionary in benefits. Regional public goods fall between national or local public goods and global public goods. The non-rival property implies that the consumption of the good by individuals within a single nation does not lower or mitigates the consumption of the same good by other individuals within a different country, because one country's consumption does not lower the supply of the good. The non-exclusionary property suggests that it is difficult or impossible to prevent any country within the region from enjoying the benefits of the good. Usually non-exclusion of benefits means that an effective price cannot be charged for the use of the good. Goods that satisfy both properties are called pure RPGs and include services such as public prononouncements and best practice laws and procedures that are not provided by state governments, but by RRNs.[3]

If the good is either partially non-rival or partially non-exclusionary, or lacks any one of the two properties, then the good is considered to be an impure RPG. Congestion effects might be important where scale economies are involved in the production of the RPG. For instance the use of satellite in telecoms in communicating across border will be subject to interference and static if there is congestion, and the good is no longer available to everyone. Satellite communication, however are exclusive since users pay a charge per unit for its use. Regulatory networks RPGs that are excludable include benchmarking data. Capacity building is an example of an impure RPG with some rivalry of benefits.

Many infrastructure related RPGs are not completely non-rival or non-exclusionary, but the publicness of goods tends to be dynamic, so that a regional good with public properties could emerge as a private good in the future. Likewise a regional good that was once considered to be private may emerge as a RPG probably by design or due to the fact that technologies change overtime, and influences how the good is produced. (See Kaul, 2003).

The dynamic nature of public goods forces us to contemplate the definition of RPGs not based only on consumption properties (non-rival and non-exclusionary benefits), but based on production technology as well. Regional public goods may be considered to be public goods because their production requires collective action, and are thus dependent on regional cooperative efforts.[4] The role of RRNs becomes very important when this view of RPGs is considered. As Kaul (2004) suggested, in cases like this the role of national governments may be only to provide incentives and support to individual nations in aligning their private interests with the regions overall policy goals and objectives. Based on this view, regional public goods may be considered as regional collective action goods that are put in place by state or non-state actors (in the form of regulatory networks) to encourage certain behavior by countries.

In summary, public goods are RPGs if their benefits or costs extend beyond the borders of countries within certain geographical locations.

## 4. Facilitating Infrastructure through Regional Regulatory Networks

Encouraging infrastructure development through the use of RRNs can be viewed as synonymous with facilitating regional collective action, since RRNs usually take the form of groups consisting of representative member nations that must come to agreements on issues including the nature of the provision of infrastructure, information sharing, capacity building, and the development of broad policy objectives. Because the provision of the RPG is dependent on cross-country cooperation, collective actions will create challenges in the provision of the good that are not limited to a single country or group as with private goods. Challenges associated with the provision of RPGs emerge on both on the production as well as the resource ends, primarily because the provision of a good that has cross-border benefits (and costs) – whether among states or private actors- usually involve actors with national self interests instead of regional goals and objectives. Countries with greater influence will promote policies that benefit their country the most. Likewise collective action production processes encourage free-rider problems. The production of the RPG may therefore be undersupplied as some countries that consume the non-exclusive good will chose not to contribute to its production in the hope that other countries will. RRNs could respond to the free-rider challenge by instituting mandatory participation in cases where the free-rider can be distinguished from situations where contributions are low or non-existent because recipients (countries) place very little value on the good that is being produced. Differential preferences across countries could make this distinction unclear. At the knowledge sharing workshop in Cairo in 2005, participants identified some inhibitors to the proper functioning of regional networks. Among the challenges identified were conflict of interest among network members, unequal power relations, lack of commitment, unequal representation, hidden agenda, high

---

[3] See Berg and Horrall, 2007 for definitions of various infrastructure related RPGs.

[4] The Inter-American Development Bank (2005) defines a RPG as any good, commodity, service, systems of rules or policy that is public in nature and that generates shared benefits for the participating countries and whose production is a result of collective action by the participating countries. The regional dimension of the

membership fees, language barriers, the inability of networks to adjust to the changing environments, the lack of encouragement from local institutions to form new or to support existing networks, inadequate resources and lack of know-how and strategies to approach potential networks or partners. (Global Development Network, 2005). This section is devoted to analyzing the resource concerns (financial and technical) that are represented on this list of inhibitors, but many of the other challenges mentioned will also be discussed since the source of the problem may lie with the inadequate supply of resources.

## 4.1. Production (Technical) Challenges

As indicated before, RPGs are public in part because they have to be produced through regional collaborations or cooperations. RPGs are collectively produced because they cannot be (or are not) adequately produced through domestic policy alone. Individual country network participants recognize the need for international level cooperation even though such cooperation can give rise to productive incentive problems. The production methods or technologies that are involved in the creation of infrastructure related RPGs can be described primarily by the weaker link, summation, weighted sum, threshold, and better shot, aggregation technologies. (Berg and Horrall, 2007). Each of these production processes possess advantages, but there are also challenges for the adequate production of infrastructure related RPGs. Table 1 outlines the characteristics and production related problems of infrastructure related RPGs.

With the *weaker link* technology, all participating countries contribute to the production of the RPG, but the quantity and quality of the public good depends mostly on the smallest level of country contribution, with contributions that exceed the minimum level adding to the impact of the RPG at a diminishing rate.[5] Individual country contribution is therefore likely to vary within a region as some countries may have different willingness or ability to contribute (financially or otherwise). In other words, there might be a weak link that does not pull its weight. Heterogeneous contribution levels result because the benefits of a RPG will vary across countries, depending on country specific situations. For instance, as Berg and Horrall, 2007 pointed out, unanimous public pronouncements will depend heavily on the views of countries with the least interest in the topic. This kind of adverse selection results in the quality of the good being biased towards the country with the least interest, unless a way is found to cushion this bias. The most eager contributors to the RPG are not the countries with the strongest infrastructure or the ones with the most resources, but rather countries that are most wanting in the knowledge, the regimes, the standards and the rules that are required to challenge common or cross-border issues in infrastructure. It is possible for shortcomings in the participation of one country to be cushioned by the contribution of others when the weaker link technology is relevant, but usually only in part, because increasing amounts of contributions have successively smaller and smaller impacts on the production of the RPG. The low quality contributors crowds out high quality contributions, because there is a high cost associated effort. Countries that invest heavily into contributions are not likely to see the

---

definition is important, because if countries do not cooperate to produce the good, then it is not an RPG even if the benefits of the goods are shared once it is produced.

[5] The definitions of aggregation technologies used are based on Sandler, 2006.

returns on their investments, because the quality and quantity of the good that is produced jointly is to a large extent only as good as the weakest link. That is, countries with the least to contribute will have the most influence in determining the output. In cases where adverse selection is particularly serious, regional regulatory intermediaries can be beneficial and are needed to create guidelines and provide appropriate incentives for stronger participants to assist weaker links. The use of an independent intermediary (RRN) may reduce the influence of self-seeking behaviors of politicians as well as non-state actors interventions, especially when the provision of the RPG involves intergovernmental participation.

Each country or participant is required to contribute to the production of the RPG with the *summation* technology. The total amount of the RPG produced using this process therefore equals the sum of the contributors from all the participating countries. (Sandler, 2003). The strength of the summation technology depends completely on cooperation to increase the benefits of the solution to a problem or opportunity, so that the collaborative efforts will yield greater benefits to every country. (IDB, 2005). As an example, recent national network news can be disseminated across countries that would augment the experience of regulators in other countries to understand the implications of new rulings (Berg and Horrall, 2007). In the case of communications, the quality of a regional informational infrastructure will depend on the quality and quantity of basic networks in participating countries. Many developing countries, however, lack the some of the basic components that are essential for the working of such an infrastructure, and so the working of the regional infrastructure is inhibited. In cases like this, it might be necessary for regional intermediaries to intervene and provide support to countries in which the infrastructure or resources that are needed to provide the RPG. It could be beneficial to all countries if the technically stronger countries assist the weaker countries as a means of improving the overall quality of the regional good. (Kaul, 2004) The right incentives have to be in place for the stronger providers to assist the weak links in the production process. The quality of the RPG might be lower than the average expectation if the countries self-selected (adverse selection), so that only countries with the most need for the RPG participates in its production, because their perceived benefits from collaboration are large relative to those perceived by countries with greater technical (and financial) resources. The latter group may choose not to participate in the production of the RPG, causing its quality to be lower than if the group of participating countries included the stronger players as well as the weaker links.

The *weighted sum* process of producing RPGs occurs where the total amount of the RPG produced is the weighted sum of each country's contribution to the good. All countries have to make a contribution financially or in the form of specific behavior in order for the good to be produced. For instance, regulations or laws, that allows the collection of quality data from one country which would serve as a model for other nations, thereby giving data from that country greater weight than data for other countries. For instance, in southern Africa models of telecommunication production policies are usually adapted from South Africa thereby giving processes utilized by that country more weight. This happens mainly because South Africa in the past decade tend to be ahead of many of the other nations in Sub-Saharan Africa in terms of telecommunications infrastructure development, but as a result South Africa gets to have a stronger voice that countries that are struggling more with infrastructure development issues.

**Table 1. Characteristics and Production Related Problems of Regulatory Network Regional Public Goods**

| Pure RPG | Impure RPG (excludability) | Club Good (impure with some rivalry of benefits) | Aggregation Technology | Production Problem |
|---|---|---|---|---|
| | | Events and Meetings | *Weighted-Sum* | *Moral hazard* |
| | Benchmarking Data | | *Threshold* | *Free-Rider* |
| Public Pronouncements | | | *Weaker Link* | *Adverse Selection* |
| | Stakeholder Material | Capacity-Building/ Training | *Better-Shot* | *Prisoners' Dilemma* |
| Best Practice Laws, Procedures & Rules | | | *Better-Shot* | *Prisoners' Dilemma* |
| Network News | | | *Summation* | *Adverse Selection* |
| Studies | | | *Weighted-Sum* | *Moral Hazard* |

Because of the difference in contributions across nations that occur when the weighted sum production process is utilized, differential impacts of country contributions results, which could negatively impact incentives to contribute or participate. Incentive problems can be perceived in situations where data rich countries may have little interest in participating because incremental benefits from obtaining data from other countries may be minimal.[6] Countries that are unlikely to derive enough benefit from the RPG are less likely to contribute to its provision. Issues of moral hazard may emerge for RPGs that utilize the weighted sum production process. Countries with the most resources or are not as susceptible may not be interested in participating because they have less need for the RPG. Other countries may not take the appropriate amount of action because they know that if their contribution is small or suboptimal, it will carry very little weight. States may be less careful in terms of capacity development, or data gathering for instance, which results in a lower likelihood that the product will be produced in adequate amounts or quality, since the offending states realize that they will be protected by superior contributions which carry more weight, and which in the case of some RPGs such as rules and regulations, may be adopted. That is, if practices developed by individual nations are not appropriate or adequate, the country can simply adopt best practice models developed by a dominant country. This option may reduce the incentive for countries to pull their weight. Even the dominant nation may have a low incentive to "take care", because the benefit from such care is shared, while the incremental cost is borne largely by country alone, when its contribution to the RPG is weighted more heavily than the

---

[6] Positive incentives to contribute are also possible. In the case of data sharing, the dominant country could enjoy substantial benefits since standards used by the data-rich country tend to be the model that is adopted. The dominant country, therefore incurs no addition costs in order to modify its old model or to adopt an entirely new template.

contributions of others. In addition the problem of moral hazard may be compounded in the absence of significant heterogeneity of contributions. If country contributions are similar, there may be no dominant contribution and the RPG produced may be inferior. There is therefore a tradeoff because the existence of a dominant player can change the incentives of other nations to contribute, while the absence of a primary contributor may result in lower public standards.

In some instances, the total supply of the infrastructure related RPG relies mostly on the strongest (largest) contribution, with the importance of each remaining contribution diminishing with its size. This process of RPG production is known as the *better shot* production technology. As long a one country provides best practice examples, others could benefit by adding incrementally and thereby benefit from information sharing. The best practice may be in the form of general network policies or approach that works well. Problems arise with this production process, because the uniqueness of countries even within particular regions could reduce the usefulness of best practice methods and some countries may not benefit from adopting generic approaches. The fear that the one policy fits all approach my not work in individual countries could discourage countries from departing from the current norm. Countries may be unwilling to invest in improvements that could generate payoffs that are too small. Where incentives are sufficiently strong, RRNs could facilitate opportunities for countries to pool resources regionally, while encouraging the strongest participant to provide the good. The structure of the better shot aggregation technology encourages smaller contributions from nations with less resources, or know-how, which especially from a resource standpoint improves the opportunity for a country with limited contribution to gain as much benefit from the RPG as another country with significant contribution gains. The problem, however, is that the diminishing importance of smaller nations contributions can result in their voices not being heard. As Kaul et al, (2003) puts it, the decision making process of some (globally inclusive) public goods exclude some of the people that are affected by a good's spillovers, or where they are included, their full participation is not ensured in the decision making process. If there is insufficient representation from these "weaker" nations, they may be ignored and insufficient attention will be given to issues that are relevant to them, hence reducing the effectiveness of the RPG, and at the same time create a lack of incentive for cooperation from the underrepresented nations. In other words, the better shot aggregator may result in a variation of the prisoners' dilemma, in that the outcome from regional collaboration may not be Pareto efficient. A Pareto improvement to the outcome using this technology is possible, since smaller contributors could benefit more from the RPG if their voices could be heard, without diminishing the benefit of the RPG to the region.

The other primary production technology for infrastructure related RPGs is the *threshold aggregation* method. Where this method of production is utilized, the benefits of the RPG can only be achieved after a certain level of RPG supply is achieved. For instance, if cost data that is used for cost-based tariff setting for telephone and internet service is not of a certain quality (for example, if it is not disaggregated enough), them the data is relatively useless. The information must achieve a certain standing for it to be useful. The challenge with this production technique as it relates to data collection has to do with the difficulty in making data collection standards, such as common definitions, auditing procedures and validity checks. The interdependence that is inherent in the threshold technology is particularly susceptible to free-rider problem: One country might decide that the amount contributed by other countries is sufficient and that it would therefore be unnecessary for that country to contribute anything at all towards the RPG. The most desirable outcome (threshold) is unstable, because each country has an incentive to cheat. A small contributor may have the incentive not to put the effort into contributing what they may perceive as irrelevant to achieving the desired threshold, or unsubstantial relative to the contribution of others. Otherwise, some countries might have the incentive to lower individual contributions, as a means of lowering their costs, and maximizing their benefits when the desired level or quality of the RPG is achieved. If enough countries adopt this strategy, then the outcome is as in the prisoners' dilemma where the final achievement is smaller that what is desired or truly achievable.

## 4.2. Financial Challenges

Funding public goods that are shared by different countries is difficult because countries place different valuations on the benefits of cross-national cooperation. This makes the influence of Regional Development Banks (RDBs) and international organizations important catalysts of collective action (Ferroni, 2002). Empirically, seed money for the creation of infrastructure related organizations (particularly in developing regions) are generally funded through international and regional development assistance, whose financing are usually voluntary.[7] For instance South Asian Telecommunications Regulators' Council and Telecommunications Regulators Association of Southern Africa are just two infrastructure related organizations whose developments were encouraged by the ITU. The World Bank also provides financial assistance to numerous RRNs including the South Asia Forum for Infrastructure Regulation, the African Forum for Utility Regulators, and the Association of Water and Sanitation Regulatory Entities of the Americas. Regional funding agencies include the South African Development Community and the European Commission. There is also evidence of funding through public private partnerships, such as the Public Private Infrastructure Advisory Facility, usually in conjunction with the World Bank, national governments contributions and private sources. Regional funding agencies especially are uniquely positioned to provide funding for RPGs, because such agencies tend to have longstanding relationships with the countries. The regional development funding agencies specialize in fostering or supporting regional cooperation, and therefore have the experience that would help to generate and transfer adequate funding of RPGs production. (IDB, 2005). Financial problems associated with RPG production include the obvious lack of money to finance projects, but could also exist due to issues with the institutions that provide funding, the approach required for payment of benefits of the RPG and the divergence of means across countries.

### *Limited National Financial Resources*
The stimulus for the development of regional regulatory networks that provide infrastructure related RPGs often come from outside agencies because one significant

---

[7] Ferroni, 2002 argues that countries' recognition that national advantages could be maximized by using the right combination of national and regional policies is not sufficient to overcome the challenges of collective action. Regional development banks are therefore important organization that can act as catalysts of collective action that is needed in the production of RPGs.

obstacle to facilitating cross-border cooperation is the lack of financial resources. Many developing countries do not have the where-with-all in terms of economic resources to acquire the systems that are needed to improve network infrastructure in their country. The success of collaborations is hindered by the inability to establish a funding source that will sustain network operations. In addition, while the benefits of sharing are obvious, many countries do not have the money to acquire the necessary inputs to take advantage of opportunities. As Riggs, 1996 argues, electronic networks are connecting on a daily basis with each other throughout the world, and with advancing technology it has become possible to create effective information infrastructures in developing countries to become part of a global information infrastructure. Such an information infrastructure, however, requires not just software and hardware, but physical, financial, and human resources as well data or knowledge. Some developing countries lack at least one of these components and could therefore benefit from sharing resources with countries that are close by. The lack of financial resources is evident in other national sectors as well and inhibits the benefits that can be gained from cross-border collaborations.

## *Funding Institutions*

Because funding for RPGs are primarily voluntary, uncertainties are created in the budgets of international organizations in the absence of cross-border financial intermediaries who could ensure financial commitment to contributions. The conflict of interest that exists between the regionally desirable contribution to the RPG and the optimal commitment of individual countries circumvents the implementation of a Pareto optimal solution to the financing of RPGs. As a result there is a need for an independent institution that can enforce contributions and issue sanctions to violators. The use of institutions with authority to enforce payments is fairly straightforward in the case of national public goods, because government institutions are typically in place that has the authority (using the laws of the country) to back up enforcements. (Gerber, and Wichardt, 2007). With cross-border public goods, however enforcements are more complicated, because even where international intermediaries exist, there are often no common laws that exist to back up enforcements.

Individual regions are, however beginning to take the initiative to address financial concerns. In 2004 for example, the board of executive directors of the Inter-American Development Bank (IDB) approved the Initiative for the Promotion of Regional Public Goods. The bank made up to $10 million dollars in resources to finance projects that support the development of RPGs in Latin America and the Caribbean. By taking this initiative the IDB is embracing the notion that rather than depending on international organizations, regional cooperation can be effectively address some of its problems. Many concerns and opportunities shared by countries within a region can addressed more effectively at the regional level, through cross-border cooperation in the production of public goods. RPGs are however sometimes under supplied because there is much difficulty in obtaining financial and institutional support for the regional effort. Countries are concerned about the possibility of subsiding free-riders, since the RPG will benefit everyone regardless of their support. This requires the intervention of an appropriate regional organization to fund (and to encourage dialogue and action) at least at the early stages of a RPG production. (IDB, 2005).

## *Payment Approach*

Financing of RPGs may be via loans or grant-based funding from regional and international agencies. Often countries are required to pay on the basis of their ability to do so, and in other circumstances RPGs may be financed from private contributions. Grant-based funding and loans are means through which the development banks can contribute to the provision of the RPG. Grants can be problematic when the RPG being produced generates free-riding. Non-paying parties cannot be excluded from the benefits of the public good so much care has to be taken with their allocation governance and management (Ferroni, 2002) to help mitigate moral hazard concerns. The problem with the use of loans by individual countries to finance RPGs is that the benefits from the investment that the loan financed are shared by all participating countries, while the borrowing country is responsible for the costs of the loan. The divergence between the beneficiary of the loan and who bears the burden of the loan can discourage countries from utilizing this method of financing.

Where contributions to the budget of cross border institutions are based on a country's ability to pay, counties with very limited financial resources may enjoy the benefits of the RPG for free. A more rigorous financial agreement that includes specific user fees, that essentially converts the RPG to a club status in some cases is essential to ensure financial stability in the provision of the public good. In recent years increased strain on water resources in Africa motivated actions that would protect and sustain the quality and quantity of water sources such as the river Nile. Approaches for water protection involve services that are provided by regulatory networks with jurisdiction within countries along the Nile. Protecting the water source could generate public goods that are not whose benefits are unevenly spread across countries regionally. (Jäkerskog, A, J. Granit, A Risberg and W. Yu, 2007). For instance, policies that improve downstream irrigation potential or improve downstream drinking water obviously give greater benefits to downstream countries, compared to countries that are located further upstream the Nile. This raises the issue of whether the downstream countries should pay more or whether financial contributions should be based on ability to pay, since the primary beneficiaries of the RPG probably cannot afford to pay fully for the value of the benefits derived. That is, the economic status of the benefiting country could mitigate the financing approach that benefiting countries should be responsible for the bulk of the costs associated with the RPG's production. Gerber and Wichardt, 2007 proposes a two step process to address the financial challenge in public good provision so that even in cases where payments are based on the ability to pay, the incentive to free ride would be removed. In this model, beneficiaries must commit to the RPG by paying a deposit to a neutral institution prior to the country making its contribution for its share of the public good. As an incentive to contribute, participants would be refunded their deposits (minus minimal fees) once their financial contribution to the RPG has been made. Of course the deposits would have to be sufficiently large to ensure that the depositing country desire its return.

Funding from international and local investors is not uncommon in the production of RPGs. Often the private investor has some individual objectives that could be helped by ensuring a quality RPG. According to Kaul et al, 2002, responsibilities for providing and financing public goods in the areas of communication and transportation are often based on a mix of public and private financing. For instance, between 1998 and 2003, cellular service in the SADC grew from 2.9 million connections to 21.5 million connections. This significant growth in cellular service was funded by a consortium of domestic and foreign investors. At the same time, growth in communications services including the internet, data communications and the provision of customer premises equipment is largely a result of private financing. (Goulden and Msimang 1996)

### *Incomes*

If income levels differ significantly across countries within regions, that at the same time have heterogeneous preferences, the RPG produced could be biased toward the preferences of countries with greater financial means. Richer countries are usually the ones that have the financial means to support their policy preferences for instance and to contribute financially towards policy reform in developing countries. The louder voice of a country due to its financial advantages can result in the adoption of policies that are not based on regional consensus.

## 5. CONCLUDING OBSERVATIONS

The increased porousness of our borders and the challenges faced by many countries in the pursuit of infrastructure reform and development have fostered a new way of thinking that is regional, rather than completely domestic in nature. However, understanding RPGs and using them to the advantage of regions is very daunting, so problems associated with the production and delivery of infrastructure through the sharing of resources and regional collaborations is growing in importance across the world. The major impediments associated with the direct provision and supply of RPGs is similar to the ones that result in the public provision of domestic or local public goods. These problems are incentive related and include adverse selection, moral hazard, free-rider issues and the prisoners' dilemma. Production challenges of this nature are complex as they involve both national and regional concerns. Another important obstacle to the provision of RPGs is the availability of financial support for the production of the RPG. Financial concerns have much to do with the inability of some countries to contribute to production. Funding sources exist from regional and international organizations, national governments as well as from the private sector, but adequate cross-border cooperation still remains a challenge that must be addressed.

The presence of an independent intermediary such a regional regulatory network is essential to facilitate the production and the financing of infrastructure related RPGs. RRNs in their role as facilitators is itself a public good and therefore RRNs may be viewed as RPGs as well. Capacity building remains an issue for newly formed networks with limited resources. The need for regional cooperation in the sharing of information, knowledge and experience is essential to develop the quality of RRNs and to improve the provision of the RPGs that they provide.

The technical and financial problems associated with infrastructure related RPGs, and the need for well trained and experienced RRNs are only a subset of the vast array of concerns that must be addressed. Chief among them is the fact that successful regional cooperation requires the development of a clear strategic plan-much like a white paper- that must be followed in setting the direction of regional policy. National incentives must be aligned with regional objectives in order to improve the success of RPGs provision. If countries believe that regional cooperation will generate benefits that could not be achieved autonomously by that country, then countries will the incentive to take advantage of regional initiatives that could increase benefits and reduce the costs of cross-border interactions. Understanding the nature of RPGs and the challenges that they create is a first step in developing solutions that would take full advantage of the use of regional cooperation to build capacity and to reform, modernize and further develop infrastructure within regions.

## REFERENCES

Berg, Sanford, and Jacqueline Horrall (Forthcoming, 2007). *Review of International Organizations.*

Brown, Ashley C., Jon Stern, Bernard Tenenbaum and Defne Gencer (2006). *Handbook for Evaluating Infrastructure Regulatory Systems*, Washington, D.C.: World Bank. xx-397.

Ferroni, Marco (2002). "Regional Public Goods: The Comparative Edge of Regional Development Banks," unpublished manuscript. February. 1-21.

February1http://www.iie.com/publications/papers/ferroni0202.pdf

Gerber, Anke, and Philipp Wichardt (2007). "Providing Public Goods Without Strong Sanctioning Institutions." Governance and the Efficiency of Economic Systems Discusion Paper No. 194, February. 1-10.

Global Development Network (2005). "Knowledge Sharing for Development: Africa Regional Program. Workshop Report." Global Development Network, Cairo, Egypt, February 27-28. 1-35. http://www.gdnet.org/pdf2/knowledge_sharing_workshop/workshop_report.pdf

Goulden, Brian and Mandla Msimang (2005). "Collaboration in ICT Regulation in the Southern Africa Development Community: A Regional Approach to Capacity Building." Center on Regulation and Competition Working Paper Series, Paper no. 98, March. 1-17. http://www.competition-regulation.org.uk/publications/working_papers/wp98.pdf.

Inter-American Development Bank (2005). Regional Public Goods: Call for Proposals 2005. http://www.iadb.org.

Jäkerskog, A, J. Granit, A Risberg and W. Yu. (2007). "Transboundary Water Management. as a Regional Public Good. Financing Development. – An Example from the Nile Basin." Report No. 20. Stockholm International Water Institute, Stockholm. http://www.siwi.org/downloads/Reports/Nile_Basin_Report_07.pdf

Kaul Inge, Pedro Conceição, Katell Le Goulven, and Ronald U. Mendoza (2003). "How to Improve the Provision of Global Public Goods" in *Providing Global Public Goods: Managing Globalization*, ed. Kaul Inge, Pedro Conceição, Katell Le Goulven, and Ronald U. Mendoza, New York: Oxford University Press.

Kaul, Inge, Katell Le Goulven and Mirjam Schnupf (2002). "Financing Global Public Goods: Policy Experience and Future Challenges" in *Global Public Goods Financing: New Tools for New Challenges .A Policy Dialogue*. Ed. Inge Kaul, Katell Le Goulven and Mirjam Schnupf. United Nations Development Programhttp://www.undp.org/ods/ffd-monterrey.html.

Kaul, Inge, Isabelle Grunberg and Marc A. Stern (1999). "Defining Global Public Goods" in *Global Public Goods: International Cooperation in the $21^{st}$ Century*, ed. Kaul Inge, Isabelle Grunberg and Marc Stern, New York. Oxford University Press.

Kaul, Inge (2004). "Financing Global Public Goods: Trends and Challenges." Discussion Draft.http://www.sdnp.undp.org/gpgn/pdfs/Financing_GPGs.pdf

_____ (2003). "Public Goods, A positive Analysis." Office of Development Studies, October. 1-23. http://www.sdnp.undp.org/gpgn/pdfs/Financing_GPGs.pdf

Sandler, Todd (2006). "Regional Public Goods and International Organizations," *Review of International Organizations*," 1: 5-25.

_____ (2003) "Assessing the Optimal Provision of Public Goods: In Search of the Holy Grail" in *Providing Global Public Goods: Managing Globalization*, ed. Kaul Inge, Pedro Conceição, Katell Le Goulven, and Ronald U. Mendoza, New York: Oxford University Press.

Riggs, Donald E (1996). "Building a Global Information Policy: Issues, Challenges, and Opportunities." http://web.simmons.edu/~chen/nit/NIT'96/96-231-Riggs.

Van der Linden, G. H.P.B. (2005) "Asian Regional Consultation on the Work of the International Task Force on Global Public Goods. Welcome Remarks." Asian Development Bank, Mandaluyong City. February. 1-4. http://www.adb.org/Documents/Speaches/2005/ms2005015.asp

*Chapter 6*

# VALUE CREATION IN ACQUISITIONS: PRIVATIZATION OF TURK TELEKOM[*]

### *Fırat Demir*[†]
Department of Economics, University of Oklahoma, Hester Hall, 729 Elm Avenue, Norman, OK 73019, USA

### *Vahap B. Uysal*[‡]
Michael F. Price College of Business, University of Oklahoma
307 West Brooks, Norman, OK 73019, USA

## ABSTRACT

This paper examines value creation and distribution in acquisitions in the context of Turk Telekom privatization. We find that operational synergies, in addition to traditional sources of synergies, play an important role in value creation. Specifically, re-allocation of resources from inefficient business units to efficient ones contributes to value creation. We also document that the privatization method as well as the post-privatization regulatory framework are significant determinants of the success or failure of privatization programs in developing countries especially with regard to increasing competition, efficiency and value transfer to the public.

**Keywords:** Mergers and Acquisitions, Privatization, Value Creation, Efficiency

## 1. INTRODUCTION

Studies on the mergers and acquisitions (M&A) conclude that acquirers do not create value for their shareholders (Andrade, Mitchell & Stafford, 2001). Although these studies are

---

[*] We thank William L. Megginson for insightful discussions on telecommunications privatizations.
[†] E-mail: fdemir@ou.edu; Tel: 405-325-5844
[‡] E-mail: uysal@ou.edu; (405) 325 5672

able to identify which acquirers create value in M&A, they seldom offer evidence on how value is created in M&A. In a recent meta-analysis, King et al.(2004) find that the most commonly used constructs in the M&A literature fail to predict which acquirers are likely to obtain gains in newly announced M&A. Therefore, further theoretical work and the development of new constructs are required to identify factors that contribute to value creation in M&A (Andrade, Mitchell & Stafford, 2001; Zollo & Singh, 2004).

Examining privatizations of State Economic Enterprises (SEE) offers a novel approach to examine value creations in M&A. SEEs offer typical examples for agency problems created by conflicting interests of managers and the owner (i.e. public). Managers and boards of SEEs are politically appointed and serve in the best interest of political authority. To the extent that interests of political authority are not aligned with the public, the political authority may execute policies in SEEs that may favor a small fraction of political supporters at the expense of the whole society. For example, the political authority may employ more people in SEEs than the optimal level (Megginson, 2005). Using SEEs as an employment agency temporarily extends the voter base for the political authority while creating permanent uneconomical SEEs subsidized by the tax-payer citizens of the society. This is especially important in developing countries where SEEs constitute a large fraction of the economy and democracy is not well-developed. Furthermore, the privatization wave together with financial liberalization has been one of the primary reasons behind the steep increase in M&A type Foreign Direct Investment (FDI) in developing countries during the 1990s and 2000s.[1] In terms of the sectoral distribution of M&A, we see that the highest concentration is in services with 60% of total flows followed by manufacturing taking 38%. Among services, the highest share is occupied by transportation, storage and communications sectors, followed by financial services (UNCTAD, 2007). Thus, examining privatizations in services in developing countries offer a natural experiment to identify sources of value creation in M&A.

Specifically, the Turkish experience with liberalization and privatization is of particular importance for researchers and policy makers. Turkey started liberalization and privatization policies two decades earlier than most other developing countries. Thus, its experiences carry important lessons for countries that now go through similar structural changes. In terms of its weight in the international markets, Turkey is one of the major emerging markets and received 23 cents out of every dollar invested in middle income countries in the form of portfolio investment in 2000. The total amount of short term capital flows reached $120 billion during 1990-2005.

The Turkish example also underlines the challenges posed by a fragmented and polarized political system with a legacy of crony capitalism and corruption with weak prudential regulation and accountability. Ercan and Onis (2000) also pointed out that the intra-bureaucratic conflicts within as well as between different segments of the state (e.g. the Constitutional Court) or the bureaucracy (e.g. the Treasury or the Ministry of Finance) created additional obstacles. Therefore, privatization and deregulation experiences of Turkey do not only shed light on improving privatized SEE, but also offers insights on improvements in the overall domestic product and service markets.

In this study, we examine the Turkish privatization experience in the context of sale of Turk Telekom (TT), state-owned telecommunications service provider of Turkey. We find

---

[1] Approximately, 28% (or 35% excl. China) of total FDI inflows during 1990-2004 were in the form of M&A. (UNCTAD, 2007).

that gains to acquirer came from tax shields, better incentive alignment of new TT management, financial synergies and increased productivity in operations. Although prior financial studies on M&A document the non-operational channels of value creation (tax shields, incentive alignment of managers and financial synergies), problems associated with measuring operating synergies prevented them to detect improvement in operational performance related to M&A (Grinblatt and Titman, 2001). This study documents that one of the channels of operational synergies is the re-allocation of labor force from inefficient business units to value-creating ones. The change in the business and legal environments, which eliminated uncertainty in the *ex-post* privatization period, also increased *ex-ante* bidder participation.

Although gains to acquirer are quiet distinct, the benefits for the public are mixed. Privatization of TT associated with lower tax collections for the public. There were also steep increases in local distance calls since telecommunications commission was not able to exercise its powers right after the sale was finalized. On the other hand, breaking up the satellite operations from TT paved the way for competition. Furthermore, privatization of TT conveyed credible signal about the liberalization policies and growth opportunities in Turkey, which, in turn, are likely to improve FDI.

This paper also sheds light on failures and successes of privatization of state-owned telecommunications service providers. A growing body of literature indicates that privatized telecommunications providers show significant improvements in financial and operating performance (Ros, 1999; Wallsten, 2000; Boyland and Nicoletti, 2000; Bortolotti et al., 2002). In this respect, the most important factors in achieving efficiency gains in developing countries are found to be the increasing competition and effective independent regulation and supervision of privatized enterprises (Fink, Mattoo and Rathindran, 2002; Wallsten, 2001; Megginson, 2005).[2] Due to data availability on the effective date of privatization, researchers tend to focus on the role of privatization. Long-time span and country-specific experience of deregulation posit barriers for researchers and call for in-depth studies to disentangle effects of privatization and deregulation. Since the privatization of Turk Telekom has been on the agenda of numerous administrations and each followed a different strategy to implement the privatization, the differences in privatization policies allow us to compare and to evaluate deregulation policies and their role in the success of privatization. We document that political consensus and commitment play an important role in the implementation of TT privatization.

This paper is presented as follows: Section 2 summarizes the liberalization and privatization experience of Turkey. Section 3 describes the privatization process of TT. Section 4 examines the post-privatization benefits/costs for the public and the acquirer. Section 5 concludes.

---

[2] For a review of privatization experience in developing countries see Parker and Kirkpatrick (2005).

## 2. A Brief History of Economic Liberalism and Privatization in Turkey[3]

The year 1980 constituted a corner stone in the modern economic history of Turkey marking the date of switching from a closed-inward oriented economic structure based on Import Substitution to an outward-oriented path of development. One of the common features of the pre-liberalization era under Import Substitution Industrialization regime was the fact that during these years, the private accumulation process highly depended on policy and politics rather than markets.[4] Entrepreneurs became increasingly dependent on the state and bureaucracy and on the subsidies they provided rather than exploiting the opportunities created by the market. The political and economic environment inevitably created vast opportunities for rent seeking as the business people competed fiercely for the special set of incentives provided by the state. The system, as a result, encouraged and generated serious moral hazard problems and rent-seeking behavior on a systematic basis in both public and private spheres.

Consequently, following the major balance of payments crisis in the late 70s, Turkey, under the directions of international financial community, tried to undertake a profound switch in the philosophy of state structure regarding its role in economic affairs. The new economic model aimed at reducing the size of the public sector involvement in the real sector through its operations of State Economic Enterprises (SEE) as well as at reducing the degree of intervention in the organization of the market activities. As a part of this broad program, in the course of the 1980s, Turkey went through a step-by-step liberalization in its economy. Liberalization of the foreign trade regime, removal of exchange rate controls, adoption of special policies to attract foreign direct investment (FDI), liberalization of interest rates, privatization of SEEs, were some of the components of the new economic program. In the final stage of this program, the August of 1989 witnessed a complete transition to an extremely liberal capital account regime.

Despite the initial success of the reform programs especially with regard to increasing export performance and public finances, the results were disappointing on several accounts. There were low fixed investment rates, high and chronic inflation, increasing GDP volatility, and financial instability that showed itself in three consecutive crises in 1994, 2000, and 2001. Public finances, on the other hand, went nearly bankrupt partly thanks to double and sometimes triple digit real interest rates that averaged at 9.4% during 1991-2005 with annual average peak being at 23.8% in 2002.

Consequently, the share of total public sector borrowing requirement (PSBR) reached 16.4% while the share of interest payments in consolidated budget increased to 25% of GNP by the year 2001 from around 4.7% and 0.5% in 1975 respectively. The policy of keeping interest rates high to encourage short-term capital inflows have resulted in a rapid build up of domestic debt with deteriorating burden on the budget in the form of high interest payments. The share of interest expenditures in the consolidated budget increased from 3% in 1975 to 51%, 47% of which was on domestic debt in 2001. The share of principal and interest payments to total tax revenues increased from around 12% in 1980 to a record number of 236% of which 203% was on domestic debt in 2001. To service public debt, the central

---

[3] This section is partly based on Demir (2004).

government channeled 103% of tax revenues - about 52% of its total expenditures- to interest payments in 2001 (which was only 4.2% in 1980). In other words, tax revenues have become insufficient to pay even for the interest payments themselves alone.

The growing interest burden on the consolidated budget has been tried to be financed through investment cuts, current expenditure reductions (most of which were personnel expenditures), and privatization of SEEs. The share of current expenditures in the consolidated budget decreased from around 52% in 1975 to 25% in 2001 while the share of investment expenditures dropped from 20% to 5% for the same period.

On the privatization front, the actual date of implementation of privatization plans goes back to 1986. The initial rationalization for such a radical change in a heavily state-controlled economy[5] was the productive and operating efficiency arguments as well as the moral hazard problems. However, over the course of time (and mostly thanks to large PSBR) the main motivation for privatization has become to finance public sector deficits rather than improving the efficiency and competitiveness of these firms and the market in which they operate.[6] In other words, revenue generation has steadily replaced the efficiency gain arguments.

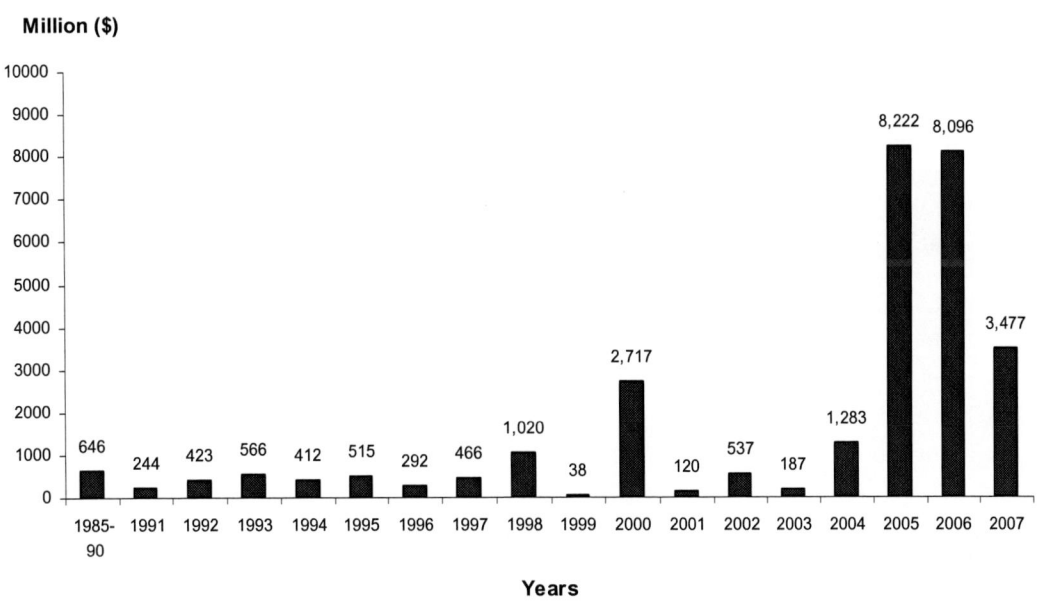

Source: For figures 1- 4, Privatization Administration.

Figure 1. Annual Privatization Implementations.

On the other hand, the privatization experience of Turkey did not follow a straight line resulting from the ongoing struggle between the pro and anti-privatization groups and

---

[4] For a detailed analysis of this relationship between the state and the business class, see Bugra (1994).
[5] The value added to GDP ratio of SEEs fell from 8.5 in 1985 to 2.6 in 2005. During the same period the total employment at SEEs dropped from 635,000 to 247,000.

institutions (see Ercan and Onis, 2000). Nevertheless, despite the fact that the move towards privatization had started much earlier than most developing countries, it has been much limited in scope and in its success.[7] Accordingly, between its starting date in 1986 and 1999, total proceeds from privatization program has been only $4.7 billion in current prices (Figure 1).

The privatization program gained pace mostly due to the rising public sector indebtedness and consecutive financial crises in 1994, 2000, and 2001 after which privatization of large and often profitable and marketable SEEs became a top priority of successive governments. (such as the *Petrol Ofisi,* petroleum products and distribution agencies-POAS, Turkish Telecom Inc, GSM Licensing, *Seka*, the paper, cellulose and pulp plants, and the Turkish Airlines). The program further accelerated its pace especially after 2000-2001 that mark the most serious economic crisis in its history (Figure 1).[8] In fact, the policy makers capitalized on the crisis and managed to persuaded the shocked public to accelerate the privatization of previously controversial and politically challenging SEEs such as TT. The appointment of the World Bank vice president, Kemal Dervis, as the minister of economy to calm the markets and to ease access to foreign financing also helped speed up the privatization process without much political opposition. Since 1985, state shares in 244 companies, 103 establishment, 22 incomplete plants, 8 toll motorways, 2 Bosporus bridges, 1 service unit and 393 real estates and 6 ports have been taken into the privatization portfolio (Privatization Administration). In total, 210 companies have been privatised. Overall, the public sector has withdrawn completely from cement, animal feed production, milk-diary products, forest products, handling and catering services and petroleum distribution sectors. Moreover, more than 50% of the public sector shares have been privatized in tourism, iron and steel, textile, sea freight and meat processing sectors as well as most of the ports and petroleum refinery sector.

The initial objectives of privatization programme in Turkey were identified by a commissioned master plan by Morgan Guaranty Bank in 1986. In the study 14 objectives were discussed in the order of importance including targets like improving market mechanisms, improving efficiency and productivity, contributing to the development of capital markets and contributing to the widening of ownership distribution of firms, promoting competition, reducing SEEs burden on Treasury, eliminating the monopoly pricing and indirect taxation by SEEs, attracting modern technology and management techniques. Strengthening the strategic connections in international arena through alliances with foreign investors was also included among the objectives. The last on the list was generating revenues for government. However, thanks to growing PSBR this last objective has quickly become the primary motivation for privatization.

In terms of methods of privatization in Turkey, it has been done mostly through three modes of sales techniques that are "block sales", "public offers for floatation", and "direct sales of assets and premises of the SEEs and their subsidiaries" (Karatas, 2001). However, despite the initial recommendations, governments mostly used the block sale method in privatizations partly because of the lack of sufficient depth and the liquidity of capital

---

[6] This is also visible form the radical cuts in the investment expenditures of SEEs to levels that do not cover even the depreciation. Accordingly, the investment spending to operating revenues ratio steadily fell from 13% in 1989 to 3% in 2004.

[7] In fact, Turkey started implementing all the articles of Washington consensus before it was a consensus in 1989.

markets. The first method that account for 64% of all privatizations between 1986 and 2007 (Figure 2) is the direct sale of the majority of the public shares to a single buyer and is generally ranked as the least successful one for increasing competition and efficiency. In fact, with block sales either new monopolies are created or existing public monopoly is transferred to private monopoly. The second method, "privatization via public offering" accounts for only 18% of all privatizations during this period and had only marginal effect on capital market development or increasing ownership distribution of firms that were ranked among the top five objectives in the initial master plan.

The net account of privatization proceeds (net of expenses) is documented in Table 1 that gives the total balance of privatization income over 1986-2004. During this period, net privatization revenues reached 10.351 billion dollars while expenses amounted to 4.5 billion between bringing net gain from privatization to 5.8 billion dollars. Out of this 5.8 billion, 3.4 billion is transferred to the treasury; that is 33% of net revenues of $10.351 billion.

### Table 1. Privatization Income and Expenses ($1,000)*

|  | 1986-2002 | 2003 | 2004 | TOTAL | % |
|---|---|---|---|---|---|
| **Revenues** |  |  |  |  |  |
| Gross Privatization Revenues | 7,225,914 | 255,054 | 1,133,743 | 8,614,710 | 60 |
| Block Sale | 3,433,525 | 37,835 | 364,042 | 3,835,402 | 27 |
| Asset Sale | 591,740 | 129,615 | 453,850 | 1,175,205 | 8 |
| Public Offering | 1,595,578 | 0 | 37,011 | 1,632,589 | 11 |
| International Offering | 1,004,471 | 0 | 125,488 | 1,129,959 | 8 |
| I.S.E. Sale | 597,033 | 87,603 | 153,352 | 837,988 | 6 |
| Incompleted Asset Sale | 3,568 | 0 | 0 | 3,568 | 0 |
| Dividend Income | 2,068,137 | 63,122 | 223,291 | 2,354,550 | 16 |
| Issued Principle Bills | 1,454,169 | 95,914 | 1,160,364 | 2,710,447 | 19 |
| Principle Payments | 57,553 | 15,911 | 3,900 | 77,364 | 1 |
| External Loan and Grant | 259,865 | 39,342 | 97,786 | 396,994 | 3 |
| Other | 134,103 | 17,878 | 14,473 | 166,454 | 1 |
| Total | 11,199,740 | 487,222 | 2,633,557 | 14,320,519 | 100 |
| **Expenditures** |  |  |  |  |  |
| Transfer To The Companies | 318,593 | 0 | 0 | 318,593 | 2 |
| Consulting and Auditing | 52,721 | 2,931 | 1,268 | 56,920 | 0 |
| Advertisement | 52,079 | 1,878 | 6,796 | 60,753 | 0 |
| Capital Increase | 3,648,750 | 234,640 | 369,102 | 4,252,492 | 31 |
| I.S.E. Repurchases | 134,243 | 0 | 0 | 134,243 | 1 |
| Social Assistance Supplements | 190,905 | 13,087 | 54,209 | 258,201 | 2 |
| Credit To The Companies | 1,117,707 | 138,182 | 41,162 | 1,297,052 | 9 |
| Loan Payments | 1,915,590 | 91,244 | 1,930,622 | 3,937,457 | 28 |
| Others | 124,363 | 12,147 | 25,308 | 161,819 | 1 |
| Transfer To Treasury | 3,403,846 | 0 | 0 | 3,403,846 | 25 |
| Total | 10,958,798 | 494,111 | 2,428,467 | 13,881,376 | 100 |
| **Net balance** | 240,942 | -6,889 | 205,090 | 439,143 |  |

Source: For Tables 1 - 4 Privatization Administration.
* Including previous years' installments collections.

---

[8] The 2001 crisis led to a 9.5% contraction in GNP, 18% fall in private sector real labor cost, and a sharp increase in unemployment rate. Accordingly, the non-agricultural unemployment rate increased to 13% and 15% in 2001 and 2002 respectively from 9% in 2000.

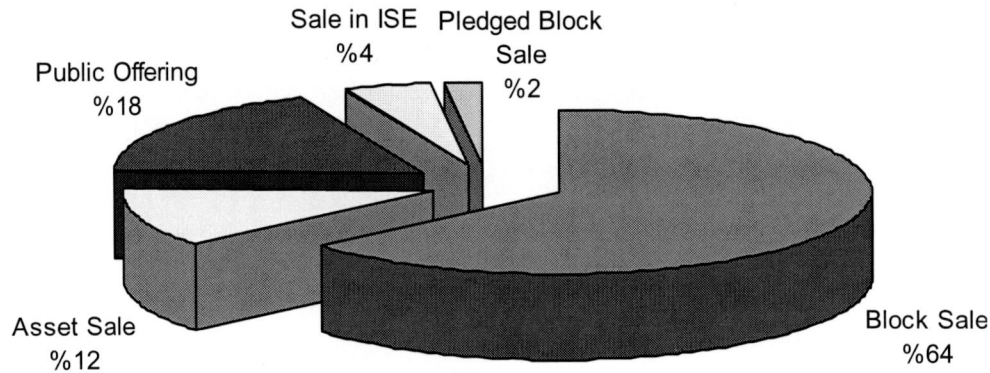

Figure 2. Privatization Methods, 1986-2007.

## 3. PRIVATIZATION OF TURK TELEKOM

### 3.1. A Brief History of Turk Telekom

Turk Telekom has been part of the General Directorate of Postal, Telegraph and Telephone that was created in 1924 with the article 406 that granted it a state monopoly status in postal, telephone and telegraph services. The operational activities of TT included establishing, managing and operating telephone centrals, local and long distance fixed telephone lines, cable systems, data networks, communication networks, GSM, cable TV etc.

In the aftermath of the major economic restructuring along free market based liberal economy in early 80s, TT initiated a radical change in its operations and increased its investment rates radically. As a result, the share of telecommunications investments in GDP rose from less than 0.3% in early 80s to around 1% at the end of 1980s. Yet, following 1994, the investment rates fell drastically to less than 0.3% partly thanks to the 1994 financial crisis and growing need for financing budget deficits, and partly because of its privatization decision and its subsequently delayed implementation.[9] Given that TT was put on the list of firms to be privatized, investment rates fell and its growth lacked behind other telecommunication firms in the world. Accordingly, the telecommunications investment as a share of revenues displayed a steady decline from 39% in 1993 to 9% in 2004. In contrast the same figures for Belgium, and Italy were 24% in 2004 (World Bank, 2007).

As a result, Turkey was lagging behind other European countries in IT development especially since the second half of 1990s. Broadband subscribers per 1,000 people, for example, were only 22 as opposed to 250 in Netherlands, or 155 in France in 2005. International internet bandwidth was 8861mbps as opposed to 200,000mbps in France or 566,056mbps in Germany. Internet users per 1,000 people were 222 as opposed to 430 in France or 455 in Germany in 2005 (World Bank, 2007).

---

[9] For example, as a part of IMF agreement in the aftermath of 1994 financial crisis, Turkey agreed to cut SEEs investment spending. As a result, TT cut back 25% of its planned investment in 2004. Consequently, the fixed line connection que increased from 654,528 in 2004 to 785,328 in 2005 (Yilmaz, 1999).

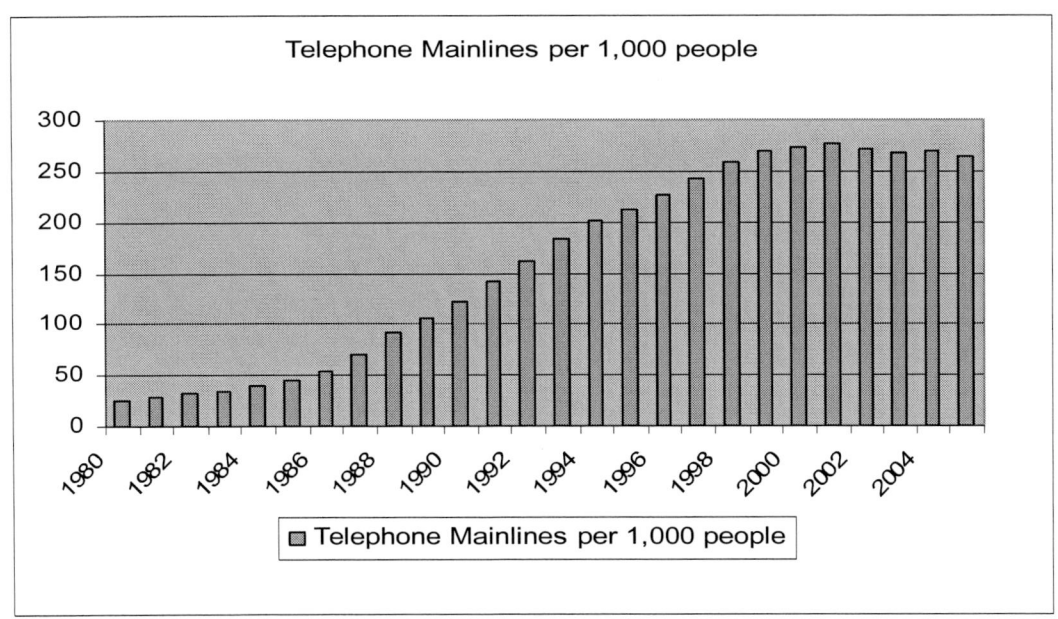

Figure 3. Telephone Mainlines between 1980 and 2005.

On the other hand, despite low averages, telecommunications sector is one of the fastest growing sectors in Turkey and its growth rate is one of the highest in Europe. Internet users, for example, have increased by 55% on average every year between 1995 and 2005. This growth rate is higher than any other European country. Similarly mobile phone subscription has increased around 50% a year since 1995 and reached 605 per 1000 people in 2005 from 7 in 1995. The total number of telephone subscribers (1.3 million in 1981) increased to 12.4 million in 1994 and 62.6 million in 2005. Telephone mainlines per 1,000 people was 29 in 1981, 202 in 1994 and 264 in 2005 (Figure 3). The annual growth rate of subscribers has been around 16% on average between 1980 and 2005 as opposed to 7% average growth rate in Europe (World bank, 2007).

Likewise, telecommunications revenues have shown a steady increase unlike in other European countries. While telecommunication revenues as a share of GDP fell from around 67% in Belgium, 12% in France, 4% in Germany and Netherlands in 1996 to around 2-3% in 2004, it has steadily increased in Turkey from around 1-2% to 4% of GDP during the same period.

## 3.2. Privatization of Turk Telekom

Historically, the telecommunications, and in particular telephone services, have been seen as strategically important for nation states and has been a prime example for natural monopolies in all introductory economics textbooks. As a result, they have been either owned directly by the state as a monopoly or in case they are private are granted monopoly rights.

Yet, with the information revolution during the 1990s the sector is no longer seen as a primary candidate for natural monopoly status and several countries started experimenting with its deconsolidation and liberalization. For example, most of EU countries fully liberalized their telecommunications markets by 1998 (Pennings et al., 2005). According to Li et al. [2001] while only 2% of telecommunications firms in 167 countries were privatized in 1980, it increased to 42% by 1998. Furthermore, given the revolutionary advances in IT technology and increasing competition in the domestic and international markets, telecommunication companies increasingly used mergers and acquisitions as a way of expanding their resources, know-how and market access (Chan-Olmsted and Jamison, 2001; Pennings et al. 2005; Waverman & Trillas, 2002).

On the other hand, despite radical increases in world-wide M&As in telecommunications sectors, we also see that the foreign ownership shares differ widely between developed and developing countries. For example, while the foreign ownership shares remain at 5.6% in Spain and Sweden, 12% in Netherlands and 2% in Germany, it is 65% in Bulgaria, 61% in Croatia, 60% in Lithuania, 59.5% in Hungary, 54% in Romania, and 47.5% in Poland. Megginson (1995) attributes the higher foreign ownership rates in developing countries to insufficient capital and expertise in domestic markets.

During the privatization wave of the 1990s and 2000s we also see a clear difference in the privatization methods of telecommunication firms between developed and developing countries. While public offerings through stock markets are the primary method of privatization in developed countries, we see the block sales as the main method in developing ones. For example, the percentage of public shares privatized through public offerings is 100% in the UK, 83% in Italy, 79.9 in Spain and 55% in Germany. In contrast, it was 0% in Turkey, Mexico and Tunisia. This difference may seem to result from the fact that stock markets in developing countries are not well-developed with neither the depth nor the liquidity required for stock offerings of large state-owned telecommunications service providers. Yet, given that one of the primary objectives of privatization programs in developing countries is to increase capital market deepening, this can be seen as a failure of the privatization implementation itself.

Turkey, as most other developing countries, adopted block sales as the main privatization method for TT. The first step towards the privatization of TT was in September 1993 with changes in the Telegraph and Telephone law 406 under the amendment of law 509. The telephone service is then split from the state telecommunication monopoly PTT (postal, telephone and telegraph) and converted to a joint-stock company under the ownership of the Undersecretariat of Treasury of Republic of Turkey. However, this amendment was cancelled only one month later in October 1993. In June 1994 with law 4000 it was re-established as a separate entity from PTT. In May 1995, law 4107 authorized the sale of 49% of the company .Also with this law, the GSM market was opened to competition. On June 14, 1995, it is included in the privatization list and transferred to the Privatization Administration. However, in March 1996 the Constitutional Court cancelled significant parts of the law closing the door for its privatization. In August 1996, law 4161 was enacted paving the way once again for its privatization[10].

In the end, it took 4 years for succsessive governments to pass a law that was in accordance with the objections of the Court. During this period the privatization plans also went through radical changes. For example, the law 4107 that was accepted in the parliament in 1995 (and later cancelled by the constitutional court) stated that 51% of the shares would stay under public ownership, 10% would be transferred to Postal Administration, 5% would

---

[10] See Aybar et al. (2001) for a detailed chronology of the successive rounds of amendments and legal battles.

be sold to the TT employees and only the remaining 34% would be privatized (Yilmaz, 1999). Yet even then several barriers remained such as the conversion fo TT balance sheets to international standards, seperation of TT assets from those of Postal Administartion, and the asset valuation.

Another important benchmark in the liberalization and both deregulation and re-regulation of the telecommunications sector in Turkey was in January 2000 with the passing of law 4502 that made the legal structure compatible with the European norms. One of the most visible changes brought forward with this law was the separation of policy making, regulation, and management functions of telecommunications sector. As a result, a new independent regulatory agency, Telecommunications Agency, has been created and became functional starting from August 2000 to regulate and supervise the sector. With this law the role of state in telecommunication sector has been redefined.

Since 1995 Turk Telekom has been put on auction twice for 20% and 33.5% of its shares in June 2000 and December 2000 respectively without success because of market conditions and lack of interest from international players, legal complications, and perhaps more importantly lack of political consensus.

Following the failure in the auctions, in May 2001, the parliament revised the existing regulations and passed article 4673. The new article, among others included the following changes: a) Creation of Golden Share system to give the public sector a final saying in decision making in order to protect national interests and security; b) The authorization to privatize all shares except the Golden shares; c) Limiting the ownership share of foreign firms with 45%; d) Authorizing the Telecommunications Agency to give operating permits and licences; e) Public offering of 5% of shares to TT employees and general public.

However, prior to its final privatization in 2005, several other changes and cancellations to the initial law were implemented in July 2004 including: a) lifting of restrictions on foreign ownership and restricting the coverage of golden shares and allowing privatization of more than 50% of shares; b) legal changes regarding bad loans of TT; c) Enabling the sale of 100% of TT; c) Allowing the transfer of its employees to other SEEs once the public share falls below 50%; d) Elimination of several special tax obligations of TT including contributions to earth quake fund, civil defence fund, national productivity centre fund, and Turkish Standards Institute; e) the satellite operations have been taken out of TT to function as a separate public entity.

The final stage towards privatization started with the auction on December 25, 2004 and ended on June 24, 2005. A total 4 groups made bids for operating licences of TT for 21 years and the results were announced on July 1, 2005. Oger Telecoms Joint Venture Group (composed of Saudi Oger and Telecom Italia) offered the highest bid (6.550 million dollars). The official transfer of ownership took place on November 14, 2005.

Overall, similar to other privatization cases in Turkey, the lack of political consensus and commitment, and highly fragmented and polarized political and bureaucratic structure of Turkish politics delayed the implementation of privitization of TT from 1994 till 2005. From the date of its privatization decision to 2006, 18 cases against its privatization have been brough to the courts (Yildirim, 2006). In the end, partly resulting from the legal complications, lack of political consensus, and inconsistent changes in the privatization laws, and partly because of the changes in international conjucture (i.e. Eastern European privatizations were completed before TT) the final sale was realized at a much lower value than the initial estimates of 27-40 billion dollars by Morgan Stanley and Lehman Brothers.

## 4. POST-PRIVATIZATION ANALYSIS

At the time of its privatization in 2005, Turk Telekom was the 13th largest telecommunication firm in the world with no accrued social security or corporate tax liabilities. It had 21,153,000 telephone line capacity, 80,000 paid public phones, 19,000,000 fixed phone lines, 35,000,000 km. copper cable lines, 100,000 km. fiber optic cable, and 3,000 telecom distributors. TT owned 40% of cell phone company AVEA (which was the 3$^{rd}$ largest mobile phone operator in Turkey with 6.1 million subscribers and 795 million dollar net sales at the time of privatization).[11] TT Net, (TT's internet service subsidiary), had 1,668,000 subscribers with 100% digitalization rate in exchanges and transmissions. TT's net sales in 2004 were $4.6 billion with a net income of 1.86 billion. In this section, we will examine the post-privatization performance of TT from the target (i.e. public) and acquirer perspectives.

### 4.1. Gains to the Target

Since the public was both the seller and customer of TT, this section does not only examine the target premium (as in traditional M&A studies), but also studies benefits/costs of TT privatization for the public.

#### *a. Target Premium*

After a transparent auction process, Oger Telecoms Joint Venture Group took control of 55% of TT shares for 6 billion 550 million dollars in 6 annual instalments. The first instalment that was due at the time of sale was $1.637 billion dollars and the following 5 instalments were $ 983 million each year. Since the net profit of TT in 2004 was 1 billion 700 million dollars, the annual instalments were not even equal to its net yearly profit. Furthermore, due to windfall oil gains in Gulf countries, Oger Inc. was able to secure cheaper credit lines in these countries and used them to pay off the remaining instalments as of March 2007, 4 years ahead of the schedule. Collectively, these may suggest that TT was undervalued. However, TT was sold through a transparent international auction when the economic indicators in Turkey were quiet strong. Thus, the undervaluation, if exists, was not due to the supply side of the transaction, but rather might be attributed to demand side such as poor financial performance of international telecommunications service providers. One potential solution for better target premium might be postponing the privatization. Although there is no clear indication for better pricing, there are problems associated with postponement. First, postponement may have adverse effect on the economy as a whole. Since the privatization of TT had been on the agenda for more than a decade and had been cancelled three times, another cancellation may cast serious doubts on the liberalization policies in Turkey. This, in turn, might decrease FDI which was very sensitive to changes in economic policies. Second, participating auctions is costly for bidders. Top management teams of these firms divert their attention from existing businesses and allocate their time on potential acquisition projects. These projects also have opportunity cost of forgoing positive

---

[11] After the privatization, TT acquired an additional 40.59% of Avea from Telecom Italia and increased its ownership to 81.1% of the firm.

net-present value (NPV) projects. Thus, postponing privatization when the annual GDP growth rate was 6% might discourage potential bidders from participating future auctions which, consequently, may result in further decrease in privatization value.

## b. Pricing of Local and Long-distance Calls

At the moment, 34% of the telecom bills are from taxes (15% VAT, 15% special communication tax, 1% communication tax) and therefore it is very unlikely to see a fall in general price level after privatization. In fact, TT increased its local calling charges by 27% where it has a virtual monopoly while reducing the long distance rates by 80% where it faces competition from other long distance providers. Given that 85% of total call volume is local, the net effect on final calling rates is a 10.95% increase. This corresponds to around 800 million dollar increase in its sales in 2007. However, as of July 30, 2007, telecommunications oversight committee allowed competition for local calls. Although increase in competition will be likely to decrease rates for local calls, the time period between the effective privatization and competition dates (2 years) served as an exclusivity period for the acquirer.

## c. Labor Transfers

There was a sharp reduction in the number of employees between 2005 and 2006 (Figure 4). The new management of Turk Telekom reduced the work force by more than 10,000 employees in addition to another 5,000 that were eliminated prior to privatization. Based on the privatization agreement, laid-off TT workers were employed by other SEEs. On the one hand, the transfer of former TT employees is likely to increase the current expenditure of the state and will be extra burden on the budget. On the other hand, this transfer can be interpreted as re-allocating resources. To the extent that the reallocation improves productivity of existing workers, TT is likely to provide more efficient services for the customers.

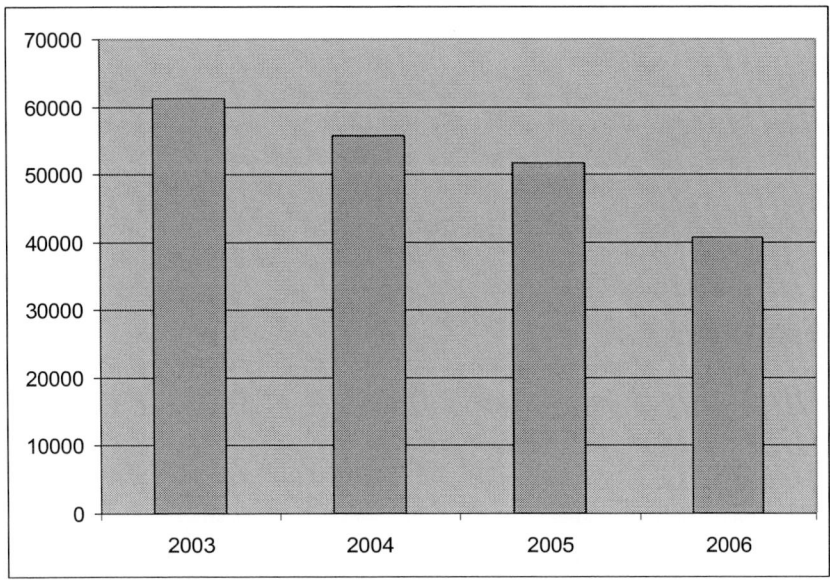

Figure 4. Number of Employees in TT.

### d. Transfer of Funds to the Treasury

Prior to privatization, TT transferred around $4 billion to public funds in 2004 and $3.8 billion in 2003. Thus, privatization agreement significantly reduced this large sum of transfer by eliminating TT's tax obligations to extra-budgetary funds and semi-public foundations as well as through reducing the corporate tax rate from 30 to 20% right after privatization of TT in 2005. Furthermore, the state will give $500 million for investments in rural areas in each year. In addition to state subsidies, the loss of funds and foregone taxes may contribute to imbalances in the state budget in the future. Furthermore, it is very likely that the new owner of TT will purchase equipments and other inputs it needs from one of its foreign subsidiaries in a country with lower tax rates. This may have two major implications: First, as a result of differential pricing or intra firm pricing strategies, it may further reduce its tax obligations. Second, as a result of falling demand from TT (that is their main customer), local industries supplying such equipments and inputs may be financially distressed with serious ramifications for employment creation and technological know-how development.

### e. Progress in FDI

Privatization of large telecommunications SEEs catch the attention of large international investors seeking for new growth opportunities. Specifically, these investors are very hesitant to make investments in developing countries which are prone to political and economic instability. Under the instable conditions, investors cannot precisely forecast long term benefits/costs and are unable to calculate the liquidity of their investments. To the extent that developing countries provide stable economic and political environments, international investors can be expected to seize the growth opportunities in these countries. Thus, execution of TT privatization is likely to bring more FDI towards Turkey. Although there is no consensus in the literature regarding the net employment effect of FDI, the increasing FDI inflows, especially in the form of Greenfield FDI, can be expected to help solve the unemployment problem in the Turkish economy.

## 4.2. Sources of Value Creation for the Acquirer

Mergers and acquisitions create value for the shareholders to the extent that the merged firm have larger after-tax cash flows than the sum of the after-tax cash flows of the individual firms before the merger. The gains to acquirer can be achieved through four channels:

### a. Tax Shields

Taxes constitute large fraction of value transfer from shareholders to government. To the extent of magnitude of tax savings, shareholders capture firms' cash flows. Since interest payments are tax-deductible, debt-increasing strategies allow large tax savings. Acquisitions associated with debt increases create tax savings. The acquirers can achieve these tax savings through financing acquisitions primarily with debt. They can also increase the debt level following acquisition if the total debt capacity of the merged firm is greater than the sum of individual debt capacities of acquirer and its target. This is more likely to be the case if target is under-leveraged prior to being acquired. Merged company can also have greater debt capacity if it is better diversified than the individual firms and thus has lower financial distress cost. Firms with low financial distress costs have larger optimum debt levels which in turn, create excess debt capacity. In addition to debt-driven tax savings, government can offer tax breaks for acquisitions in industries that are essential for well-functioning economies. Taxes played an important role in value creation for TT. Prior to privatization, TT had no debt which allowed the prospective acquirer to create tax shields and to create value through tax savings. Although tax savings generate value for acquirer shareholders, they are not in the best interest of the public. Tax savings of TT may contribute to imbalances in public account deficit.

## b. Operating Synergies

Operational synergies are attained when the merged firm is more profitable than the separate firms. This is only possible through improving efficiency (revenues) or cutting costs. The merged firm can increase efficiency through effectively transferring resources from one division to another. Firms can reduce cost through combining facilities, sales forces, or disposing underutilized equipments.

Table 2 shows that following privatization net sales dropped dramatically from $5.500 million in 2005 to $4.431 million in 2006. However, net profit and EBITDA do not show significant changes in the pre and post privatization periods. Collectively, these suggest that there are no short-run gains or losses in the privation of Turk Telekom.

### Table 2. Financial Indicators for TT, 2004-2006

|  | 2004 | | 2005 | | 2006 | |
| --- | --- | --- | --- | --- | --- | --- |
|  | (YTL) | ($) | (YTL) | ($) | (YTL) | ($) |
| Net Sales | 6.581 | 4.606 | 7.411 | 5.5 | 5.963 | 4.431 |
| EBITDA | 3.351 | 2.345 | 4.007 | 2.974 | 3.093 | 2.298 |
| Net Profit | 2.516 | 1.866 | 3.242 | 2.405 | 3.2 | 2.378 |
| Investments | 0.526 | 0.368 | 0.471 | 0.349 | 0.284 | 0.211 |

### Table 3. Financial Ratios for TT, 2004-2006

|  | 2004 | | 2005 | | 2006 | |
| --- | --- | --- | --- | --- | --- | --- |
|  | (YTL) | ($) | (YTL) | ($) | (YTL) | ($) |
| Net Profit/ Net Sales | 0.38 | 0.41 | 0.44 | 0.44 | 0.54 | 0.54 |
| Net Profit/ # of employees | 45094.45 | 33444.46 | 62663.08 | 46485.11 | 78587.39 | 58400.26 |

### Table 4. Number of Employees and Customers in AVEA, 2004-2006

|  | **2004** | **2005** | **2006** |
| --- | --- | --- | --- |
| Number of employees | 1,012 | 1,169 | 1,335 |
| Pay-as-you go Customers | 2,994,902 | 2,694,737 | 4,458,794 |
| Subscribed Customers | 1,825,693 | 3,421,095 | 3,005,306 |
| Total Customers | 4,820,595 | 6,115,832 | 7,464,100 |

## Table 5. Financial Indicators of AVEA

|  | 2004 | | 2005 | | 2006 | |
| --- | --- | --- | --- | --- | --- | --- |
|  | (YTL) | ($) | (YTL) | ($) | (YTL) | ($) |
| Net Sales | 691 | 515 | 1.066 | 795 | 938 | 659 |
| *EBITDA* | (169) | (126) | 26 | 19 | 154 | 108 |
| Net Profit | (804) | (599) | (783) | (584) | 99 | 69 |
| Investments | 215 | 160 | 498 | 371 | 349 | 233 |

Meanwhile, investments decreased from $349 million in 2005 to $211 million in 2006. Since TT also reduced work force following privatization (Figure 4), these findings collectively suggest that Turk Telekom sustained short-run profitability through cost reduction.

Although there was no significant increases in the profits in the short-run, TT attained superior production efficiencies. For example, profit margin increased from 0.41 in 2004 to 0.54 in 2006 (Table 3). Similarly, Net profit per employee also improved between 2004 and 2006 due to labor force reduction following privatization. These findings substantiate the view that the new TT management improved the productivity.

Extending our analysis on AVEA, the mobile phone subsidiary of Turk Telekom, provides further insight on the strategies of the new management. The labor force of AVEA increased from 1012 in 2004 to 1335 in 2006. This change corresponds to 30% increase in the number of employees of AVEA. This finding, along with the work force reduction in TT, suggests that the new management shifts (labor) resources from unprofitable divisions to efficient divisions. Furthermore, there is change in the target customers of the firm. AVEA focuses on pay-as-you-go customers and almost doubled its revenues, whereas its revenues from subscribed customers decreased by 10%. This is consistent with the improvement in the efficiency of the TT following privatization. Consequently, net profit increases from negative $584 million in 2005 to a positive $69 million in 2006.

To the extent that the new management in TT transfer resources from inefficient divisions to efficient ones, improvements in efficiency will create value for acquirers. This may also translate into better service quality for the public. Furthermore, acquirers (OGER Inc and Italia telecommunications) are also likely to cut the costs of TT. The acquirers have already operations in telecommunications. Thus, they can consolidate activities and outsource services through geographical segments in which the firms operate. This will also allow TT to create value through reduction in costs.

### c. Better Incentive Alignments for the Top Management Team

Separation of ownership and management leads to agency problems between managers and shareholders. To the extent that managers are entrenched and do not behave in the best interest of shareholders, they manage assets of the company to benefit them personally. This, in turn, results in inefficiencies and undervaluation of firm's assets. Acquirers that detect these problems can create value through better management practices in target companies.

One of the major problems associated with SEEs is the poor management incentives. The appointments of the top management team in TT were political which motivates managers to behave in the best interest of the ruling party at the expense of shareholders (public).

Cumulative findings of previous studies on privatization indicate performance improvements following the change of ownership in SEEs. Consistent with this view, compensation of the new management is tied to firm performance which aligns the interest of managers with those of the firm.

### d. Financial Synergies

Acquisitions can create value through combining value companies (cash-rich but low growth) with growth companies (high growth, but cash-poor). Companies with high growth potential may not finance projects. This is especially important for capital intensive investments. To the extent that positive NPV projects receive funding they would not have otherwise received, acquisitions of growth companies by value companies create value.

Following 1994, investments in TT declined rapidly partly due to economic crisis in Turkey and partly because of its privatization decision. Therefore, its growth lacked behind other telecommunication firms in the world. Thus, prior to privatization, TT had significant growth potential with insufficient financial slack for investments. Meanwhile, Oger Inc, located in Saudi Arabia and the lead firm in the acquisition consortium, had access to low-cost capital due to windfall gains in oil exports and had few growth opportunities in existing business units. Collectively, acquisition of high-growth TT by cash-rich Oger Inc. may create substantial financial synergies.

## 5. CONCLUSION

In this study, we examine the Turkish privatization experience in the context of sale of Turk Telekom (TT), state-owned telecommunications service provider of Turkey. In addition to traditional synergy sources including tax shields, better incentive alignment of new TT management and financial synergies, we document the role of operational synergies in value creation in M&A. Specifically, the re-allocation of labor force from inefficient business units to value-creating ones contributed to operational synergies. This finding suggests that efficient allocation of merged company resources is one of the channels to create value in M&A and improves our understanding of how value created in privatization of telecommunication SEEs, in particular and in M&A, in general.

In addition to value creation, we also examine how value is transferred to the public. We find that privatization of TT was associated with lower tax collections and higher rates for local calls. On the other hand, breaking up the satellite operations from TT paved the way for competition. Furthermore, privatization of TT served as a credible signal about the liberalization policies and growth opportunities in Turkey and is likely to improve FDI.

This study also documents the importance of business and legal framework in privatization. The lack of political consensus and commitment, and highly fragmented and polarized political and bureaucratic structure of Turkish politics delayed legal foundation necessary for implementation of privitization of TT for more than a decade. The privatization of TT was implemented only after the majority party took decisive steps to pass laws in accordance with privatization and liberalization policies between 2001 and 2004.

# REFERENCES

Andrade, G., Mitchell, M. & Stafford, E. (2001) New evidence and perspectives on mergers. *Journal of Economic Perspectives, 15*, 103-120.

Aybar, C.B., Guney, E.S. & Suel, H. (2001) *Privatization And Regulation In Turkish Telecommunications: A Preliminary Assessment SNHU International Business Program Working Paper Series* No. 2001-02. Available at SSRN: http://ssrn.com/abstract=278089 or DOI: 10.2139/ssrn.278089

Bortolotti, B., D'Souza, J., Fantini, M. & Megginson, W.L. (2002) Privatization and the sources of performance improvement in the global telecommunications industry, *Telecommunications Policy, 26*, 243-268

Boyland, O. & Nicoletti, G. (2000) *Regulation, market structure and performance in telecommunications.* Paris: Organization for Economic Cooperation and Development

Bugra, A. (1994), State and Business in modern Turkey. Albany, NY; State University of New York Press.

Chan-Olmsted, S. & Jamison, M. (2001). Rivalry through alliances: Competitive strategy in the global telecommunications market, *European Management Journal* 19(3):317-331.

communications: An introduction, Editorial, Telecommunications Policy, 26:219-224.

Demir, F. (2004), A Failure Story: Politics and Financial Liberalization in Turkey, Revisiting the Revolving Door Hypothesis." *World Development, 32*(5):851-869.

Ercan, M. & Onis, Z. (2000), *Politics within the state: Institutions and Dilemmas of Turkish Privatization in Comparative Perspective*, Bogazici University.

Fink, C., Mattoo, A. & Rathindran, R. (2002)An Assessment of Telecommunications Reform in Developing Countries, *World Bank Policy Research Working Paper* 2909, Washington, DC: World Bank.

Gutierrez, L.H. & Berg, B.(2000) Telecommunications Liberalization and Regulatory Governance: Lessons from Latin America, *Telecommunications Policy 24*, pp.865–84.

King DR, Dalton D.R., Daily, C.M. & Covin J.G. (2004) Meta-analyses of post-acquisition performance: Indications of unidentified moderators. *Strategic Management Journal* 25, 187-200

Li, W., Qiang, C. Z-W. & Xu, L.C. (2001) The Political Economy of Privatization and Competition: Cross-Country Evidence from the Telecommunications Sector, *Development Research Group mimeo,* Washington, DC: World Bank.

Megginson, W.L.(2005) *The Financial Economics of Privatization.* New York: Oxford University Press

Parker, D. and Kirkpatrick, C. (2005) Privatisation in Developing Countries: A Review of the Evidence and the Policy Lessons, *Journal of Development Studies, 41*, 513 - 541

Pennings, J., Kranenburg H.V. & Hagedoorn, J., (2005) Past, Present and Future of the Telecommunications Industry, in *The Aging New Economy: The Growth and Dynamics of New Media Firms.* Editor: Cinzia Dalzotto.

Privatization Administration, http://www.oib.gov.tr/index.htm

Ros,A.J.(1999) Does Ownership or Competition Matter? The Effects of Telecommunications Reform on Network Expansion and Efficiency, *Journal of Regulatory Economics* 15, 65–92.

Unctad (United Nations Conference on Trade and Development) (2007), *Foreign Direct Investment Online Database,* Extracted on 7.9.2007.

Wallsten, S.J. (2000) *Telecommunications Privatization in Developing Countries: The Real Effects of Exclusivity Periods,* mimeo, Stanford University and the World Bank.

Wallsten, S.J., 2001, An Econometric Analysis of Telecom Competition, Privatization and Regulation in Africa and Latin America, *Journal of Industrial Economics, 69,* 1–19.

Waverman, L. & Trillas, F. (2002). Corporate control and industry structure in global communications: An Introduction, *Telecommunications Policy 26,* 219-224

World bank (2007), *World Development Indicators Online database,* World Bank.

Yildirim, B. (2006), *Türkiye Büyük Millet Meclisi Genel Kurul Tutanağı* 22. Dönem 4. Yasama Yılı 61. Birleşim February 09.

Yilmaz, K. (1999) Turk Telekomunikasyon Sektorunde Reform: Ozellestirme, Duzenleme ve Serbestlesme. (Reform in the Turkish telecommunications Sector: Privatization, Regulation, and Liberalization) in *Atiyas, İzak, (Ed.), Devletin Düzenleyici Rolü (State and Regulation),* Turkish Economic and Social Research Foundation (TESEV) Publication, August 2000

Zollo M. & Singh H. (2004) Deliberate learning in corporate acquisitions: Post-acquisition strategies and integration capability in US bank mergers. *Strategic Management Journal* 25, 1233-1256

Reviewed by Omar S. Dahi, *Assistant Professor of Economics,* School of Social Science, Hampshire College, Amherst, MA 01002.

Chapter 7

# CROSS MEDIA OWNERSHIP: AN ANALYSIS OF REGULATIONS AND PRACTICES IN AUSTRALIA, HONG KONG AND SINGAPORE

*T.Y. Lau[1], Katie Look[1], David Atkin[2] and Carolyn A. Lin[2]*

[1]University of Washington, Seattle, WA, USA
[2]University of Connecticut, Storrs, CT, USA

## ABSTRACT

This study compares the regulations and practices of cross media ownership in Australia, Hong Kong, and Singapore, utilizing a conceptual framework to explore relationships between the government, service providers and consumers. Government regulatory policies are then analyzed in the context of recent technological trends pushing media convergence. Study findings suggest that market size does not affect policy-making and that private ownership is the key determinant of policy outcomes. The relative merit of various regulatory approaches (i.e. "government- guided" vs. "market-oriented") are discussed, alongside implications of cross media ownership changes wrought by technology.

The pace of global media consolidation is intensifying, driven by dramatic changes in business models, technology and regulation. Pelton (2003) notes that the telecommunication sector has grown 500% among developing countries since the 1970s, contributing to an emerging $4 trillion I.C.E. (information, communication and entertainment) sector that's helping fuel economic growth. Anowkwa, Lin, and Salwen (2003) suggest that this growth results from globalization, technical as well as economic convergence, privatization, and deregulation. The latter is driven by on Western business models, particularly those stressing the importance of free markets and media deregulation. Perhaps the most significant experiment in media deregulation can be found in the Telecommunication Act of 1996, notable for its liberalization of media ownership rules and removal of line-of-service entry barriers between broadcast, cable, local and long distance industries in the U.S.

Long championed by free-market conservatives, media deregulation is aimed at removing strictures on cross media ownership (e.g., Baer, 1995). Symptomatic of the paternalistic models under which most global media structures have been based, such restrictions might take the form of an enterprise in a specific market being prevented from owning shares--or establishing and operating another legally separate enterprise--in an adjacent market (Sato, 1998). In the communications sector, this includes ownership between telecommunications and cable television, telecommunications and broadcasting services, cable television and broadcasting services and within television services (see, e.g., Atkin, Hallock, & Lau, 2007; Bates, Jones, & Washington, 2002; Baxter, 2005; Gillet, & Vogelsang, 1999; Pelton, 2003). The present study compares the regulations and cross media ownership dynamics in three cosmopolitan Asian markets: Australia, Hong Kong, and Singapore.

Governments adopted cross media ownership rules decades ago when media outlets were distinct (e.g., Auferheide, 1999; Atkin, 1999; Coutreau, 2003; FCC, 2004; Garcia-Murillo, 2001; Horwitz, 1991). Now, with market structures changing and media converging, conservatives believe that current rules are in desperate need of reform (e.g., Baer, 1995; Baxter, 2002; Ireland, 2002; Rosenberg, 2001; Lowenstein, 1997; Telecommunications Reform, 1996). Others are less sanguine about such deregulation, suggesting that relaxation of cross media ownership rules could imperil viewpoint diversity, localism, competition, community interests (e.g., Atkin, 2002; Doyle, 2002; Horwitz, 1991; Kahn, Tardiff, & Weitzman, 1999; McChesney, 1997; Parenti, 1992; Ross, 1996).

The present analysis employs a conceptual framework outlining a triangular relationship between the government, service providers and consumers to explore the differences and similarities in each country's regulatory approach. In particular, the actions or inactions taken by each of these Asian governments will be evaluated in the context of a comparative framework. The analysis seeks to determine relationships between media convergence and media ownership policies, including the influence of these emerging market structures on media conduct and performance.

## CHANGING MARKET STRUCTURES-ESTABLISHING A BASELINE COMPARISON WITH THE U.S.

Dating back to Siebert, Peterson and Schramm's (1963) seminal "Four Theories of the Press," the U.S. has been cited as a model of media freedom, a fact which helps explain the country's leadership in media deregulation. In particular, the Telecommunication Act of 1996 (110, Stat. 56, 1996)—which eliminated most cross-ownership restrictions across broadcast, cable, and telephone media--represents the most radical experiment in media deregulation to date (e.g., Atkin, 1999; Atkin et al., 2006; Auferheide, 1999). This measure has been followed by relaxation of ownership caps in electronic media, championed by the Bush administration (e.g., Atkin & Lau, 2006; Dreazen, 2002), and generally supported by the courts (Belson, 2005; Dreazen & Solomon, 2001; Gellhorn, 2004).

As the chief beneficiaries of media deregulation, American industry executives support consolidation and claim there is no evidence that, for instance, commonly owned newspapers and broadcast stations pose any threat of anticompetitive behavior (Belson, 2005; Dreazen, 2001). Citing the advantage of scale economies, conservatives maintain that co-owned stations have extensive newsgathering resources and strong community ties that facilitate more in-depth news coverage than other media outlets in the community (Strum, 2003). Others point to the decline of scarcity-based rationales for regulation, given the proliferation of new wired and wireless channels like the Internet (e.g., Chan-Olmsted & Kang, 2003; Lee & Chan-Olmsted, 2004). In addition, an unprecedented and ever-growing level of media and programming diversity can be found in today's media marketplace (e.g., Albarran & Dimmick, 1996; Labaton, 2005).

Free market advocates thus claim that it makes no sense to subject the broadcasting industry to rules developed when broadcast TV was the chief source of programming, as current rules prevent companies from competing effectively against other forms of media (e.g., Baer, 1995). Larry Pressler (R-S.D.), cosponsor of the Telecommunication Act, argues there are plenty of outlets to protect viewpoint diversity, which will be lost if rules are not relaxed and media outlets can no longer compete (see Telecommunications Reform, 1996). To wit, the best protection against television becoming an increasingly marginalized source of information in today's market place is less regulation and more competition (Ireland, 2002).

Yet, as became evident in the uproar over the FCC's attempt to eliminate media ownership caps—vacated by an appeals court in *Prometheus Broadcasting v. FCC*--not all industry players are in favor of deregulation (see Belson, 2005; Labaton, 2005). Liberal commentators (e.g., McChesney, 1997; Parenti, 1992) maintain that elimination of ownership rules would increase the profits of media conglomerates, undermine competition, reduce the already limited diversity and reduce quality of local media content. Industry hyperbole about the raft of new satellite networks belies the fact that broadcast networks and cable providers own 90% of the most popular cable channels (e.g., Abelman & Atkin, 2002), as new media outlets do not necessarily imply new media content (Albarran & Dimmick, 1996). According to this argument, the public needs multiple, competing, diverse and independent sources of information and entertainment (Crouteau, 2003).

**Table 1. A profile of the U.S. media environment as a baseline for comparison**

| United States | 1960 | 1980 | 2000 |
|---|---|---|---|
| Households with TV | 45 million | 77.8 million | 100.8 million |
| Daily Newspapers | 1,700 | 1,745 | 1,480 |
| Newspaper Readers | 58 million | 62.2 million | 58.8 million |
| Cable Subscribers | 750,000 | 19.2 million | 68.5 million |
| Radio Stations | 4,086 | 9,278 | 12,615 |
| Broadcast TV Stations | 537 | 1,011 | 1,616 |
| Websites | 0 | 0 | 30 million |

Source: (FCC, 2003).

Emulating the U.S. model above, the Asian markets under consideration here were also undergoing radical transformation in their media environments, changes that we explore in the section to follow.

Commentators on both sides of the spectrum acknowledge, however, that the modern media world is dramatically different from that of a quarter-century ago (e.g., Baer, 1995;

Doyle, 2002). As new technologies have emerged with the progression of time, access to the numerous media outlets (television, radio, newspaper, cable, satellite, Internet) has shifted. As shown in Table 1, in the United States from 1960 to 2000, the number of radio outlets grew by 142%, independent radio station owners grew by 74%, television outlets grew by 217% and the number of independent television station owners grew by 150%. Newspapers declined by 9% while the number of newspaper owners remained unchanged (FCC, 2004).

Alongside the growth in traditional media, new media systems mushroomed as analog systems shifted to digital transmission (Pelton, 2003). After experiencing a 1000% growth rate during the 1990s, the Internet claimed over 30 million websites in the new millennium (FCC, 2003); they were joined by 68.5 million cable subscribers and over 20 million DBS subscribers (Labaton, 2005).

## MEDIA CONVERGENCE

Broadcasting and telecommunication authorities globally are trying to find a workable model to balance the impact of media convergence and cross media ownership. New regulatory schemes must consider different network platforms, particularly each's ability to carry similar voice, video and data services (e.g., Bates et al., 2002; Lin & Atkin, 2002). Pelton (2003) maintains that previously disparate wired and wireless modalities--which traditionally faced limitations and were only used for a single application—are now swapping or merging applications. In particular, digitization is expanding the functions of these technologies, allowing all forms of content to be handled over same network in the same manner. The reduced cost of bandwidth allows consumers to access more media outlets, while the concept of spectrum scarcity is fast becoming a constraint of the past.

## COMPARING AUSTRALIA, HONG KONG AND SINGAPORE

In the parlance of Siebert et al.'s (1963) Four Theories of the Press, the present cases were useful because they provide a set of comparison points where South Asian/Pacific countries can be arrayed along a political continuum. Although this framework has been refined over the years, alongside the media terrain that it profiles, the fundamental tenets of the media systems paradigm remain in place; that is, the relationship between a country's media system and its political economy is reciprocal. In this case, the contexts range in scope from a Western style libertarian/social responsibility model (e.g., Australia), to an authoritarian system moving towards a more open, social responsibility model (Singapore) and a social responsibility system that may be moving in an opposite direction as an authoritarian central government asserts its dominion (Hong Kong).

The latter two cases could be classified under one reconceptualization of Siebert's model, developmental theory, which states that press freedom is valued among several developing nations in Asia and elsewhere. But, as Chang and Tai (2003, p. 28) observe, "(A)s a result of their colonial pasts, unstable social and political structures, and inadequate economic infrastructures, these developing nations stress that they cannot afford the luxury of Western-style press freedom." When comparing cross media ownership regulations between countries such as these, it's useful to consider such evaluative criteria as: size of the market, size of advertising revenues, government initiative and leadership, service providers' financial strength and active consumer participation.

Australia, Hong Kong and Singapore were selected for the study because of their similarities and differences in the afore-mentioned factors (see Table 3). Similarities among all three countries include open economies that allow for open competition, a government that promotes competition, and a market rich in media outlets. Singapore and Hong Kong represent countries with higher per-capita GDPs, a factor which usually correlates well with the level of infrastructure development. Given the small size and high population densities of both territories, as well as their relative wealth, Singapore and Hong Kong are well-positioned to continually implement state-of-the-art technologies (Vallath, 2000). A comparison the three Asian countries reveals that Australia's media market is the largest, with over 600 radio stations and 100 broadcast stations (see Table 2).

Hong Kong and Singapore's markets are similar, however, in terms of population. At the beginning of 2003, Hong Kong had 841 registered publications, four free to air channels, over 100 cable TV channels and 14 radio channels (Hong Kong SAR Government, 2004). Singapore is served by over 1.5 million imported publications, 5,500 foreign newspapers and publications, 72 foreign news agencies, seven free-to-air channels, 17 local FM stations and about 40 cable TV channels (MDA, 2004). Besides the variances in market size, other differences can be found in terms of ownership and regulation. While Australia and Hong Kong's media are privately owned by individuals, Singapore's media infrastructure is in transition from government to private ownership. The Australian government thus champions a "hands off" approach, while Singapore offers a more paternalistic regulatory regime. Hong Kong's media were relatively free during the waning days of British colonial rule, but— moving in a direction opposite that of Singapore-- have been increasingly subject to government regulation and inference since the arrival of the Chinese communist administration.

**Table 2. A profile of media infrastructures in Australia, Hong Kong and Singapore**

| Country | Radio Broadcast Stations (AM/FM) | Television Broadcast Stations | Newspapers | Internet Service Providers | Internet Users (2002) |
|---|---|---|---|---|---|
| Australia | AM 262, FM 345 | 104 | N/a | 667 | 10.63 million |
| Hong Kong | AM 7  FM 7 | 4 | 53 | 201 | 4.35 million |
| Singapore | AM 0, FM 18 | 7 (2004) | 10 dailies | 9 | 2.31 million |
| United States | AM 5,804, FM 6,161 | 1,714 | 10,855 | 7000 | 165.75 million |

Sources: (Central Intelligence Agency, FCC, U.S. Census Bureau, Singapore Media Development Authority, Singapore Infomap, Australian Bureau of Statistics, Hong Kong Office of the Telecommunications Authority, Hong Kong SAR Government Information Centre, various years.

**Table 3. The political economy of Australia, Hong Kong and Singapore**

|  | Australia | Hong Kong | Singapore | Similarities |
|---|---|---|---|---|
| Size of Market (population) | 19.7 million | 7.3 million | 4.6 million | Open economies which allow open competition |
| Ownership | Private | Private | Previously government owned | Rich in Media |
| Government Concept | Minimum interference | Minimum interference | Strong interference | Government promotes competition |

## CONCEPTUAL FRAMEWORK: GOVERNMENT, SERVICE PROVIDERS AND CONSUMERS

The present study explores a conceptual framework outlining a triangular relationship between the government, service providers and consumers. First, we assume that these three major players will interact in their perception and impact of cross media ownership. In more paternalistic regulatory models, governments will take a leading role in deciding if new guidelines are necessary to redefine cross media ownership in response to changing market and technology convergence environments. Moreover, service providers may push for regulatory changes owing to self-interest, changing market needs and operation modes.

Despite recent moves towards media privatization across the globe (e.g., Ogan, 2007; Tunstall, 2008) government remains the major service providers in several countries. For example, Singapore's television broadcasting services were owned by the government before undergoing privatization in 2000. However, in the triangular relationship with regulators and media, consumers are typically the "silent voices" unless there are outspoken interest groups to speak for their rights in this cross-media policy making process (see Krasnow, Longley, & Terry, 1992). Therefore, by exploring the conduct of these policy determiners, the present study examines two key questions: (1) how do the criterion nations act to ensure a level playing field for all service providers to operate in the market, and (2) how do regulators act in these domains to ensure that the public's best interests are served?

## CURRENT STATUS OF CROSS-MEDIA OWNERSHIP

The global media landscape continues to be transformed by globalization, privatization, deregulation, and even a decline in Western domination (e.g., Ogan, 2007). In order to analyze policy implications in Australia, Hong Kong and Singapore, it's useful to briefly profile the current status of cross media ownership for each country in the respective sections to follow.

## Australia

The cross-media ownership rules in the Australian Broadcasting Services Act (BSA) of 1992 prohibits a person from being in a position to exercise control over any combination of commercial television station, commercial radio station and a newspaper in the same license area (ABA, 2004). Australia's cross-media rules were created with the intent of ensuring diversity in the sources of information and opinion and a plurality in the ownership of media. These rules are seen to act as an absolute barrier to media firms taking advantage of economies of scale and scope, which might arise from owning different types of media in the same market. The Broadcasting Services Amendment Bill of 2002 was introduced into Parliament to reform cross-media ownership restrictions, although it did not pass. The initiative abolishes or relaxes cross-media ownership rules contained in the BSA--encouraging greater competition and use of new technologies--while providing strict safeguards to ensure diversity of opinion and at least minimum levels of local news and information (OECD, 2003). The bill was passed into law during November of 2006, with several provisions related to local diversity taking effect in 2007 (http://www.acma.gov, 2007). In a further attempt to streamline regulatory oversight in light of media convergence, the Australia Communications and Media Authority took over the role of the Australian Broadcast Authority (ABA) in July of 2005. Australia thus stands poised to emulate the deregulatory initiatives of her Anglo-American counterparts (Tunstall, 2008), although the freedoms granted by her social responsibility traditions than those afforded her counterparts on the Asian mainland.

## Hong Kong

In Hong Kong, authorities are overseeing a media infrastructure that hews generally to the tenets of social responsibility—instilled under British colonial rule—but must now ultimately answer to the more authoritarian leanings of the mainland. Consistent with this more paternalistic approach, a newspaper proprietor or sound broadcasting licensee shall not exercise control of a Domestic Free and a Domestic Pay license except with prior approval (HKBA, 2004). In the year 2000, Li Ka-Shing required an approval from the Broadcasting Authority since he occupied top positions at both Cyberworks--a Hong Kong Internet company with a pay TV subsidiary (HKTVOD)--and Metro Broadcast Corporation, a radio broadcaster. The Broadcasting Authority said that neither HKTVOD nor Metro were dominant players in either of markets and went on to characterize each market as distinctly separate. In addition, the companies agreed there would be no cross-management, cross-control or cross-subsidization between firms (Creed, 2000). This market-driven model stands in contrast to the more paternalistic regulatory approach found in Singapore, which is reviewed in turn.

## Singapore

Despite her capitalist leanings, Singaporean media evolution during the late 20$^{th}$ century reflected the authoritarian leanings of its dictator, Lee Kwan Yew. More recently, Singapore could be classified in the 'tutelary' media system alongside other Western-leaning proto-democracies like Tiawan (e.g., Lin & Salwen, 1986). Here the Code of Practice for Market Conduct in the Provision of Mass Media Services, introduced in 2003, prohibits a broadcasting licensee, newspaper company and any other person that provides Mass Media Services or Ancillary Media Services from entering into a Consolidation that is likely to unreasonably restrict competition in any Mass Media Services Market or sub-market in Singapore (SMDA, 2004). The Media Development Authority (MDA) determines cross-media licenses on a case by case basis. Factors for consideration in consolidation include the market in which the applicant is entering, the market participants, the level of concentration in the market, the structure of the market, the likelihood that existing market participants or new entrants would respond to consolidated entity, and efficiencies that would likely result from the consolidation.

To further analyze the cross-media ownership discussion, the perspectives of the government, service providers and consumers are provided below. Table 4 provides a summary of stakeholder perspectives across each country.

**Table 4. A summary of stakeholder perspectives in Australia, Hong Kong, and Singapore**

|  | Australia | Hong Kong | Singapore |
|---|---|---|---|
| **Government** | Deregulate – Current restrictions are outdated and should be reformed to adapt to converging media. | Regulate – Restrictions are necessary to minimize conflict of interest, build-up of monopoly of the media and editorial uniformity. | Both – Consolidation can create pro-competitive effects but can also create monopolistic power. |
| **Service Providers** | Deregulate – Technology gives Australians access to many voices and current rules restrict a company's ability to grow. | n/a | Both – Minimal government regulation is necessary for healthy development of the industry, especially during times of tremendous technological changes, but regulations are necessary when one firm dominates. |
| **Consumers** | Regulate – Consolidation of media would concentrate power, restrict variety of opinion, reduce competition and diminish local content. | n/a | n/a |

## PERSPECTIVES OF KEY POLICY DETERMINERS

### Australia

The Australian government is in favor of reforming the current cross media restrictions to improve media companies' access to capital, facilitate investment in new technologies, enable media companies to grow and expand in the new content-driven converging global media environment, and ensure that customers have access to high quality media offerings (ABA, 2004). Current restrictions on cross-media ownership constrain Australia's media sector within outdated structures while, around the world, media businesses are being driven by the imperative of delivering readily adaptable content across multiple platforms. At the same time, convergence within the communications sector is arguably making these restrictions increasingly redundant.

The Cross-media ownership law enacted in 1992 was designed to meet public policy concerns, but the media market has since changed materially. Fairfax Holdings, one of Australia's top publishers, continues to advocate a liberalization of crossownership strictures that limited the company's ability to grow and restricted access to its share register. The launch of pay television and the exponential growth in the Internet gives Australians access to many more sources (News Interactive, 2004). Telstra, Australia's leader in telecom services, believes that cross media ownership will open up opportunities for companies that could be potential allies, or rivals (Telstra, 2003).

Consumers in Australia are primarily concerned about declining levels of local and regional news and information programs on both TV and radio. Local services play an important role in developing community identity, and ensuring that important information is relayed in a timely fashion (DCITA, 2004). Although the 2006 crossownershp law continues to restrict 3-media (radio, TV, paper) combinations locally, concerns remain that concentration of media would concentrate power, restrict variety of opinion, reduce competition and diminish local content. While consumers agree with the government and services providers that it is necessary to amend the Act to consider the new forms of media that have emerged in recent years (e.g., subscription television), and to account for the convergence of telecommunications, broadcasting and the internet, diversity will not be preserved by removing the current restrictions. This is particularly evident given that Australia's media infrastructure remains dominated by a few large corporations (Sheehan, 2002). New initiatives like Bridge Network's venture with telecommunications provider Telstra promise to enhance program choice, by sending multiple video channels to mobile receivers (Keshishoglou, 2004). But, drawing from the American model, true competition between phone and video providers may be years in the making (e.g., Atkin et al., 2006) and have only a minimal impact on content diversity.

### Hong Kong

Although Hong Kong enjoys a rather unique status as a former British colony, it does provide an interesting experiment in the cohabitation of a traditionally open media system moving under the suzerainty of a command-style regime. Chan (2003, p. 214) notes that "In the wake of the reunification with China, the media played a role in monitoring the performance and the policies of the Chinese ruling government." He concludes (p. 214) that, despite this openness, "(M)any journalists are worried that the government will eventually enact laws to restrict their coverage of politically sensitive issues."

The Hong Kong Government believes restrictions are necessary to minimize conflict of interest, build-up of monopoly of the media and editorial uniformity (HKBA, 2003). Such policies flow from a rather paternalistic regulatory regime, albeit one under which the Chinese communists strive to avoid the appearance of a heavy hand. Chan's (2003) analysis of the Hong Kong model thus concludes that "Broadcast enjoys political freedom and could, in theory, carry programs critical of government policies" (p. 211). For that reason, media and audience interests seek to preserve the strains of press freedom from Colonial days which—while not completely open—did encourage the development of a multiparty press. To the extent that such deregulation can serve as a model for the mainland, Hong Kong media may well facilitate the adoption of freedom and democracy on the mainland. And while the mainland has allowed limited involvement from such foreign media potentates as Rupert Murdoch and Bill Gates, the Communist Party is likely to strictly limit any foreign ownership. This approach is in line with the tenets of the *developmental press theory* (Anowkwa et al., 2003). To wit, Chang and Tai (2003) note that, "If unchecked, press freedom tends to rock the boat, disrupt national harmony, or hinder economic development (p. 28)."

By comparison, in the United States, consumer advocates fear that loosening cross-media rules would lead toward greater media consolidation, and result in less independent, less diverse points of view on our airwaves. Media is central to democracy and the public interest should supercede all others (Inslee, 2003). In addition, new media may not be a substitute for traditional media. Emerging web and cable news outlets reach different kinds of audiences—typically one that's more upscale--with a different kind of news (e.g., Lin & Atkin, 2007). Television is not an exact substitute for reading the newspaper (e.g., Stepp, 2001). In that regard, consumers are worried about the growing power of big media conglomerates to snuff-out independent viewpoints and charge exorbitant prices for basic services.

## Singapore

The Singapore Media Development Authority believes that consolidation can have pro-competitive effects, such as creating economies of scale and scope. However, as the rationale underlying much antitrust regulation suggests, consolidation may also harm competition (see Gellhorn, 2004). For example, mergers could create an entity that has market power or facilitate unlawful collusion amongst competing entities. The MDA will reject any application for consolidation if it is likely to unreasonably restrict competition in the Mass Media Services Market (MDA 2003).

In Singapore, service providers such as Starhub Cable Vision support competition in the mass media industry, especially in light of revolutionary changes caused by convergence. As one commentator notes, this change encompasses "(d)irect interconnections between competing wire line phone companies and interconnections between different media that never used to compete at all. Television is leaving the air in favor of the wires; the telephone is leaving the wires in favor of the air. Copper and coax, wired and wireless, terrestrial and satellite...universally interconnected standard for the transmission of everything – voice, data, video, the lot...The dynamics of the new technology have forced regulatory change" (Starhub Cable Vision, 2003, p. 1). This convergence of sending modalities has been termed the "Pelton merge" (Pelton, 2003). Deregulation advocates maintain that minimal government regulation is essential to the healthy development of the industry, especially when the industry is undergoing tremendous technological changes which in turn are producing an industry- transforming convergence in each sector. Regulation should, according to this view, be imposed only when a firm has achieved market dominance.

## CONCLUSION

The issue of how a merger will affect diversity and pluralism is prevalent in all perspectives examined here. Although scholarly studies on the impact of consolidation media diversity are often inclusive (e.g., Dimmick et al., 2000), particularly in the context of political voice and pluralism, it's useful to consider some of the ramifications of ownership deregulation. It is axiomatic that a diversity of choice is critical for assuring congruence between consumer preferences and the available media information, while at the same time assisting in the maintenance and development of national or community cultures. Viewpoint diversity is also important for preserving democracy by ensuring that a multiplicity of voices heard in the "marketplace of ideas," a marketplace that's threatened by media consolidation (Agrawal, 2006; McChesney, 1997). Under the tenets of libertarian theory, such openness represents the best guarantor of the public interest and helps ensure that both private and public decision makers are held accountable for their actions.

Unfortunately, government initiatives to encourage diversity through radical deregulation—as exemplified by the U.S. since the 1980s—often encourage the growth of private monopolies that imperil access to the press and freedom of speech (e.g., Auferheide, 1999). Since the three countries examined all promote competition, it is possible that companies will have an incentive to offer differentiated content and increase diversity to avoid competing on price. Conversely, a reduction in diversity could induce greater reliance on price, and benefit consumers by offering higher quality at a lower price. However in countries such a Singapore, where diversity and quality are tightly regulated, it is possible that no changes will take place. The question of how deregulation will affect diversity and pluralism thus remains unresolved in such contexts.

In sum, the present study of cross media ownership policies in Australia, Hong Kong and Singapore yields the following observations:

- The size of a market does not make a difference in the policy making process of cross media ownership, and
- Privatization yielding a high level of private ownership represents an important facilitator of media infrastructure development.

It remains an open question as to which policy vehicle can best achieve these goals in the national contexts explored here. Singapore follows a "government guided" policy while Australia and Hong Kong adhere to "market-oriented" initiatives, and it's difficult to determine which can most effectively reconcile the often competing goals of market efficiency and public service. This situation remains fluid in the context of the nascent, unfolding era of convergence in which various stakeholders now find themselves.

What's clear is that cross media ownership dynamics continue to evolve as media technologies develop and converge. In terms of the policy of cross media ownership, based on the Asian contexts considered here, two options present themselves: either maintain or change the status quo. In terms of the former, perhaps the most extreme model for deregulation can be found in the U.S., which moving towards a pure marketplace approach based on media self-regulation. Given America's strong leadership role in pushing for open markets and less paternal regulatory models, other countries will be looking closely to see whether initiatives to vacate decades-old media cross-ownership and audience cap strictures can succeed.

The answer to this question will depend, in large measure, on the extent to which American-style self-regulation can ensure that there is a leveled playing field and the publics' best interests are served. It will be important, in later work, to explore the extent to which policymakers draw from Western examples, and how they would evaluate the success of American style deregulation over the past decade. Later work should also consider cases in markets in economies that are less developed than those explored here. As the market and media convergence environments continues to evolve, cross media ownership may assume either a reactive or proactive approach.

## REFERENCES

Abelman, R. & Atkin, D. (2002). *The televiewing audience: The art and science of watching television.* Cresskill, NJ: Hampton.

Agrawal, B. (2006). *Television in South Asia.* New York: University Press.

Albarran, A., & Dimmick, J. (1996). Concentration and economies of multiformity in the communication industries, *Journal of Media Economics, 9*, 41-50.

Anowkwa, K., Lin, C., & Salwen, M. (2003). *International Communication: Theory and cases.* (New York: Wadsworth).

Atkin, D. (1999). Videodialtone reconsidered: Prospects for competition in the wake of the Telecommunication Act of 1996. *Communication Law and Policy Journal, 4*, 35-58.

Atkin, D. (2002). Convergence across media. In C.A. Lin and D. Atkin (Eds.). *Communication technology and society: Audience adoption and uses* (pp. 23-41). Cresskill, NJ: Hampton.

Atkin, D., Hallock, J. & Lau, T.Y. (2007). Local & long distance telephony. In A.G. Grant and J. Meadows, *Communication technology update* (pp. 273-283). London: Elsevier.

Atkin, D., Lau, T.Y. & Lin, C.A. (2006). Still on hold: Prospects for competition in the wake of the Telecommunication Act of 1996 on its 10 year anniversary. *Telecommunications Policy, 30*, 80-95.

Auferheide, P. (1999). *Communications policy and the public interest: The telecommunications Act of 1996.* New York: Guildford.

Australian Bureau of Statistics. Accessed on 22 February 2004 at http://www.abs.gov.au/websitedbs/D3310114.NSF/home/Statistics.

Australian Broadcasting Authority (2004). Limitation on Control – Cross Media Limitations. Accessed on 22 February 2004 at http://www.aba.gov.au/ownership/limitations/x_media.htm.

Australian Communications and Media Authority. Media ownership and control. http://www.acma.gov.au/WEB/STANDARD//pc=IND_REG_MEDIA (accessed June 20, 2007).

Baer, W.S. (1995). Telecommunications infrastructure competition: The costs of delay. *Telecommunication Policy, 19*, 351-370.

Bates, B., Jones, K.A., & Washington, K.D., (2002). Not your plain old telephone: New services and new impacts. In C. Lin and D. Atkin (Eds.), *Communication technology and society: Audience adoption and uses* (pp. 91-124). Cresskill, NJ: Hampton.

Baxter, J. (2002). Executives defend cross-media ownership. *Vancouver Sun.* PG E9. Accessed on 9 December 2002 through LexisNexis.

Belson, K. (2005, Jan. 28). Dial M for merger. *The New York Times*, C1, 4.

Central Intelligence Agency. The World Factbook. Accessed on 22 February 2004 at http://www.cia.gov/cia/publications/factbook/index.html.

Chan, J.M. (2003). Mass communication and development in Hong Kong. In K. Anowkea, C. Lin, & M. Salwen (eds.). *International communication: Cases and issues* (pp. 210-214). Belmont, CA: Wadsworth.

Chan-Olmsted, S. & Kang, J.W. (2003). Theorizing the strategic architecture of a broadband television industry. *Journal of Media Economics, 16*, 3-16.

Chang, T.K. & Tai, Z. (2003). Freedom of press in the eyes of the dragon. In. K. Anowkwa, C. Lin, & M. Salwen (Eds.). *International communication: Cases and issues* (pp. 24-46). Belmont, CA: Wadsworth.

Coutreau, D. (2003). Remarks to the FCC Broadcast Ownership En Banc. Accessed on 22 February 2004 at http://www.fcc.gov.

Creed, A. (2002) "CyberWorks clears Hong Kong cross-media ownership laws. *Newsbytes.* http://www.newsbytes.com.

DCITA (2004). Background on broadcasting services Amendment Bill 2002. Accessed on 22 February 2004 at http://www.dcita.gov.au/Printer_Friendly/0,00_1-2_10-3_492-4_114882-LIVE_1,00.html.

DCITA (2004). Modernizing Australia's media ownership laws. Accessed on 22 February 2004 at http://www.dcita.gov.au/Printer_Friendly/0,00_1-2_10-3_492-4_103841-LIVE_1,00.html.

Dimmick, J. (2000). *The theory of niche.* Mahwah, NJ: LEA.

Doyle, G. (2002). What's 'new' about the future of communications? *Media, Culture & Society.* Vol. 24, pg. 714-724.

Dreazen, Y. (2002, July 15). FCC's Powell says telecom 'crisis' may allow a Bell to buy Worldcom, *Wall Street Journal*, p. A4.

Dreazen, Y. & Solomon, D. (2001, March 5). Court overturns FCC's ownership caps, in victory for AT&T, AOL, cable firms. *Wall Street Journal*, p. A3.

Federal Communications Commission (2003). FCC 03-127. Accessed at on 27 March 2004 at http://www.fcc.gov/mb/policy/

Federal Communications Commission (2004). Office of the Broadcast License Policy. Accessed on 22 February 2004 at http://www.fcc.gov/mb/broadcast_policy.

Federal Communications Commission. (2003). FCC sets limits on media concentration. Accessed on 22 February 2004 at http://www.fcc.gov/ownership/documents.html.

Garcia-Murillo, M. (2001). FCC organizational structure and regulatory convergence. *School of Information Studies, Syracuse University.* Accessed on 22 February 2004 through Proquest.

Gellhorn, E. (2004). *Antitrust law and economics.* St. Paul, MN: West Publishing.

Gillett, S.E. & Vogelsang, I. (1999). *Competition, regulation and convergence: Current trends in telecommunications policy research.* Minneapolis, MN: West.

Horwitz, R. (1991). *The irony of telecommunication reform.* London: Oxford.

Hong Kong Broadcasting Authority (2004). Who may obtain broadcasting licenses. Accessed on 22 February 2004 at http://www.hkba.org.hk.english//licences/who.html.

Hong Kong Office of the Telecommunications Authority. Accessed on 22 February 2004 at http://www.ofta.gov.hk/datastat/main.html.

Hong Kong SAR Government. (2004). Hong Kong: The facts. Accessed on 26 March 2004 at http://www.info.gov.hk/hkfacts/media.pdf.

Inslee, J. (2003). Media mergers endanger democracy, diversity of news. *The Seattle Times.* Accessed 11 March 2003 at http://www.seattletimes.com.

Ireland, J. (2002) Prepared Statement for the FCC Official Hearing. Accessed on 22 February 2004 at http://www.fcc.gov.

Kahn, A.E., Tardiff, T.J., & Weisman, D.L. (1999). The telecommunications act at three years: An economic evaluation of its implementation by the Federal Communications Commission. *Information and Economics Policy, 11*, 319-340.

Keshishoglou, J.E. (2004). Television content in transition: Today's entertainment landscape is moving by leaps and bounds, leaving consumers bewildered. *Media Asia, 31, 218-223.*

Krasnow, E., Longley, C., & Terry, H. (1992). *The politics of broadcast regulation.* New York: St. Martins.

Labaton, S. (2005, Jan. 28). U.S. backs off relaxing rules for big media. *New York Times*, pp. C1, C2.

Lee, C. & Chan-Olmsted, S. (2004). Competitive advantage of broadband Internet: A comparative study between South Korea and the United States. *Telecommunications Policy, 28,* 648-677.

Lin, C.A. & Atkin, D. (2002). *Communication technology and society : Audience adoption and uses.* Cresskill, NJ: Hampton.

Lowenstein, R. (1997, Feb. 27). Antitrust enforcers drop the ideology, focus on economics, *Wall Street Journal*, pp. A1, 4.

McChesney, R. (1997). *Corporate media and the threat to democracy.* (New York: Pine Forge Press).

Parenti, M. (1992). *Make believe media.* New York: Wadsworth.

Pelton, J. (2003). International telecommunications. In K. Anowkwa, C. Lin, & M. Salwen, *International Communication: Theory and cases* (pp. 267-284). Belmont, CA: Wadsworth.

News Interactive (2004). Media ownership laws not needed: Fairfax. Accessed on 27 March 2004 at http://www.news.com.au/common/story_page/0,4057,2973761%5E1702,00.html

OECD Secretary General. (2003). *Media mergers.* Organization for Economic Co-operation and Development (OECD).

Rosenberg, M. (2001). Dispatch from the telecom battlefield, *Utility Business*, May, p. 1.

Ross, S.D. (1996). When the wires cross: Ensuring diversity in the era of video dialtone, *Communication Law and Policy, 1*, 65-97.

Sato, K. (1998). *Cross-ownership and convergence: Policy issues.* Organization for Economic Co-operation and Development (OECD). Working Papers, Vol. VI.

Sheehan, P. (2002). Media ownership and control: The next step. *Newsweekly.* Accessed on 9 March 2003 at http://www.newsweekly.com.au.

Siebert, F., Peterson, T., & Schramm, W. (1963). *Four theories of the press.* Urbana: University of Illinois Press.

Singapore Infomat. Accessed on 23 February 2004 at http://www.sg.snapshot/snap_media.asp.

Singapore Media Development Authority (2004). Accessed on 23 February 2004 at http://www.mda.gov.sg/media.

Starhub Cable Vision. (2003). Comments on the Proposed Code of Practice for Market Conduct in the Provision of Mass Media Services. Accessed on 27 March 2004 at http://www.mda.gov.sg/MDA/documents/Starhub%20Cable%20Vision.pdf.

Stepp, C. (2001). Whatever happened to competition? American Journalism Review. Accessed on 10 March 2003 at http://www.ajr.org.

Strum, J. (2003). Statement to the FCC. Accessed on 22 February 2004 at http://www.fcc.gov.

"Telstra Eyes Cross-Media Opportunities" (2003). *Business Times.* Accessed on 9 March 2003 at http://adtimes.nstp.com/my/archive/2002/mac11f.htm.

Telecommunications Act of 1996, 104 Pub. L. 104, 110 Stat. 56, 111 (1996) (codified as amended in 47 C.F.R. S. 73.3555).

Telecommunications Reform (1996). *Congressional Digest, 75*, 3-26.

Vallath, C. (2000). The technologies of convergence. In Mark Hukill, Ryota Ono, and Chandrsekhar Vallath ed., *Electronic Communication Convergence: Policy Challenges in Asia* (pp. 33-37). Thousand Oaks, CA. Sage Publications, Inc.

Umino, A. (2004). *Broadband Audio-Visual Services: Market Developments in OECD Countries.* OECD.

U.S. Census Bureau. Statistical Abstract of the United States: 2003. Accessed on 23 February 2004 at http://www.census.gov/prod/www.statistical-abstract-03.html.

In: Telecommunications Research Trends
Editors: H. F. Ulrich, E. P. Lehrmann, pp. 143-157
ISBN: 978-1-60456-158-6
© 2008 Nova Science Publishers, Inc.

*Chapter 8*

# NIGERIA: REVIVING A FORMER MONOPOLY IN A RAPIDLY EVOLVING MARKET

### *Chuka Onwumechili*
Department of Communications at Bowie State University,
Bowie, MD, 20715, USA

## ABSTRACT

Nigeria's telephone market has changed remarkably in the last few years with NITEL, the erstwhile monopoly, rapidly losing market share to vibrant competitors. Prior to 1992, NITEL was the sole provider of telephone services to Nigeria. However, NITEL was unsuccessful and regulatory changes in 1992 led to policies allowing the entrance of private providers into the market.

A decade later, the market reached a watershed when the regulator, the Nigerian Communications Commission (NCC), auctioned GSM licenses that dramatically changed the market. The new licenses quickly re-shaped the market environment by generating rapid increases in customer subscription and introduction of value added services. NITEL, clearly unable to compete, attempted to stem the rising tide by creating interconnection bottlenecks. The attempts failed and NITEL's market share went into a free fall. The government later sold NITEL to private investors – Transcorp. Ltd. – but there remains a tough road ahead for the erstwhile monopoly.

This chapter addresses the issues identified above by reviewing the Nigerian telephone market before 1992 and discussing what followed market liberalization. A key focus is an analysis of NITEL's options in a market, which continues to evolve.

## INTRODUCTION

Nigerian Telecommunications Limited (NITEL) has the challenge of facing a changing telephone market and having to re-invent itself in the process. The challenge is a monumental one considering that NITEL had been a public monopoly for virtually all its history and it now has to compete in a liberalized market against heavily capitalized competitors who have entered the Nigerian market. This paper focuses on providing an analysis of options available to NITEL as it seeks to compete effectively in the market and avoid liquidation in the face of fierce competition.

One must note that NITEL finds itself in a competitive market largely because of its own failure to provide service satisfactorily to Nigerians since 1886 when telephone service first arrived in Nigeria. Therefore, this paper will begin with a discussion of key reasons why NITEL failed in its charge to provide effective service to Nigerians and the consequences for the country. In addition, it reviews how Nigeria sought solutions through several policy instruments including Decree 75 and the Nigerian Communications Act (NCA) of 2003, which helped to liberalize the market and introduce private competition. The paper also discusses NITEL's initial reactions to competition and the impact of such reaction on the viability of NITEL and its readiness for competition. Finally, it analyzes the options available to the company in today's market.

## NITEL's Failure in Providing Service

NITEL, previously known as Posts and Telecommunications (P & T) until the merger between P & T and Nigerian External Telecommunications (NET) in 1985, had a checkered history, which featured undelivered promises. It was similar to other state owned enterprises (SOEs) such as the then Electricity Corporation of Nigeria (ECN) and the Water Corporation which repeatedly failed to deliver on the promises of universal basic services to Nigerians.

Though telephone service arrived in Nigeria as early as 1886, service was slow in reaching a large number of Nigerians. The initial network completed before 1962 connected the "colonial office in London with Lagos and the commercial centers in the country with local authority offices" (Ajayi, Salawu, and Raji, 1999, p. 163). Line expansion was notoriously slow, particularly after the country achieved political independence from Britain. Ajayi, Salawu, and Raji (1999) attributed the difficulties to several problems associated with the budget. A major problem was that the level of funds budgeted for the project was much lower than what NITEL needed for such massive development of lines. There was an attempt to ameliorate this by introducing five-year plans where the number of lines progressively expands. Unfortunately, this was not effective as unforeseen circumstances often disrupted such plans. A second problem was that released funds fell far short of budgeted amounts, which further stunted line expansion. In addition, there were catastrophes that retarded expansion plans. The most remarkable catastrophe was the Nigerian civil war, which took place from 1967 to 1970. The federal government shelved all plans for line expansion during the period and instead diverted the funds to support war efforts. It is notable that this situation did not end in 1970 because the effect of the war stretched to a post-war period of reconstruction and repairs to existing lines. Another disrupter was the introduction of new technology in the 1980s. At the time, NITEL undertook an ambitious project of digitizing existing lines. During the period, line expansion became insignificant and attention shifted to digitization. Remarkably, the country's population was growing exponentially while the slow expansion of lines was taking place and, thus, the teledensity for Nigeria remained well below 1:100 as demonstrated in Figures 1 & 2 below.

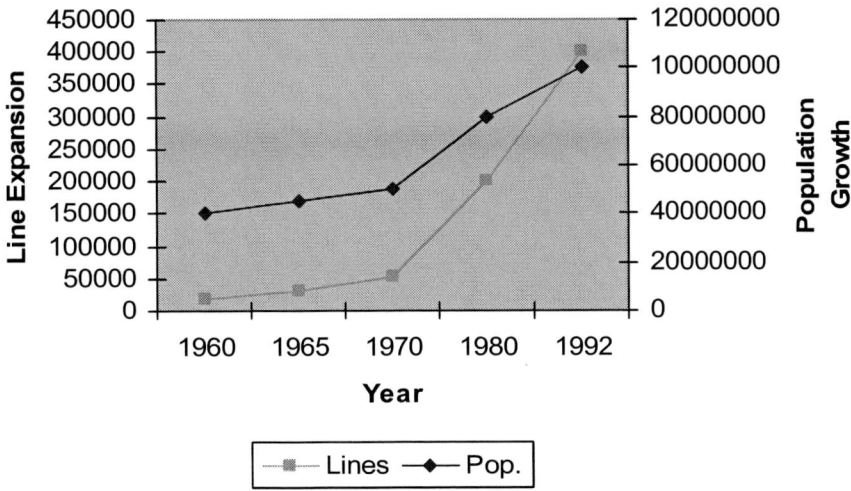

Figure 1. Population and Telephone Line Expansion.

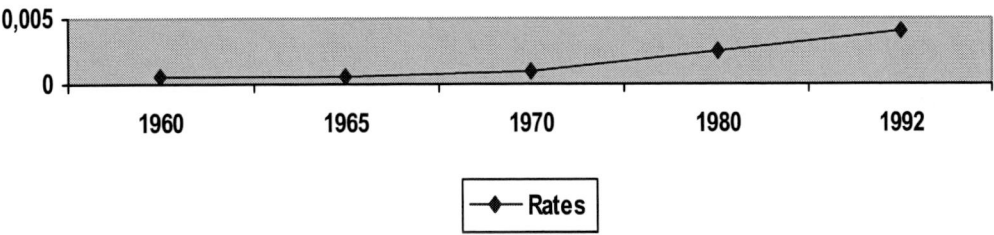

Figure 2. Teledensity 1960-1992.

NITEL, in spite of its poor services, regularly increased its domestic tariffs some times by as much as 600 – 800% (Aragba-Akpore, 1999). However, much of the ensuing revenue went to the benefit of the top managers as profits declined over the years (Nigeria, 1998).

An analysis of the problems described above leads us to several conclusions that were antecedents to the problems of NITEL and they provide us with underlying explanations. These antecedents, described in the subsequent paragraphs, are as follows: the practice of prebendalism and rent-seeking behavior, government bureaucracy, lack of competitive services, and the perception of the services as elite.

Government in Nigeria is widely seen as "distant" and exploitable for private gain. This explains the excessive rent-seeking behavior where employees (1) demand for gratuities in order to provide services, and (2) develop an elaborate system of patronage in procurement. Joseph (1987) described this as *prebendalism* and he is credited with the first use of the term *prebendalism* to describe the sense of entitlement that Nigerians express about state revenue and their focus on exploiting the offices of the state. In essence, government workers, politicians, and members of ethnic groups feel a right to share government funds, properties, and positions amongst themselves and their kin groups. Therefore, a state office becomes similar to a *catholic prebend*, which is the right of chapter members of a church to share in the revenue of the church. This has led to the plundering of government funds and property by employees and politicians. This attitude is prevalent in Nigeria among employees of

NITEL, a government-owned company. NITEL workers have helped plunder the company for private gain and have not taken company goals seriously. Ayogu (1999) wrote: "employees were dishonest, routinely converting payments for billings, carrying on temporary unauthorized line switching, and colluded in the theft as well as in vandalizing the agency's equipments" (p. 12).

The government bureaucracy also helps to worsen the situation. The bureaucracy often creates red tape, delaying and stunting expansion and other types of telephone services. Telephone installation delays could reach "somewhere between 8 and 10 years ..." (Obadare, 2005, p. 13). In addition, Ayogu (1999) identified a number of red tapism that emerged in the bureaucratic activities of NITEL. He listed among others: " ... long delays in obtaining government approval of investments, a 52-week procurement cycle for replacement parts ...." (p. 13).

The lack of competition discouraged NITEL from changing its practices. There was little incentive for NITEL to improve its services because it was a monopoly and could do as it saw fit. At least, this was entirely true in the residential market. However, in the business market, the failures of NITEL did lead to the rise of limited competition. For instance, the government allowed the state-owned Nigerian National Petroleum Corporation (NNPC) to set up its own telecommunications network that bypassed NITEL and a few national banks also received concessions to develop limited networks employing telecommunications transmission technologies such as VSAT.

More telling is the fact that telephone services, at the time, were dedicated to the elite alone. The then Minister of Communications, Colonel David Mark, in the 1980s was cited as noting that telephones were not for "every Tom, Dick, and Harry" as he sought to explain NITEL's failure to provide universal access while he was Minister[1]. Ayogu (1999) stated: "Crucially, it (telephone) has also been a strong element in social stratification as the possession of a telephone was supposed to be an index of where one belonged on the social ladder" (p. 12). The pattern of telephone ownership in Nigeria had long encouraged this phenomenon. Telephones in Nigeria were always accessible based on social class even before Nigeria's political independence from Britain. At that time, telephones were mostly available to the British colonialists and in the colonial administrative offices. After independence, the pattern remained with the administrative offices and a few elites laying claim to ownership of telephone sets. The government offices and corporations owned half of the 400,000 telephones that were operational in 2000 (Ndukwe, 2001). Wealthy Nigerians owned the rest. The cost of the service did not help according to Obadare (2005) who claimed that the cost of telephones (without adding the exorbitant installation charges) was as much as $800! This was at a time when the average annual salary in Nigeria was less than $400.

## NITEL's Poor Performance

The above antecedents explain NITEL's slowness in providing services, particularly in expanding telephone lines. However, the antecedents go beyond explaining the slowness in line expansion. Instead, they also lead us to the explanation of poor customer service,

---

[1] Colonel David Mark was the Minister for Communications under the military dictatorship of General Ibrahim Babangida. Today, David Mark is the Senate President of Nigeria.

overstaffing, the unprofitability of NITEL during the period under discussion and an increasing demand for services.

Onwumechili and Okereke-Arungwa (2003) studied NITEL's customer service by surveying 286 persons from across Nigeria. They tested three hypotheses that focus on NITEL's services and the company's prospects in a competitive market. There were statistical support for the hypotheses, which predicted customer's poor perception of NITEL and its services. The authors concluded, "NITEL is headed for a certain decline on its stranglehold as the dominant provider of telephone service in Nigeria" (p. 69). That prediction has come to fruition. Below summarizes the authors' views of NITEL following the research study:

> ... NITEL's service to its existing customers was increasingly troubling. There were numerous uncompleted calls, delays on repairs, and inappropriate billing, among numerous other complaints. (p. 66)

In addition, there are numerous other references in the literature to NITEL's poor customer service (NCC, 2003-2004 and Ayogu, 1999). Customer cut-offs in mid-sentence are among the numerous problems. A notable story occurred in 2001 when Chief Bode Akindele, Chairman of an investment company that had just completed a deal to purchase NITEL, was cut off in mid-sentence while placing a call to the United States (Ujah and Adeshida, 2001). That symbolized the fact that the company's bungling did not respect the elite or upper class but was endemic. Call completion rates were poor at only 30% in 1990 (Ayogu, 1999 and Aragba-Akpore, 1996) and its wireless subsidiary, M-TEL, was "frequently experiencing service congestion" (Ayogu, 1999). In spite of all the problems, NITEL had no compulsion to improve customer service since its clients did not have access to an alternative service at the time.

Overstaffing was also a problem. Ayogu (1999) wrote that NITEL had "60 employees per 1000 main exchange lines in comparison to 0.2 for New York Telephone (Nynex) ..." (p. 13). More telling, in spite of its large number of employees, were: (1) the long delays for repairs, and (2) NITEL's 327 faults per 100 main lines were, by far, the highest in Africa at the time (Laffont and Meleu, 1997). In essence, productivity is low for the high number of employees. Moreover, there was little training and employees received few or no tools to work efficiently. Yet, NITEL had to pay workers, adding to NITEL's expenditures in comparison to its low revenue.

NITEL's finances were precarious and we have already mentioned several reasons for this. The existence of prebendalism, low line expansion in spite of increasing demand for telephone service, and overstaffing all combined to eat away NITEL's revenue. Ayogu (1999) adds: "NITEL posted operating losses all the way from incorporation to commercialization" (p. 13). Government data shows, for instance, that losses for 1990 and 1991 alone were $18 million and $1.5 million respectively (Nigeria, 1998). As we shall see later in this paper, NITEL's precarious financial situation became an albatross around its neck preventing investors, or at least making them hesitant, from purchasing the company.

The demand for NITEL's services continued to rise exponentially during the pre-liberalization period. One is tempted to attribute this to satisfaction with the company's services. However, studies note that there were, instead, significant levels of service dissatisfaction (Onwumechili & Okereke-Arungwa, 2003). A deeper analysis shows that the rise in the demand for service was symptomatically a demand for general telephone service and not necessarily a specific demand for NITEL's service. Unfortunately, NITEL's long delays in service installation also meant that supply of service did not keep pace with the rapid rise in demand for services.

## ARRIVAL OF A LIBERALIZED MARKET

Though NITEL's troubles were evident, before 1992, it had very little to fear. It was still the monopoly with no alternative service sources for Nigerians. Thus, there was very little incentive for the company to change its behavior as noted previously. An initial government effort towards change was to corporatize NITEL in 1985. Corporatization meant a reduction in government's subsidy of NITEL and a means for the company to move towards commercialization instead of siphoning government funds. However, as we saw with the negative revenue of 1990 and 1991, the finances of the company continued to deteriorate instead of improve.

However, NITEL was to face its first challenge in 1992 after the government promulgated Decree No. 75 liberalizing the telecommunications market and establishing a market regulator i.e the Nigerian Communications Commission (NCC). Though the decree opened the market for the entrance of private competitors, it did not provide the NCC with the authority to regulate NITEL and NITEL was to use this loophole to frustrate activities in the market. Ayogu (1999) wrote: "... the decree establishing NCC left NITEL out of NCC's regulatory domain (and) ... NITEL was often not responsive to efforts from NCC to secure compliance" (p. 20).

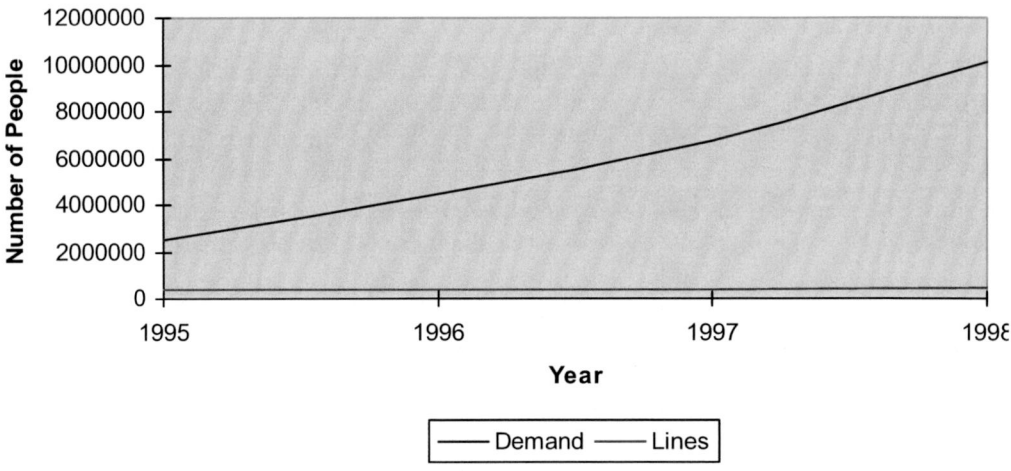

Figure 3. Comparing Demand with Supply of Lines.

NITEL remained the dominant provider during the early years of privatization (1992 – 2001) through a strategy of stifling competition and the NCC's inability to generate a viable competitor for NITEL. More troubling is the fact that the wait list continued to grow rapidly and the number of service lines remained virtually constant. The wait list for subscribers was 2.5 million names in 1995 and then 10 million in 1998 (Aragba-Akpore, 1999). Figure 3 shows a graph identifying the growing gap between the demand for and the supply of service lines.[2] NITEL's practices were unfortunate and eventually the outcome was a catastrophe.

NITEL's strategic fulcrum of stifling competition centered on: (1) delaying competitors' attempts to interconnect to NITEL's lines and (2) setting of high interconnection fees for competitors (Moshiro, 2004 and Ayogu, 1999). These tactics were effective because the early licensees in the market did not have adequate capital to compete effectively. For instance, they did not have the funds to develop their own facilities. Thus, most of them had little choice but to rely on NITEL's facilities. This was the major and critical difference between NITEL's competitors during the early years of liberalization and the company's competitors post-2001. Competitors during the early years filed several to the NCC complaints against NITEL but the NCC had little authority over NITEL and, thus, NITEL's activities continued without abating. Competitors also complained that the interconnection charges were outrageous. They were more expensive than what NITEL charges its own customers to make calls and, thus, this predatory practice had the effect of running competitors aground, as they could not offer service to customers at a competitive price without going under.

The NCC, then under Ogbonna Iromantu, was unable to find solutions. Instead, NITEL's hands strengthened because of Iromantu's policy of granting licenses without verifying the capabilities and resources of applicants. In fact, many of the licensees during the period never provided service (Onwumechili, 2003 and Kubeyinje, 1998). Instead, they procured licenses with the hope that they could later resell the licenses for profit. Moreover, several of them obtained the licenses because of their relationships with people in powerful positions in government. In essence, there were very few serious and capable telecommunications companies in the market at the time. To make matters worse, NITEL was unchallenged as no provider received license to offer nationwide service. Instead, the licensees such as the early private telephone operators (PTOs), who offered fixed services and the fixed wireless access (FWA) services, were limited to regional geographical service areas.

However, NITEL's inattentiveness to strategies that would make the company an effective and efficient service provider in a competitive market eventually backfired. The Iromantu phase ended with the military return to the barracks in 1999. A new democratic government proceeded quickly to change the market environment. There were three major actions by the democratic government that significantly impacted the market as well as NITEL's status in that market: (a) An immediate revocation of licenses awarded under the Iromantu era (Onwumechili, 2003), (b) the appointment of Ernest Ndukwe to head the NCC (Ojo, 2005 and Esselaar, Stavrou, & O'Riordan, 2004), and ( c) the enactment of the Nigerian Communications Act (NCA) of 2003.

The revocation of licenses caused an uproar and litigations followed. However, the revocation decision was effective. For one, it immediately rid the market of speculators whose only interest was to resell the licenses and not to provide service or develop the market. It also cleared out individuals who received licenses because of their relationships with certain persons in the government.

Esselaar, Stavrou, and O'Riordan (2004) credit Ndukwe with transforming the NCC from "an inefficient bureaucracy to a governmental organization run with private sector efficiencies" (p. 10). Ernest Ndukwe, who replaced Iromantu, consulted widely and acted

---

[2] Data of number of lines in service is sourced from the United Nations Statistics Division which is available online through Http:// unstats.un.org/unsd/. Data on demand is from Aragba-Akpore (1999).

rapidly. He decided early on two things: (1) that new market entrants would have adequate capital with the ability to develop their own facilities, and (2) that new entrants must show experience in working in the field. Those two key requirements irrevocably changed the course of the telecommunications market in Nigeria and posed the first major challenge to NITEL's monopoly. The entrance of heavily capitalized competitors brought companies that had the funds to rapidly build their own networks and, thus, largely bypass NITEL. Secondly, these companies had experiences with developing large networked services elsewhere before entering the Nigerian market. This assisted them in quickly developing the Nigerian market and thereby challenging NITEL's monopoly status.

Furthermore, the NCA of 2003 firmly brought NITEL under the regulatory ambit of the NCC (Ndukwe, 2006). For instance, it clearly mandates the NCC to set interconnection rates where providers are unable to do so. For instance, NITEL has set up the current interconnection rates for providers. These rates are 9 cents for wireless services and 4 cents for fixed services (Moshiro, 2004). Those rates are dramatic reductions from the average rates of 13 cents and 10 cents respectively (NCC, 2003-2004). Section 97.2a-c of the Act also authorizes the NCC to intervene when it deems interconnection agreements inconsistent with the Act's provision or when it considers such agreement to be against the public's interest (NCA, 2003). Prior to the Act, NITEL had regularly delayed interconnection and at times provided such interconnection at an exorbitant and predatory rate in an anti-competitive move. Moshiro (2004) confirms this through the following statement: "The PTOs experienced difficulties in interconnecting with NITEL's network, particularly before NITEL was subjected to oversight by the NCC" (p. 9).

## Liberalized Market: Market Evolution and Impact on NITEL

The post-Iromantu era clearly denotes a major change in the telecommunications market, particularly in the areas of competition and regulation. Invariably, those changes became significant for NITEL. They were to change NITEL forever to the point where NITEL's survival is now in the balance.

The key impact was on market competitiveness. Ndukwe's arrival in 2000 made an immediate impact within one year. The NCC, under his leadership, oversaw the first major auction of three GSM licenses involving a screened list of bidders selected in terms of capitalization strength and technical capability. MTN Nigeria, a subsidiary of MTN South Africa; Econet propped by Zimbabwean interests; and Communications Investments Limited (CIL)[3], mostly owned by Nigerians won the licenses at a cost of $285 million each. NITEL received an additional license. The first shock was that companies would pay such a large amount to enter the Nigerian market. Analysts previously believed that the market had few people who could afford to pay for telecommunications services but the over 10 million people on the wait list for NITEL services since the late 1990s was enough to convince investors that there was money in the market. Though NITEL paid the fees, its market

---

[3] The NCC denied Communications Investment Limited (CIL) license after the latter failed to make payments on agreed deadline. CIL's CEO, Mr. Mike Adenuga, later established Globacom receiving a Second National Operator (SNO) license allowing it to offer both fixed and GSM services.

behavior did not change and it continued to believe that its tactics in the early phase of market liberalization would continue to be effective against new but well capitalized competitors.

This logic, however, collapsed. NITEL's tactics proved annoying to the new competitors but it failed to deter them. MTN and Econet quickly built up subscribers and by 2003, GSM telephones surpassed NITEL's fixed lines as the service of choice. MTN and Econet, at the beginning, had problems interconnecting with NITEL but they went ahead to build their own facilities and as time went on, the importance of NITEL's facilities diminished. Moreover, these new entrants became very aggressive and NITEL could not cope. For instance, the new companies introduced several value-added services, variety of service plans, and per second billing. The aggressive competition also forced prices down while the quality of service improved all to the consternation of NITEL. The result was NITEL and MTEL's rapid loss of subscribers to new competitors and the resultant decline in market share (NCC, 2005 and NCC, 2003-2004). Figures 4 & 5 show market share changes impacted by the entrance of adequately capitalized competitors. The new competition helped Nigeria to surpass the modest goals of the government's Vision 2010 committee, which is to achieve a teledensity of only 2:100, increase fixed lines to 4 million and wireless lines to 3 million. See Table 1 for current industry achievements compared to Vision 2010 goals (Vision 2010, 1997).

However, for NITEL that story was different. NITEL did not only suffer declining market share but its revenue plummeted and liabilities rose rapidly to the point that liquidation had become a serious consideration for the one-time monopoly.

The revenue loss was not surprising because NITEL's operating lines fell dramatically over the years. Most subscribers also left NITEL to sign up with one of the new GSM providers because of poor customer service. By 2004, the number of subscribers to NITEL's fixed line service fell by half and subscribers to its wireless service (MTEL) fell to less than 50,000 (Aragba-Akpore and Okwe, 2005). These contributed to the declining revenues. Unfortunately, NITEL also failed to implement any significant project initiative to attract new customers. Its attempt to expand its lines, under Pentascope management in 2005, failed to attract the necessary funding.

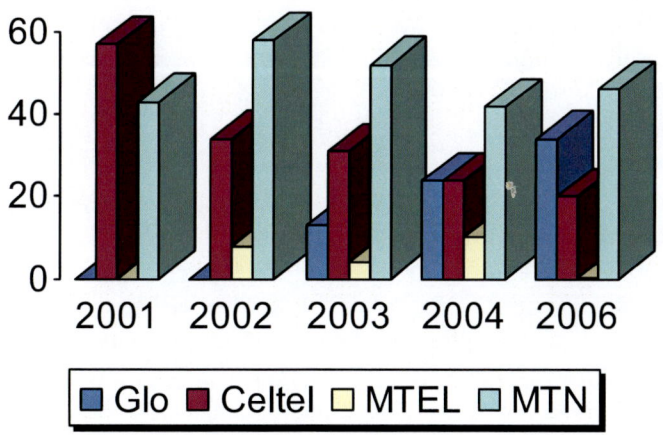

Figure 4. Market Shares of Wiress Companies.

## Table 1. Comparing Recent Industry Data with Vision 2010 Goals

|  | Vision 2010 Goals | 2006 Industry Data |
|---|---|---|
| **Teledensity (Fixed/Wireless)** | 0.02 | 5.25 |
| **Number of Fixed Lines** | 4.0 million | 1.5 million |
| **Number of Mobile Lines** | 3.0 million | 25.1 million |

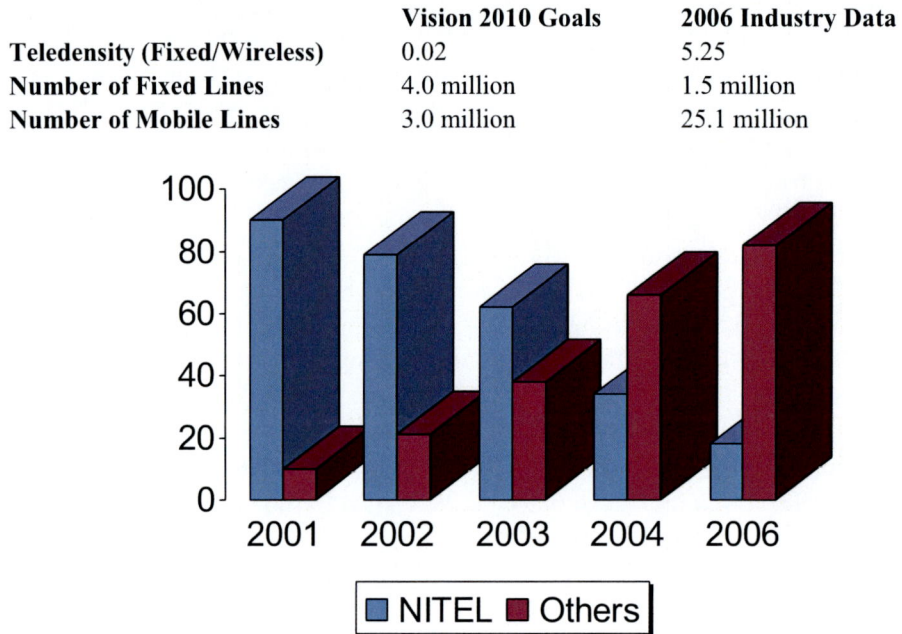

Figure 5. Market Shares for Fixed Line Companies.

More troubling was the increasing liability that NITEL accumulated. For instance, NITEL borrowed large amounts of funds, particularly to pay for its $285 million GSM license. Moreover, the prebendalism that existed prior to liberalization was still present and its effects compounded the company's problems. Overstaffing also persisted. NITEL's labor fought vigorously against attempts to reduce its numbers to a more manageable and productive work force.

In essence, NITEL's reactions to an evolving competitive market failed and the impact on NITEL's survivability was increasingly negative. Its move to slow interconnection failed against its competitors, it was losing revenue, and its liabilities were increasing rapidly (NCC, 2003-2004). It was under this situation that the Bureau of Public Enterprises (BPE) moved to sell NITEL. However, even that attempt met stiff resistance. BPE had long slated NITEL for sale with a plan to dispose 40% of the company to a single private investor, 20% to the public and the government retaining the balance of 40%. However, with NITEL's value predictably falling because of two major problems – declining market share and a depressed global market for telecommunications – BPE moved to dispose NITEL for less than an offer of $1.317 billion that it received in 2002 from Investors International (London) Limited (ILL)[4]. In addition, it changed the sale structure of the company to 51% investor share in an attempt to sell the company. Unfortunately, NITEL's labor fought long and hard against several sale attempts, which further forced the company's sale price downwards. Eventually, BPE was fortunate to sell NITEL for $750 million to a local group of investors under the name

---

[4] BPE revoked NITEL's sale to IIL after the latter failed to pay the balance of the sale fee on deadline. The BPE placed NITEL back in the market at the time.

Transcorp Limited in late 2006. NITEL faces the challenge of restructuring itself in preparation for a competitive market. The problem of prebendalism is now likely in the past but a sale to a private group of investors does not immediately save the company. It still must work to survive and compete having fallen so far from a market monopoly to a position of a minor player in the telecommunications market.

## WHAT ARE NITEL'S OPTIONS

Transcorp Ltd has encountered several problems since its purchase of NITEL in 2006. It has found it difficult to raise substantial funds to reposition NITEL in the market. Worse still, the purchase rules have hampered attempts by Transcorp to attract other major investors in its bid to raise additional funds. Furthermore, Transcorp failed to raise projected funds in a recent Initial Public Offering (IPO) that included attempts to attract interest overseas. Then its technical partner, British Telecommunications (BT), withdrew from the partnership with Transcorp citing financial problems. However, in spite of these bleak events, Transcorp remains in a position to turn around NITEL. A trump card is that several Transcorp principals are close to the Nigerian government and, thus, have access to persons who make policy changes that have tremendous impact on the telecommunications market.

The key question is: What can Transcorp do to turnaround NITEL? In the subsequent paragraphs, I discuss the options available to NITEL. These options include restructuring management, addressing technology issues, improving customer service, and finding the appropriate market niche for the company.

Restructuring management addresses the company's lingering problem. Over the years, the depth of prebendalism and anti-competitive strategies formed the bedrock of the organization. Those practices led to the situation where NITEL is facing the real threat of liquidation. Therefore, Transcorp must initiate restructuring the company with a focus on discontinuing those practices and changing the company culture. Transcorp clearly understands a need for change. In its purchase negotiation with the government, it reached an agreement requiring government to take responsibility of laying-off workers and taking care of worker pensions. Furthermore, Transcorp's partnership with BT demonstrated its understanding of the need for organizational culture change. BT was to install a new culture of productivity while dismantling the entrenched regime of prebendalism. Transcorp must find a replacement for BT that would continue a focus on culture change.

There are reports of decaying and outdated equipment counting as part of NITEL's assets. Analysts note that it could cost as much as $1.5 billion to replace or repair equipment alone. Of course, the figure is astronomical but NITEL has no choice particularly in a market where the competitors are aggressive and have implemented innovative technology. NITEL must leap frog by sourcing funds to begin replacing and repairing the equipment.

New technology will go a long way in helping the company improve its customer service. Customer service has negatively affected NITEL for years (Onwumechili and Okereke-Arungwa, 2003). Its most recent and rapid loss of subscribers is also symptomatic of its customer service problems. New technology will help with quicker repair time and installation while the restructuring of the company's management will change the work culture in the longer term in order to assist with reaching a new employee attitude to work. All of these are important in improving customer service.

However, NITEL's biggest challenge is reintroducing itself to the Nigerian customer and repositioning against aggressive competitors. The company might change its business name for starters because the name NITEL is associated with inefficiency in the customer's mind. A new name will also signal readiness for a re-birth. Importantly, NITEL must reposition itself in the market by locating a market niche. NITEL is far from its former monopoly status and, thus, it can no longer be everything to everyone. Instead, it must locate market niches. For instance, NITEL may chose to focus its interest on the business and government markets and reap larger margins of profit. Intelecon's studies (2005a & b) found that businesses, on the average, spend more compared to individuals per week on telephone calls. There is no major provider currently dedicating service to those two niche areas. The smaller PTOs and FWAs presently provide services in those niche areas but may not be able to compete effectively against an adequately capitalized NITEL (NCC, 2003-2004). Yet, the government remains the largest employer in Nigeria and business is increasingly booming (Economist.com, undated). Increasingly there is substantial remigration to Nigeria by Nigerians who fled overseas under the economic depression prevalent during long periods of military rule. The Economist.com (undated) reports an improved economic outlook including rising GDP, growing foreign exchange, positive account balance since 2003, strengthening exports, falling debts, and decreasing imports. The government remains the largest employer in Nigeria despite growth in private sector employment since 1994 (The World Bank Group, 2002).

Furthermore, NITEL may focus attention on MTEL's wireless services since customers are presently showing preference for wireless over fixed line services (Itelecon, 2005a & b; NCC, 2003-2004; and Akinrele, 2002). There were more Nigerians subscribing to wireless telephones compared to fixed lines by 2003 (Moshiro, 2004). Itelecon's study (2005a & b) involving over 5,000 primary interviews in 144 localities reported that 94% of individual respondents and 95.5% of businesses prefer a mobile telephone to a fixed one. Moreover, there is significant profit taking in the wireless sector. Celtel[5] made over $220 million in six months in 2004 (Ojo, 2005 & Onwumechili, 2005) and MTN made $1.1 billion in six months in 2005 (Ajakaye, 2005 and Onwumechili, 2005). NITEL has an opportunity of being part of those takings if it only makes itself agile by reducing its bloated workforce and repositioning the company's business.

## CONCLUSION

NITEL is clearly at a crossroads where it could either survive or liquidate. NITEL, under its former name P & T, was a monopoly in the Nigerian telecommunications market for over a century until 1992 when the market was first liberalized using Military Decree No. 75. The market has now evolved to a highly competitive one, particularly after the President signed the Nigerian Communications Act of 2003 into law.

Now, NITEL must make the difficult choice of maintaining its behavior status quo or change. If it chooses to remain on its current path of business as usual by utilizing anticompetitive strategies, then its demise shall take place. This paper has chronicled NITEL's past and points out how budget constraints, prebendalism, government bureaucracy, lack of competition, and the perception of its service as elite all contributed to the company's failures before the market was liberalized in 1992. The results of those problems were low line expansion, increasing demand for services, poor customer service, overstaffing, and the company's unprofitability.

NITEL's initial reaction to a liberalized market in early 1992 was to continue with the status-quo while employing anti-competitive strategy of delaying or refusing to interconnect new competitors. It was effective then because (1) the market regulator – the NCC – had little authority over NITEL, (2) NITEL's new competitors were limited to far smaller geographical areas, and (3) NITEL's early competitors had very little capital to compete head-on. Thus, NITEL had little incentive to compete.

However, the enactment of the NCA 2003 coupled with the appointment of a new head of the NCC and the revocation of licenses dramatically changed the market. The new entrants during this period were large and adequately capitalized companies that could compete on facilities and technical knowledge. Furthermore, the NCA 2003 gave the NCC clear authority over NITEL. This rapidly eroded NITEL's dominance and its anti-competitive strategies failed.

This paper, however, argues that NITEL could choose to change under a new ownership following its purchase by Transcorp Limited in 2006. It lists several options for the company in a move to survive. These include restructuring the company's management, addressing technological problems, improving customer service, and locating an appropriate market niche.

## REFERENCES

Ajayi, G., Salawu, R., and T. Raji. (1999). Nigeria: After a century of telecommunications, what next? In E. Noam (Ed.), *Telecommunications in Africa* (pp. 163 – 177). New York: Oxford University Press.

Akinrele, A. (2002). Privatization and deregulation in Nigeria: A paper delivered at the workshop organized for the occasion of the visit of the Canadian Minister for International Trade and his delegation in Lagos, November 21.

Aragba-Akpore, S. and M. Okwe. (2005, February 3). Government sacks pentascope as NITEL's managers. (Available online). Http:// www.nm. onlinenigeria.com/

Aragba-Akpore, S. (1999, September 7). Six million queue for telephone lines. (Available online). Http:// www.ngrguardiannews.com/

Aragba-Akpore, S. (1996, October 15). Tale of woe in telecommunications sector. *The Guardian* (Nigeria), p. 13.

Ayogu, M. (1999). Case studies: Private sector participation in infrastructure in Uganda, Ghana, and Nigeria. (African Development Bank: Economic Research Papers No. 44). Paper prepared for the African Development Report.

Economist.com. (undated). Country briefings: Nigeria. (Available online). Http://www. economist.com.

---

[5] Celtel was formerly Econet at inception in 2001 and then VMobile.

Esselaar, S., A. Stavrou, & J. O'Riordan. (2004). VSAT case studies: Nigeria, Algeria, and Tanzania. Research report prepared on behalf of LINK Centre for the IDRC, CATIA, and GVF.

Intelecon. (2005a). Final report on expanded national demand study for universal access project (Part I – Household Survey). Submitted to NCC.

Intelecon. (2005b). Final report on expanded national demand study for universal access project (Part 2 – Businesses and Institutions Survey). Submitted to NCC.

Jerome, A. (1997). Public enterprise reform in Nigeria: Evidence from the telecommunications industry. African Economic Research Consortium Working Paper.

Joseph, R. (1987). *Democracy and prebendal politics in Nigeria: The rise and fall of the second republic.* Cambridge University Press.

Kubeyinje, K. (1998, June 1). Commission revokes licenses. (Available online). Http:// www. Bday.co.za/98

Laffont, J., & M. Meleu. (1997). A positive theory of privatization for sub-Saharan Africa. Paper presented at the plenary session on Institutions, Governance, and the Political Economy of Development in SubSaharan Africa. AERC Economic Research Workshop in Harare, Zimbabwe.

Moshiro, S. (2004). Licensing in the era of liberalization and convergence (The case study of the Federal Republic of Nigeria). Paper prepared for the International Telecommunications Union (ITU).

Nigerian Communications Act (NCA). (2003). (Available online). Http:// www. ncc.gov.ng/

Nigerian Communications Commission (NCC). (2005). Trends in telecommunications markets in Nigeria, 2003-2004. Abuja: NCC.

Nigerian Communications Commission (NCC). (2003-2004). Trends in telecommunications markets in Nigeria. Abuja: NCC.

Ndukwe, E. (2006, October 11). An overview of the Nigerian telecommunications environment and successful initiatives to promote communications development. Paper prepared for the NCC.

Ndukwe, E. (2001). Address delivered at the Year 2001 Conference and AGM of the Nigerian Society of Engineers, Enugu Branch on October 11.

Nigeria, Federal Republic. (1998). *Telecommunications development and investment opportunities in Nigeria.* Lagos: Prince of Prints Ltd.

Obadare, E. (2005). The GSM boycott: Civil society, big business and the state in Nigeria. (Civil Society Working Paper No. 23). London School of Economics and Political Science.

Ojo, A. (2005, January 1). Overview: Telecommunications, the past is another country. (Available online). Http:// www. newage-online.com/

Onwumechili, C. (2005). Reaching critical mass in Nigeria's telephone industry. *Africa Media Review, 13*(1), 23-40.

Onwumechili, C. (2003). *Reform, organizational players, and technological developments in African telecommunications: An update.* Lewiston, NY: The Edwin Mellen Press.

Onwumechili, C., and J. Okereke-Arungwa. (2003). The morning of competition: Nigeria's NITEL drags its feet with poor customer service. *Info, 5* (3), 65-71.

The World Bank Group. (2002). Pilot investment climate assessment: An assessment of the private sector in Nigeria. (Available online). Http://www. Usaid.gov/ng/

Vision 2010 Committee. (1997, September). Volume I: Main report. (Available online). Http:// www.vision2010.org/

# STRATEGIC BUNDLING IN TELECOMMUNICATIONS AND ITS ANTITRUST IMPLICATIONS FOR INTERMODAL COMPETITION[*]

### Paul R. Zimmerman[†]
U.S. Federal Communications Commission,
Washington, DC, USA

## ABSTRACT

The Telecommunications Act of 1996 and subsequent regulatory actions sought to enhance the degree of competition in the local and long-distance wireline exchange markets by, among other measures, opening the local networks of monopoly incumbent local exchange carriers (in particular, those of the Regional Bell Operating Companies or "RBOCs") to competitive entry and implementing wireless local number portability to facilitate wireless substitution. However, the recent wave of merger activity in the U.S. telecommunications industry (as well as the recent revocation of the Act's unbundling provisions) has led to rapid consolidation within both the wireline and wireless segments of the industry. At the same time, the capacity of independent wireless and Internet-based telecommunications providers to function as sources of intermodal competition against the RBOCs is often championed by policy makers and industry analysts. This chapter examines the validity of this latter contention in the context of several important characteristics of the current telecommunications industry including (but not limited to) the complementarity of wireline and wireless access, the increased tendency for subscribers to take bundled telecommunications services, and the ownership of the largest national facilities-based wireless carriers by the largest regional wireline firms. Particular attention is paid to the interaction between these factors and the use of strategic

---

[*] The views expressed in this paper are those of the author exclusively are not necessarily those of the FCC or its Chairman, Commissioners, or any other staff. All errors are my own.
[†] Email: paul.r.zimmerman@att.net Homepage: http://home.att.net/~zimmy Postal Address: Industry Analysis and Technology Division, Wireline Competition Bureau, Federal Communications Commission, 6-A165, 445 12th Street SW, Washington, DC, 20554

wireline/wireless bundling by the RBOCs to affect the development of intermodal competition (and thus consumer welfare). It is argued that RBOC wireline/wireless bundling may confer various value-added services and other benefits (*e.g.*, lower prices) to consumers in the short-run as the RBOCs employ these bundles to retain their wireline customer base. However, in the long-run, these same benefits may serve to disadvantage independent wireless (and Internet) carriers who cannot offer the same bundled offerings. To the extent that this results in the latter firms being forced to exit the market, wireline/wireless bundling may serve as means for the RBOCs to strategically retain their market power in the local wireline exchange access market while leveraging their wireline market power into the in-region wireless market. As such, the potential for independent wireless or Internet-based carriers to serve as viable intermodal competitors in the long-run appears tenuous at best. Finally, it is posited that the most viable source of intermodal competition in the local residential exchange market will stem from the entry of cable operators into the voice telephony market. As such, the long-run market structure of the telecommunications industry may evolve into a duopoly characterized by RBOCs and cable providers competing in bundles. Whether or not such a market structure is sufficient to induce "aggressive" competition between these two carriers is uncertain, which in turn highlights the need for careful antitrust and regulatory oversight in the future in order to protect the interests of telecommunications subscribers.

## INTRODUCTION

The U.S. telecommunications industry is (once again) undergoing a dramatic period of change. The revocation of the unbundling regime established in the Telecommunications Act of 1996 (hereafter "Act") in March, 2004[1] led to the exit of the two largest competitive local exchange carriers ("CLECs"), namely AT&T Corp. and MCI, from the local wireline residential telephone market and their subsequent acquisition by two of the four remaining Regional Bell Operating Companies ("RBOCs").[2] In light of these developments, federal and state telecommunications regulators are now looking to "intermodal" competition as the next panacea that will introduce viable facilities-based competition into the local wireline exchange access market. But what exactly is intermodal competition or, on a more granular level, an intermodal *competitor*? While these terms are not concretely defined, most telecommunications practitioners would likely classify an intermodal competitor as any firm that: (1) competes against the RBOCs in one or more of their core wireline business areas, and (2) relies primarily upon some *non-wireline* transmission (transport) facility to provide voice telephony services.

Two technologies appear to hold particular promise in serving as sources of intermodal competition. The first is wireless telephony. Wireless telephony uses radio spectrum as the transmission mechanism for carrying wireless voice traffic, thereby offering subscribers the convenience of mobility as well as Internet access, photography, and other advanced "next

---

[1] See *USTA II*, 359 F.3d at 564-76 (D.C. Cir. 2004).
[2] In October, 2005 SBC acquired AT&T Corp. while Verizon acquired MCI. *See* FCC (2005b) and FCC (2005d), respectively. Note that as of 2003 AT&T Corp. and MCI were also the first and second largest interexchange carriers ("IXCs"), respectively, based on national revenue and subscriber shares. *See* FCC (2005c), at Table 9.6 ("Shares of Total Toll Service Revenues") and Table 9.7 ("Residential Household Market Shares"). Note that following its acquisition of AT&T Corp. SBC changed its name to AT&T Inc.

generation" digital services. The second technology is voice-over-Internet-protocol ("VoIP") telephony. VoIP technology converts analog voice signals into digital packets at the originating end of a call, carries them over the Internet (*i.e.*, as opposed to over the wireline network), and re-converts the packets into analog signals at the terminating end of the call. Like wireless, VoIP telephony offers its subscribers access to advanced features not available in ordinary wireline connections. These features include Internet access to voice mail and the simultaneous transfer of voice and data (among others).

The development of alternative technologies to deliver voice and data telecommunications services is occurring alongside a general trend in the telecommunications industry towards consolidation. The acquisition of AT&T Wireless by Cingular in October, 2004 was the first major wireless merger to occur in the post-Act period. This particular merger raised concerns regarding intermodal competition since Cingular (the second largest national wireless carrier at the time) was wholly owned by two RBOCs (SBC and BellSouth) while AT&T Wireless (the third largest national wireless carrier) was not affiliated with a wireline parent. As such, the consummation of the Cingular/AT&T Wireless merger resulted in the largest national wireless carrier being owned and operated two major wireline carriers. And while the U.S. Federal Communications Commission ("FCC") concluded that AT&T Wireless was aggressively marketing its offerings in order to encourage mass market (*i.e.*, residential and small business) wireline subscribers to drop their wireline connections and adopt the firm's wireless service exclusively pre-merger (a practice commonly referred to as "cut-the-cord"),[3] the FCC (as well as the Antitrust Division of the U.S. Department of Justice) did not mandate any divestitures to alleviate any potential harms to the subsequent development of intermodal competition.[4]

Rapid consolidation in any industry will, of course, raise antitrust concerns regarding anticompetitive outcomes (*e.g.*, higher retail prices, fewer product choices for customers, diminished incentives for innovation, *etc.*) arising from unilateral or coordinated effects. However, at the same time, it is undeniable that the product space of the telecommunications market is transforming from an "*á la carte*" environment (*e.g.*, with subscribers choosing one firm as their local wireline exchange carrier, a second as their long-distance carrier, a third for their wireless carrier, and possibly a fourth for their Internet access provider) towards one where consumers purchase most (if not all) of their telecommunications services from a *single* firm. That is, subscribers appear to be increasingly likely to purchase "bundles" of telecommunications services. The advantages of such "one-stop-shopping" from the consumer's perspective include (but are not limited to) only having to pay one monthly bill, deal with one company's service department, and possibly receiving a discount from their subscribed carrier. As such, to the extent that the trend towards consolidation in telecommunications reflects in part carriers' attempts to offer a broader set of services to their customers, the relevant antitrust/regulatory agencies must weigh the potential benefits of consolidation (*e.g.*, cost savings that lead to lower prices) against its potential harms (*e.g.*, increased market power).

Wireless telephony is often held out as the most significant potential source of intermodal competition. Indeed, the RBOCs have on occasion cited to the presence of wireless competitors in order to receive in-region long-distance authority under Section 271 of the

---

[3] *See* FCC (2004), at para. 243.
[4] *See* Zimmerman (2006) for further discussion.

Act, local rate deregulation at the state level, and regulatory approval for their recent acquisitions the largest CLECs/IXCs. Of course, arguing that (independent) wireless carriers will function as a viable source of intermodal competition to incumbent wireline carriers relies critically on the assumption that consumers view wireline and wireless telephony as *substitutes* (*i.e.*, in deciding to subscribe to either one of the services or the other). That is, one must assume that the two services are functionally equivalent in the perception of most individuals, or that subscribers would readily switch from, say, wireline to wireless in response to a sufficiently large increase in the price of the former. However, as discussed below, current evidence suggests that these two services function (in at least some sense) as *complements* in consumption to the majority of present-day telecommunications subscribers (*i.e.*, it is more common for a consumer to use wireline *and* wireless services together rather than just one *or* the other).[5]

The purpose of this chapter is to examine the potential effects that RBOC wireline/wireless bundling strategies may have on the development of intermodal competition. Particular attention is paid to the possibility that independent *wireless* carriers will be able to constrain the market power of incumbent wireline providers given that wireless telephony is (by far) the most ubiquitous and widely-adopted alternative voice communications platform in the U.S. today.[6] In this regard, the following section reviews the extant empirical evidence as well as several academic studies regarding the substitutability versus complementarity between wireless and wireline telephony, and thus the "intensity" of intermodal competition exerted by wireless telephony to date (*i.e.*, its present capacity as a source of intermodal competition). It is argued that most of the empirical evidence/academic literature tends to suggest that wireline and wireless telephony effectively function as *complements* to most subscribers (although the tendency to cut-the-cord may increase over time).

The third section of the chapter then considers the possibility that the RBOCs will engage in "strategic bundling", *i.e.*, bundle their wireline and wireless telephony offerings (possibly at a discount) while integrating the *usage* of the two platforms. It is argued that in the short-run these strategic bundles may enhance consumer welfare since they provide various enhanced features that stem from integrating aspects of the RBOCs' wireline and wireless networks. However, in the long-run, these same bundles may serve to drive independent ("unaffiliated") intermodal competitors from the market since the latter will be unable to offer broad discounted bundles of voice, data, and video entertainment. As such, the RBOCs may employ strategic bundling in order to retain their market power in the wireline exchange access market while leveraging their wireline market power into the (in-region) wireless market (thereby mitigating the extent to which independent wireless or VoIP carriers can serve as viable intermodal competitors in the long-run). The fourth section then considers the potential future landscape of the telecommunications industry. It is posited that intermodal competition, if it is to develop at all, will likely not stem from independent wireless (or VoIP) carriers but rather from cable operators. This is because cable operators are the only firms that are likely to be able to offer similar broad-based bundles of telecommunications services that could compete against the RBOCs'. As such, the market structure of the local (residential)

---

[5] *See infra* Section 2 for further discussion.
[6] On the other hand, VoIP telephony is relatively nascent, and far few subscribers have chosen to cut-the-cord by adopting VoIP telephony exclusively relative to adopting wireless telephony exclusively. As such, this chapter will focus primarily on the potential role of wireless telephony to serve as a source of intermodal competition.

telecommunications industry may evolve into a series of regional duopolies with the RBOCs competing against cable companies in bundled service offerings. The last section concludes.

## WIRELINE AND WIRELESS TELEPHONY: SUBSTITUTES OR COMPLEMENTS?

The uptake of wireless telephony by U.S. telecommunications subscribers has increased dramatically over the past several years. Wireless industry survey data collected by the Cellular Telecommunications and Internet Association (CTIA) indicates that the number of wireless telephone subscribers has increased by approximately 344% from June 1996 to July 2004.[7] Data collected by the FCC also provide similar evidence documenting the rapid uptake in wireless service.[8] A number of explanations have been put forward to explain this trend, including (but not limited to) the obvious convenience of mobility, various enhanced features available on wireless handsets (*e.g.*, text messaging), increasing call quality, and the general decline in the price of wireless telephone service.[9]

Again, the fundamental assumption behind the notion that the presence of wireless carriers will constrain the market power of wireline incumbents (*i.e.*, that they will serve as viable intermodal competitors) is that the two services are "substitutes," *i.e.*, that for a sufficiently large increase in the price of wireline telephony subscribers would be willing to adopt wireless telephony exclusively (and *vice versa*). However, before one can discuss whatever demand interrelationships exists between the services, it is critical to understand that telecommunications services in general (and wireline and wireless access in particular) are traditionally defined along two "dimensions" of consumption. The first is the "extensive" or *access* dimension, which determines whether or not a given consumer actually takes wireline or wireless service at a given set of price and service-specific attributes. The second dimension is conditional on the first and concerns the extent to which a given customer *uses* wireline or wireless service *given* that they have chosen to take the service.

The pricing of the access dimension is typically based upon a fixed recurring charge (typically monthly). The usage dimension, on the other hand, is typically priced on a variable scale, which depends on the number and duration of the calls placed. For example, such charges include the charges levied by IXCs for each minute of long-distance calls placed by consumers on their wireline connections, as well as the per-minute charges levied by wireless carriers for usage above the number of minutes included in the "bucket" of a wireless plan

---

[7] *See* FCC (2005c), at Table 11.1 ("Measures of Mobile Wireless Telephone Subscribers").

[8] The FCC's Form 477 data (filed by providers of wireless service twice each year for each state in which they have at least 10,000 subscribers) indicates that from December 1999 (the first date these data are available) to June 2004 the number of wireless subscribers has increased by approximately 110%, while the FCC's Form 502 data (pertaining to the number of wireless telephone numbers) indicates that from December 2000 (the first date these data are available) to June 2004 the number of wireless subscribers increased by approximately 72%. In comparison, the CTIA estimates that the number of wireless subscribers increased by approximately by approximately 96% from December 1999 to June 2004, and by 55% from December 2000 to June 2004. *Id.*

[9] With respect to the latter, the value of the consumer price index for wireless telephone services has fallen by approximately 34% from December 1997 (the date the index was first published by the U.S. Bureau of Labor Statistics) to December 2004. *See* http://data.bls.gov/PDQ/outside.jsp?survey=cu.

("overage charges").[10] As such, in discussing the potential "substitutability" between wireline and wireless telephony in the context of intermodal competition, one must distinguish between the potential substitutability in *access* from the potential substitutability in *usage*.

## Substitutability vs. Complementarity in Wireline/Wireless Access

The best available empirical data on household telecommunications subscription decisions do not generally support the notion that wireline and wireless telephony are substitutes *in access* for the majority of subscribers (*i.e.*, at least not at the given set of prices and other features applicable to wireless access). That is, most subscribers *currently* subscribe to *both* sets of services (the possible reasons behind this phenomenon are considered below) rather than relying on one or the other exclusively. For instance, Tucker *et al.* (2004) consider data taken from a special supplement to the February, 2004 Current Population Survey ("CPS"). This survey asked approximately 32,000 households about the types of telephone services they employ as well as questions pertaining to the households' demographic characteristics. The survey finds that approximately 88.8% of households use either wireline telephony exclusively or use both wireline and wireless telephony. On the other hand, only 6.0% of sampled households use wireless telephony *exclusively*. The tendency for wireless-only adoption tends to be the strongest among residents of central cities, one-person households, renters, single persons, and those with "some college" level of education.

Perhaps most significantly, Tucker *et al.* (2004) find that the rate of wireless-only adoption is dominated by the youngest households. Specifically, they find that 18.0% of persons aged 15-24 use wireless service exclusively, versus 9.6% for persons aged 23-34; 5.0% for persons ages 35-54, and only 2.5% for persons ages 55 and over. Note that while the rate of wireless-only adoption is strongest among younger subscribers, even within this cohort the vast majority of households still choose to retain a wireline connection.

In its own reviews of wireline and wireless subscription patterns the FCC has also found that consumers tend to use wireless and wireline services jointly. In particular, the FCC has explicitly recognized that many consumers view the services as distinct from one another due in part to differences in their functionality, *e.g.*, wireless telephony does not equal the quality, ability to handle data-related traffic, and ubiquity associated with wireline telephony.[11] In addition, the FCC has noted that a consumer's decision to drop their wireline connection can potentially result in significant opportunity costs. For instance, home security systems (burglar alarms) often rely upon on wireline phone connection in order to connect the home security terminal to the security company's monitoring station. Thus, dropping a wireline connection could force a household to forego the benefits of security alarm protection.[12]

Academic studies examining the extent of wireline and wireless access substitution in the U.S. experience are relatively sparse. Rodini *et al.* (2003) employ the Bill Harvesting data

---

[10] In the economics literature, the terms "two-part pricing", "two-part tariffs", or "nonlinear pricing" are often used to describe the pricing structures/strategies of a firm that combines a fixed access charge with a variable usage charge.

[11] *See* FCC (2004), at para. 247 and the cites contained therein.

[12] Other potential opportunity costs associated with dropping a wireline connection include (but are not necessarily limited to) the loss of Internet connectivity via DSL broadband or dial-up, possible (adverse) credit rating effects, and traceability for emergency (911) calls made from a subscriber's residence.

from TNS Telecoms ReQuest Market Monitor® along with its survey responses. The authors estimated cross-price elasticities suggest that secondary wireline access and mobile access are (weak) substitutes for one another. Loomis and Swann (2005) examine the effect of the number of wireless subscribers on the number of lines deployed by incumbent and competitive local exchange providers (weighted by population density) with semiannual state-level panel data for the June 2000 through June 2002 period. Their estimates suggest that for every 100 additional subscribers that take wireless service 95% choose to retain the incumbents' wireline service. The remaining 5% choose to rely on wireless access exclusively. However, they find no statistically significant effect of the number of wireless subscribers on CLEC-deployed lines. Thus, the Loomis and Swann (2005) results are generally consistent with the survey results of Tucker *et al.* (2004), and provide further support for the notion that the majority of telecommunications subscribers currently use wireline and wireless service as "complements" in access.[13]

## Substitutability vs. Complementarity in Wireline/Wireless Usage

While the rate of access substitution from wireline to wireless telephony has been limited to date, the same cannot be said with respect to the rate at which subscribers are willing to substitute their minutes of usage (long distance minutes in particular) across the two platforms. Many wireless carriers offer their subscribers "free" long-distance calls as part of a wireless plan's included "bucket" of minutes (which can be used for either local or long-distance calls), as well as "free" evening, weekend, and "on-net" (*i.e.*, calls made between two end-users subscribing to the same wireless provider) local/long-distance minutes. This has led to a massive shift in the proportion of long-distance minutes placed over wireless handsets relative to wireline connections. From December 1993 to December 2003, the CTIA estimates that the overall average minutes of usage ("MOUs") per wireless subscriber per month has increased from 140 minutes (2.33 hours) to 507 minutes (8.45 hours), an increase of approximately 262%.[14] Other sources estimate the growth in wireless MOUs to be in the range of 21% to 28% between 2001 and 2002 alone.[15]

Ward and Woroch (2004) attempt to estimate the extent of wireline/wireless usage substitution in the U.S. during the ten-quarters spanning July 1999 and December 2001. The authors also rely upon the TNS Telecoms ReQuest Market Monitor® Bill Harvesting data for their analysis. They find that the average calling volumes per mobile phone more than doubled over this time period, and attribute this growth to the "free" minutes available in wireless calling plans. The authors estimate that the cross-price elasticity of wireless (wireline) prices on wireline (wireless) usage is 0.26 (0.18) for IntraLATA calls, 0.13 (0.01)

---

[13] Of course, one could argue that the tendency for joint consumption between wireline and wireline access services is simply a reflection of individual tastes [Horvath and Maldoom (2002)]. For example, high-demand users of voice telephony may tend to have both wireless and wireline service, but this does not necessarily imply that the two services are complements in the ordinary economic sense. On the other hand, many wireless calls are made to wireline connections (and *vice versa*), which might imply that higher rates of wireless penetration also increases the benefits accruing to wireline subscribers, thereby increasing the demand for wireline telephony [Gans *et al.* (forthcoming)].

[14] FCC (2005c), at Table 11.3 ("Mobile Wireless Telephone Service: Industry Service Results").

[15] FCC (2003), at para. 65

for IntraLATA-interstate calls, and 0.20 (0.11) for interstate calls.[16] Note that each of these cross-price elasticities is positive and statistically significant, suggesting that wireline and wireless services are substitutes in usage. Finally, the authors suggest that wireline usage may have been 50% higher had wireless prices not fallen as much as they did over the sample period.[17]

## THE USE OF STRATEGIC BUNDLING TO AFFECT INTERMODAL COMPETITION AND ITS IMPLICATIONS FOR CONSUMER WELFARE

Recall that the efficacy of wireless telephony to function as a viable source of intermodal competition relies upon the assumption that wireline and wireless telephony are substitutes in access. However, if anything, the discussion in the previous section suggests that (currently) the two services can be regarded as substitutes only in the *usage* dimension. However, the tendency for subscribers to substitute wireless for wireline minutes (*i.e.*, the *rate* of usage substitution) is likely to continue over time, and this trend might directly lead to an increased rate of access substitution in the long-run (*e.g.*, as the younger cohorts who tend to use wireless service more intensively and tend to drop their wireline connections more often grow older).[18] This could lead to a large decline in the overall penetration of wireline access, and thus the "cannibalization" of the RBOCs' core wireline business. As such, the RBOCs may attempt to curtail the growth of wireless-induced intermodal competition, perhaps by adopting strategies to "extend" the access complementarity between wireline and wireless telephony into the long-run.

Bundling is simply the sale of two or more separate products in a "package" (assuming that demand for each of the services comprising the bundle exists when the two goods are not bundled together).[19] A large body of economic literature considers the effects of bundling on consumer and total welfare, with the primary focus of these studies concerning the potential for bundling to serve as a potential "strategic" mechanism to implement *de facto* price discrimination or leveraging/entry-deterring practices by a multiproduct monopolist.[20] However, the bundling of various goods or services may also generate value-added services that provide additional benefits to consumers relative to the case where the goods/services can

---

[16] Note that these latter call categories are what are generally referred to as "long distance calls."

[17] Other evidence suggests that wireline toll minutes have fallen dramatically in recent years as a result of wireless usage substitution. *See, e.g.*, FCC (2005c), at Table 14.2 ("Average Residential Wireline Monthly Toll Minutes")(showing that total average residential wireline toll minutes have fallen from 143 per-month in 1995 to 71 per-month in 2003).

[18] Other factors that might lead to a greater rate of wireless-only adoption in the long-run include increasing wireless call quality and further technological advancements to wireless handsets (*e.g.*, real-time video capabilities). In addition, the development of new technologies that reduce the tendency of maintaining a wireline connection (such as wireless burglar alarm systems) could further exacerbate this trend. While this discussion is primarily concerned with RBOC incentives to mitigate wireline cannibalization from independent intermodal competitors, Zimmerman (2006) discusses why the RBOCs may possess the same incentive with respect to even their own wireless affiliates.

[19] Kobayashi (2005).

[20] *See id.* for a thorough and critical review of this literature.

only be purchased separately.[21] This effect is likely to be particularly relevant when the goods comprising the bundle are complements in demand.

Bundling may also allow firms to provision the two goods at a lower cost than they could by selling to the two goods separately (*e.g.*, through economies of scope). This effect, of course, may benefit consumers as well if firms pass-through the resulting cost savings in form of lower retail prices. As such, any discussion of bundling must recognize its potential "value-added/efficiency" effects as well as its "strategic/market power" effects. This is the approach taken below in assessing the potential impacts of incumbent wireline exchange carriers' strategic bundling of wireline and wireless telephony on intermodal competition and consumer welfare. It is argued that value-added/efficiency and strategic/market power effects in the instant case are not necessarily mutually exclusive. Specifically, while bundling wireline and wireless services may generate *short-run* consumer benefits in the form of enhanced service features and/or lower prices, in the *long-run* such bundling may serve to drive out independent intermodal (and intramodal) competitors who inherently cannot provide the same bundled offerings. To the extent that these intermodal/intramodal competitors discipline the RBOCs' exercise of market power, their exit from the industry may ultimately result in subscribers paying higher prices for the bundled offerings and/or having the quality of these services being degraded.

## Value-Added/Efficiency Effects from Bundling Wireline and Wireless Telephony

The RBOCs and their wireless affiliates have spent considerable resources in developing and marketing various wireline/wireless bundles. These bundles often involve the design of packages that "integrate" the usage of the two platforms and provide subscribers with additional value-added features that would not be realized from purchasing the two services separately (*i.e.*, either purchasing the RBOCs' wireline and its wireless affiliate's service outside of the bundle *or* purchasing one of the services from the RBOC *or* its affiliate and the other from an unaffiliated competitor). Of course, this also implies these bundles may serve to "strengthen" the complementarity between the access/usage dimensions of wireline and wireless telephony.

For example, in June 2003 Cingular and its wireline parents, SBC and BellSouth, began offering their "MinuteShare" service. This product allows SBC or BellSouth's residential wireline subscribers to share a single bucket of wireline long distance minutes and wireless local and long distance minutes offered by Cingular.[22] Another product offering by these firms that integrates the usage of wireline and wireless minutes across the two platforms is their "Fast Forward" service. With Fast Forward, an SBC or BellSouth subscriber uses a "cradle" devise to hold a Cingular wireless phone within their home or office.[23] During the time in which the wireless handset is in the cradle, all calls being made to the customer's

---

[21] Adams and Yellen (1976) classify bundling strategies as consisting of either "mixed bundling" (where the firm offers the relevant goods both a bundle *and* as separate components) or "pure bundling" (where the firm offers the relevant services only through a bundle).
[22] *See* http://www.cingular.com/about/latest_news/03_09_09.
[23] *Id.* Note that the Fast Forward offering is only compatible with *Cingular's* wireless service.

wireless number are automatically forwarded to a designated wireline phone.[24] Customers taking the Fast Forward service get unlimited incoming wireless calls/minutes forwarded to their wireline phone in the local calling area without having those minutes subtracted from the bucket of "included" minutes in their wireless plan.[25]

Other attempts at developing an "integrated" wireline/wireless platform across SBC's, BellSouth's, and Cingular's services include "simplified ordering" (allowing customers to order wireless service through SBC and BellSouth wireline sales channels), "extensive distribution channels" (allowing SBC and BellSouth call centers and Cingular retail store locations to cross-sell the firms' respective wireline and wireless products/services), and "wireless co-branding" (in which the firms promote a co-branded tag line tying the SBC and BellSouth brands to the Cingular brand in advertising and marketing efforts).[26] Clearly, these efforts may reduce a customer's search costs relative to the case where he/she obtains wireline and wireless services from different carriers. In addition, customers subscribing to SBC's or BellSouth's residential wireline service and Cingular's wireless may use a single voice mailbox system that records incoming call messages placed to either the subscriber's wireline or wireless number (thereby eliminating the need for a customer to check two different voice mailboxes and possibly having to give out two different phone numbers as contacts).[27]

Verizon and its wireless affiliate, Verizon Wireless, are also beginning to develop and market offerings that integrate the companies' respective wireline and wireless platforms. In particular, in 2004 Verizon announced its "iobi" service. This service seeks to seamlessly integrate Verizon's wireline and Internet networks with Verizon Wireless' mobile service, allowing subscribers to "manage phone calls, voice mails, calendars, address books, e-mails and more, using wireline and wireless phones, computers, laptops, and PDAs."[28] For example, a customer subscribing to iobi can receive a voice message in the form of an e-mail or text message on a PDA, or have the call directed to either the customer's wireline or wireless phone line.[29]

The above-mentioned usage-integrated wireline/wireless bundles may also confer additional benefits to subscribers. These include (but are not necessarily limited to) receiving a single bill for both services ("combined billing"), having to deal with a single call center should problems arise with either the customer's wireline or wireless service, bundled discounts (whereby the bundled price is lower than purchasing the wireline and wireless components separately), and unlimited wireless-to-wireline and/or wireline-to-wireless calls (*e.g.*, a subscriber taking a SBC/Cingular bundle could make an unlimited

---

[24] *Id.*
[25] *Id.*
[26] *See* FCC (2004), at note 581.
[27] *See* http://www.cingular.com/about/latest_news/03_09_09.
[28] *See* http://newscenter.verizon.com/proactive/newsroom/release.vtml?id=83234&PROAC.
[29] *Id.* Other enhanced features made available through iobi include real-time call management (allowing customers to choose how, where, and if they want to receive voice calls and data messages), call notifications on the customer's personal computer and the screens of other devices, programmable call-forwarding (*e.g.*, allowing calls made to the customer's Verizon wireline number to be automatically forwarded to their Verizon Wireless mobile handset and *vice versa*), interactive call and e-mail logs, conference calling, electronic contact information-sharing with automatic updating, and "click-to-dial" calling (allowing a customer to automatically initiate calls by using a computer mouse to highlight the called party's number), and "multi-modal communications" (allowing a customer to decide how a particular voice or data message is received, whether it be by e-mail, voice mail, text messaging, *etc.*). *Id.*

number of free wireline (wireless) calls as long as those calls terminated on the wireless (wireline) handset of either another SBC/Cingular bundle customer or a BellSouth/Cingular bundle customer.

Finally, the joint provision of both wireline and wireless services by the RBOCs could also generate various cognizable efficiencies (*e.g.*, cost savings or higher quality of service). These efficiencies might stem from economies of scope associated with producing wireline and wireless services jointly (as well as economies of scale within each service given the size of the RBOCs' respective customer bases). For instance, the RBOCs may realize lower costs of interconnection when terminating calls received from their wireline affiliates relative to unaffiliated wireless carriers (*e.g.*, the RBOCs' wireline and wireless networks may be jointly configured/optimized to minimize overall network costs). To the extent that these efficiencies are passed-through to consumers in the form of lower prices and/or higher quality products and services, consumer welfare is enhanced in the presences of wireline/wireless bundling.

## Strategic/Market Power Effects from Bundling Wireline and Wireless Telephony

While, as discussed above, wireline/wireless bundles (whether or not they be usage-integrated) may confer various benefits to consumers, at the same the possibility exists that these offerings will strategically advantage the RBOCs and their wireless affiliates. Specifically, the RBOCs may use such bundles and the features associated with them (*e.g.*, combined billing) to "leverage" their in-region wireline market power into the in-region wireless market. This in turn would serve to both maintain the RBOCs market power and associated revenue stream in its core business operations while potentially extending its market power into the wireless segment.

The ability of the RBOCs to gain a competitive advantage through leveraging is generated from the fact that *unaffiliated* wireless (wireline) carriers, because they do not possess a wireline (wireless) affiliate, inherently cannot offer the same enhanced features made available through (usage-integrated) wireline/wireless bundles. For example, an independent wireless carrier (*e.g.*, T-Mobile) would obviously be unable to offer its subscribers the convenience of combined billing or a discounted wireline/wireless package to their customers. Of course, an independent wireless company could enter into a joint marketing/service provisioning agreement with, say, a CLEC in order to provide a competing wireline/wireless offering, but the fact that such business arrangements are virtually non-existent suggests that the transactions costs associated with forming them may be prohibitively high. In addition, while it might be conceivable to enact regulations that require the RBOCs to offer combined billing arrangements with independent wireless carriers, it is not necessarily the case that the latter firms would be willing to enter into such transactions. This is because doing so would most likely require the independent wireless carrier to convey competitively sensitive information about their subscribers (*e.g.*, the features of the calling plan taken and the intensity of the customers' usage) to the RBOCs, and thus potentially to the RBOCs' wireless affiliates as well.

The same potential problem arises from those features of the wireline/wireless bundles that are derived *directly* from usage integration. For instance, the SBC/BellSouth/Cingular Fast Forward offering disadvantages unaffiliated wireless carriers because they simply cannot offer as similar product without having a wireline parent or affiliate. This in turn may result in Cingular possessing a significant competitive advantage relative to the unaffiliated rival. Similarly, an independent wireless firm would not be able to offer a common wireline/wireless voice mailbox, which might also give a competitive advantage to those wireless carriers that have a wireline parent. The same story holds true for all the other potential enhanced features associated with integrated wireline/wireless products discussed above. Clearly, these integrated bundles may serve as mechanisms to preserve the RBOCs local and long-distance wireline exchange access business (*i.e.*, since the integrated nature of the offerings disincents subscribers from dropping their wireline connections, thus helping to preserve the RBOCs' wireline revenues) while at the same time allowing them to leverage/extend their in-region wireline market power into the wireless segment (*i.e.*, since independent wireless carriers cannot offer the enhanced features made available through these integrated bundles).

The RBOCs have already used bundling to expand the scope of their wireline operations. Specifically, following the RBOCs' receipt of Section 271 authority (which allowed the RBOCs to offer in-region interstate long distance service on a state-by-state basis), the RBOCs began to offer bundled wireline local/long distance bundles. These bundles, such as Verizon's "Freedom" packages, resulted in the RBOCs quickly capturing a large share of the in-region long distance market. For example, as of December 31, 2004, 47% of Verizon's local residential wireline subscribers also chose Verizon as their long distance carrier.[30] At this same point in time, BellSouth reports that 48% of its mass market (residential and small business) customers subscribed to its long distance service,[31] while SBC reports that it holds approximately 52.4 million network access lines in service and 20.9 million long distance lines in service, which suggests an in-region long distance market share of approximately 40%.[32] The fact that this growth in the RBOCs' in-region long distance shares occurred *despite* the fact that their in-region intramodal competitors had *already* been able to offer local/long-distance bundles *before* the RBOCs could highlights the significant market power the RBOCs may realize from their brand names. Of course, this same factor would tend to further exacerbate the potential competitive concerns related to the RBOCs' bundling of wireline and wireless. In addition, as mentioned above the RBOCs may realize significant economies of scope in producing both wireline and wireless services (as well as economies of scale within each service). If the average costs of independent intermodal competitors or potential entrants are above those of the RBOCs' (as a result of their inherent inability to realize the same scope or scale economies), then the RBOCs' may be able to systematically undercut their competitors' prices (perhaps even setting retail prices below cost). This would also tend to drive out independent intermodal competitors from the market and/or deter potential competitive entry.

---

[30] *Verizon Communications 2004 Annual Report*, available at http://investor.verizon.com/financial/quarterly/pdf/04VZ_AR.pdf, at 20.
[31] *BellSouth 2004 Annual Report*, available at http://www.bellsouth.com/investor/pdf/annualrpt04.pdf, at 35.
[32] *SBC Communications 2004 Annual Report*, available at http://www.sbc.com/investor_relations/company_reports_and_sec_filings/SBC_2004_AR.pdf, at 5.

To the extent that the RBOCs' bundling of wireline and wireless telephony achieves the above-mentioned "strategic" objectives, the possibility exists that in the *long-run* the very same benefits that accrue to consumers from usage-integrated bundles will also serve to diminish the competitive pressure exerted by independent intramodal and intermodal competitors. This in turn would weaken and possibly remove any competitive check placed on the RBOCs in setting the prices for these integrated bundles. As such, it is less likely that the RBOCs would be willing to offer these bundles at the same price and/or with the same level of "quality" (including the number of "free" calling minutes or enhanced features contained therein) to their subscribers. Therefore, the value-added and other benefits made possible by these integrated offerings would be less likely to be passed-on to consumers.

## THE FUTURE COMPETITIVE LANDSCAPE OF THE TELECOMMUNICATIONS INDUSTRY

An important issue remaining to be addressed in light of the present discussion is how the future antitrust (*i.e.*, product and geographic) markets of the U.S. telecommunications industry might be defined. While the RBOCs' significant presence in the wireless access market and their use of strategic wireline/wireless bundling might render the efficacy of wireless-based intermodal competition as somewhat tenuous, there are other potential intermodal competitors. In particular, cable companies have begun to make significant inroads into the residential voice market.[33] These firms have already offered traditional circuit-switched telephony for some time (although the high costs of modifying the cable network to do so have largely limited the geographical scope of its deployment). The exit of many CLECs from the residential voice market (due to bankruptcy, regulatory changes, or acquisition) may give cable operators additional incentive to enter telephony and compete against the RBOCs. And, unlike resellers or CLECs that required the use of unbundled network elements to provide local exchange access service, cable companies could provide voice (and other) services entirely over their own facilities. This would entail the presence of large, entrenched facilities-based competitors with considerable brand-name recognition competing against the RBOCs in the latter's core businesses (and possibly doing so with differentiated service offerings). This is something which has been largely unachievable to date given the past technological and regulatory characteristics of the local wireline exchange market.

Cable operators are particularly well positioned to offer packet-switched voice services, *i.e.*, VoIP telephony, and have already made some inroads in competing against the RBOCs in the voice exchange access market. For example, from December, 1999 to December, 2004 the number of CLEC end-user switched coaxial cable access lines increased from approximately 0.31 million to 3.7 million, a growth rate of just over 1100%.[34] And, as of

---

[33] Cable companies are comparatively weaker in providing telecommunications services to large business subscribers. This is due to the fact that cable operators tend to have few facilities deployed in city business districts where the firms typically base their operations. For this reason, this section primarily focuses on the development of competition in the *residential* exchange access market.

[34] See FCC (2005a), at Table 5 ("Competitive Local Exchange Carrier Lines by Type of Technology"). These figures likely reflect the majority of circuit-switched cable telephony lines and some VoIP lines, but may be understated to the extent that some cable operators are not reporting the latter.

December, 2004, these lines constituted approximately 11.3% of total CLEC-deployed lines.[35]

Of course, the RBOCs will view cable operators as they would any other independent intermodal competitor, and will take efforts to prevent the cannibalization of their businesses to cable companies. Again, the most likely strategy that the RBOCs will rely upon to accomplish this objective is to offer broad bundles of usage-integrated telecommunications services.[36] Of course, to compete effectively against cable companies the RBOCs need to offer multichannel video programming services as part of their strategic bundles. Indeed, most have already begun to develop such offerings in direct response to cable competition. For instance, Verizon, SBC, and BellSouth offer satellite video entertainment service bundled with their voice service.[37] Verizon is also attempting to enter the video entertainment market within its franchise territory by deploying high-capacity fiber connections directly to end-user premises (which can also be used to offer VoIP telephony and high-speed Internet connectivity). In March, 2006 AT&T Inc. announced that it would invest $1 billion dollars over the subsequent three years in order to upgrade its fiber network in California for the explicit purpose of establishing itself as "a real competitive choice to cable companies."[38] As such, future RBOC service offerings in competition with cable companies will likely consist of a "quadruple play" of bundled (usage-integrated) wireline and/or VoIP telephony, wireless voice, data (*e.g.*, DSL), and video entertainment services.

Cable companies are also developing broad service bundles of their own. Of course, cable companies already possess the capacity to offer multichannel video entertainment services and, as mentioned above, they can offer local and long-distance service through VoIP or traditional circuit-switched telephony. Cable companies already possess the majority of the broadband access market through their cable modem offerings.[39] The "missing piece" in cable-provisioned bundles is wireless telephony. Given the ever-increasing use of mobile services among subscribers, failing to have a wireless component might greatly limit the perceived value of cable's bundles relative to the RBOCs'. However, cable companies are moving aggressively to overcome this handicap. Indeed, in late 2005 the three largest U.S

---

[35] *Id.*

[36] *See* Glenn Bischoff, *SBC, BellSouth and Cingular Team Up on Bucket Plan*, June 5, 2003, *available at* http://telephonyonline.com/news/web/telecom_sbc_bellsouth_cingular/index.html. Bischoff cites a senior telecom analyst speaking to the purpose of the MinuteShare offering: "This is preparation for that ensuing battle [against cable companies]. MinuteShare gives them a nice tight package that will be hard for the customer to churn off of..."

[37] In March, 2004 Verizon began offering DirectTV satellite television to subscribers in New York and other East Coast cities who also purchased its telephone and data services. This was done in response to Cablevision's offering of local and long-distance telephony to its Internet subscribers. Peter Grant, *Here Comes Cable…And It Wants a Big Piece of the Residential Phone Market*, WASH. POST, September 13, 2004, at R4. In 2002, SBC began to bundle EchoStar satellite television with its data services. *See* http://www.sbc.com/gen/pressroom?pid=4800 &cdvn=news&newsarticleid=7500. BellSouth also offers DirectTV satellite television as part of a bundle to some of its in-region subscribers. *See BellSouth, DirecTV Partner to Offer Satellite TV as Part of Service Package*, Aug. 28, 2003, *available at* http://www.rednova.com/news/stories/3/2003/08/27/story146.html. *See also* Almar Latour, *To Meet the Threat from Cable, SBC Rushes to Offer TV Service*, WALL ST. J., February 16, 2005, at A1 (discussing SBC's attempts to offer on-demand video entertainment services over its *own* fiber cables). Verizon has also undertaken significant efforts to build out high-speed fiber connections directly to residential end users' homes. *See* Dionne Searcey, *Verizon Hits Hurdles in Big Bet on New High-Speed Network*, WALL ST. J., March 8, 2006, at A1 for further discussion.

[38] *Communications Daily*, March 30, 2006, at 16.

cable companies (Comcast, Time Warner, and Cox) entered into an agreement with Sprint/Nextel to resell the latter's wireless service under their own brand names (along with the Sprint brand) beginning in early 2006.[40] As such, it appears clear that bundled service offerings will constitute the relevant product market in the future telecommunications industry.

Consideration must also be given to the relevant geographic market(s) that will characterize the telecommunications industry in the future. Note that cable companies are typically awarded exclusive service contracts with individual municipalities within states. The RBOCs' provide service throughout the majority of the states comprising their wireline territories as well as some out-of-region franchise areas obtained through acquisitions following the 1984 divestiture of AT&T (*e.g.*, Verizon is the incumbent carrier in some parts California, Texas, and Florida while SBC serves most of Connecticut). End-users, of course, would be unable to purchase the services of any cable operator other than the one holding the franchise rights in their region of residence (*e.g.*, they could not substitute the bundled offering of a "non-local" cable company for their RBOC's bundle).[41] Therefore, the relevant future geographic market for access will continue to be defined on a "local" basis level assuming that independent intermodal (intramodal) competitors cannot remain viable in the long-run (for the reasons discussed above).[42] This implies that there will be only two suppliers of bundled telecommunications services (*i.e.*, the RBOC and the cable operator) in many (if not most) geographic markets.

Of course, from a *policy* perspective, the ultimate question is whether two providers is sufficient to induce "competitive" outcomes in the local telecommunications market, *i.e.*, retail prices that reflect the long-run incremental cost of provisioning the various services. And while it may very well be the case that the local exchange access market is most efficiently structured as a "natural duopoly" due to its demand and technological characteristics, regulators and policy makers cannot simply ignore the potential harms that economic theory often suggests will arise with fewer firms competing in a market. For instance, policy makers will need to address the possibility of a strategic "détente" between these two large facilities-based players (*e.g.*, the RBOC and cable company "splitting" the

---

[39] Cable operators may also attempt to "integrate" the usage of their bundled services in order to reduce costs and/or mitigate subscriber churn. For instance, video telephony could be received (or Web browsing done) through a subscriber's television.

[40] Alan Breznick, Jonathan Make, Ian Martinez, and Howard Buskirk, *Three Top Cable Firms Sign Wireless Deal With Sprint Nextel*, COMMUNICATIONS DAILY, November 2, 2005, at 7. Breznick *et al.* also note the possibility of integrated cable VoIP/wireless products being developed out these arrangements ("… the cable operators and Sprint could mesh wireless and cable VoIP services into a single phone service that operators over both cellular and Wi-Fi networks, enabling subscribers to use dual-mode handsets both inside and outside the home."). It remains to be seen whether or not cable companies will attempt to acquire any of the regional or remaining national wireless carriers directly.

[41] An exception might be the purchase of VoIP telephony offered by another RBOC or a cable company operating in another area. However, to the extent that consumers prefer to obtain all their services from a single firm and/or receive incentives for doing so (*e.g.*, bundled discounts), a standalone VoIP offering by one of these firms would not be viewed as a close substitute for the "local" RBOCs'/cable company's bundled offering. As such, a "non-local" RBOC/cable company could not be considered as being within the same geographic market (nor their standalone offerings as being in the same product market).

[42] For instance, in the short-run there may be other firms without a wireline or wireless affiliate firms offering voice service over VoIP (*e.g.*, Vonage). However, the inability of such firms to provide broad bundles of various telecommunications services in a manner similar to the RBOCs and cable companies (as well as their inherent dependence on the facilities of the latter carriers to provision service) might greatly limit the extent of their operations.

market by price matching each other's offerings, or simply competing "less aggressively" in the other firm's traditional core line of business) while recognizing the potentially significant consumer benefits associated with product and service bundling/integration. And if one assumes that *structural* remedies cannot be feasibly implemented (*e.g.*, due to the technological aspects of the carriers' networks that generate economies of scale and scope), regulators must then focus on developing the appropriate *behavioral* remedies (*e.g.*, incentive regulation plans such as price caps) if, over time, consumers realize increasing service prices (*e.g.*, due to "weak" competition between the RBOCs and cable companies) and/or decreased quality of service.

## CONCLUSION

Intermodal competition from current (*e.g.*, wireless, VoIP) or pending technologies (*e.g.*, broadband over power lines) is often credited with serving as a competitive check on the behavior of incumbent wireline exchange access providers. In addition, many telecommunications industry observers argue that the influence of intermodal competition may one day render the wireline exchange access business obsolete. Indeed, it is common to hear statements to the effect that "in twenty years nobody will have a wireline phone." This chapter has shown that such sentiments are likely to be both shortsighted and naive. Even though most U.S. households now subscribe to wireless service, very few have actually chosen to drop their wireline connections entirely, and in fact the greatest propensity for doing so is reserved for the very youngest demographic class of subscribers. Of course, this might simply mean that wireless substitution will tend to increase in the future as these younger cohorts age and telecommunications users become more acclimated to using wireless (and other) forms of voice telephony.

However, the consolidation of the wireless access market and the ownership of the two largest wireless carriers by the RBOCs likely diminishes the extent to which wireless telephony can be credited as an intermodal source of competition. In the former case, consolidation may lead to independent wireless carriers competing "less aggressively" (*e.g.*, charging higher price/lower-minute wireless plans) against wireline-affiliated wireless carriers, thereby reducing the extent to which wireline subscribers view their services as a viable (cut-the-cord) substitute for wireline access. In the latter case, it is not transparent how subscriber migration from an RBOC's wireline network to the wireless network of one of its own affiliates can be credited as a meaningful competitive outcome stemming from intermodal competition. In addition, the RBOCs appear to have incentives to prevent the cannibalization of their traditional wireline business to intermodal competitors, perhaps most strongly reflected in their efforts to develop and market usage-integrated wireline/wireless bundles. The use of such strategic bundling by the RBOCs may serve to preserve both their market power in the wireline exchange access business while leveraging this market power into the wireless market (*i.e.*, by driving out independent intermodal competitors from the market in the long-run or by deterring the entry of potential intermodal competitors).

Cable companies may hold the greatest promise in serving as a source of intermodal competition against the RBOCs. Competition between these two carriers will most likely involve bundled offerings of voice, data, and video entertainment. However, it remains to be seen whether a market characterized by a series of facilities-based regional duopolies will actually result in lower retail prices for telecommunications services or whether it will simply involve the two carriers "splitting" the (monopoly) level of profits between them. Regulators at both the state and federal levels must closely monitor the progression of competition between cable operators and the RBOCs, and implement the appropriate behavioral remedies in order to protect consumers when and if such competition is deemed to be insufficiently vigorous.

## REFERENCES

Adams, W.J., & Yellen, J.L. (1976). Commodity Bundling and the Burden of Monopoly. *Quarterly Journal of Economics*, vol. 90, 475-498.

Carlton, D.W., & Waldman, M. (2002). The Strategic Use of Tying to Preserve and Create Market Power in Evolving Industries. *Rand Journal of Economics*, vol. XXXIII, 194-220.

Choi, B., Ahn, B., & Park, Y. (2003). Cross Ownership of Wireline and Wireless Communications Carriers: Synergy or Collusion?. *Information Economics and Policy*, vol. 15, 485-499.

Choi, J.P., & Stefanadis, C. (2003). *Bundling, Entry Deterrence, and Specialist Innovators*. Unpublished manuscript.

Cournot, A. (1838), *Researches into the Mathematical Principles of the Theory of Wealth* (translated by Nathaniel Bacon, 1927). New York: Macmillan.

FCC. (2003). *Implementation of Section 6002(b) of the Omnibus Budget Reconciliation Act of 1993 and Annual Report and Analysis of Competitive Market Conditions With Respect to Commercial Mobile Services*, Eight Report, WT Docket No. 02-379, July 14, 2003.

FCC. (2004). *Applications of AT&T Wireless Inc. and Cingular Wireless Corporation for Consent to Transfer Control of Licenses and Authorizations, Memorandum Opinion and Order*, WT Docket No. 04-70, October 26, 2004.

FCC. (2005a). *Local Telephone Competition: Status as of December 31, 2004*. Washington, DC: Industry Analysis and Technology Division.

FCC. (2005b). *SBC Communications Inc. and AT&T Corp. Applications for Approval of Transfer of Control, Memorandum Opinion and Order*, WC Docket No. 05-65, November 17, 2005.

FCC. (2005c). *Trends in Telephone Service*. Washington, DC: Industry Analysis and Technology Division.

FCC (2005d). *Verizon Communications Inc. and MCI, Inc. Applications for Approval of Transfer of Control, Memorandum Opinion and Order*, WC Docket No. 05-75, November 17, 2005.

Gans, J.S., King, S.P, & Wright, J. (forthcoming). Wireless Communications. In S. Majumdar, I.Vogelsang, & M. Cave (Eds.), *Handbook of Economics: Technology Evolution and the Internet* (vol. 2). Amsterdam, Netherlands: North-Holland.

Horvath, R., & Maldoom, D. (2002). *Fixed-Mobile Substitution: A Simultaneous Equations Model with Qualitative and Limited Dependent Variables*. Unpublished manuscript.

Kobayashi, B.H. (2005). Does Economics Provide a Reliable Guide to Regulating Commodity Bundling by Firms? A Survey of the Economic Literature. *Journal of Competition Law and Economics*, vol. 1, 707-746.

McAfee, R.P., McMillan, J., & Whinston, M.D. (1989). Multiproduct Monopoly, Commodity Bundling, and Correlation of Values. *Quarterly Journal of Economics*, vol. 104, 371-384.

Nalebuff, B. (2004). Bundling as an Entry Barrier. *Quarterly Journal of Economics*, vol. 119, 159-187.

Lewbel, A. (1985). Bundling of Substitutes or Complements. *International Journal of Industrial Organization*, vol. 3, 101-107.

Loomis, D.G., & Swann, C.M. (2005). Intermodal Competition in Local Telecommunications Markets. *Information Economics and Policy*, vol. 17, 97-113.

Peitz, M. (2005). Bundling May Blockade Entry. Unpublished manuscript.

Rodini, M, Ward, M.R., & Woroch, G.A. (2003). Going Mobile: Substitutability Between Fixed and Mobile Access. *Telecommunications Policy*, vol. 27, 457-476.

Schmalensee, R. (1984). Gaussian Demand and Commodity Bundling. *Journal of Business*, vol. 57, 58-73.

Ward, M.R., & Woroch, G.A. (2004). *Usage Substitution between Fixed and Mobile Telephony in the U.S.* Unpublished manuscript.

Zimmerman, P.R. (2006). *The Cingular/AT&T Wireless Merger, Wireline Affiliated Wireless Carriers, and Intermodal Competition in Telecommunications*. Unpublished manuscript.

Chapter 10

# NEW CHALLENGES IN RAMAN AMPLIFICATION FOR FIBER COMMUNICATION SYSTEMS

*P.S. André*[1,2], *A.N. Pinto*[1,3], *A.L.J. Teixeira*[1,3], *B. Neto*[1,2],
*S. Stevan Jr.*[1,3], *Donato Sperti*[1,3,4], *F. da Rocha*[1,3],
*Micaela Bernardo*[2,5], *J.L. Pinto*[1,2], *Meire Fugihara*[1,3],
*Ana Rocha*[1,2] *and M. Facão*[2]

[1]Instituto de Telecomunicações, Aveiro Portugal
[2]Departamento de Física, Universidade de Aveiro, Aveiro, Portugal
[3]Departamento de Electrónica, Telecomunicações e Informática,
Universidade de Aveiro, Aveiro, Portugal
[4]Università Degli Studi di Parma, Parma, Italy
[5]Portugal Telecom Inovação SA, Aveiro, Portugal

## ABSTRACT

Raman fiber amplifiers (RFA) are among the most promising technologies in lightwave systems. In recent years, Raman optical fiber amplifiers have been widely investigated for their advantageous features, namely the transmission fiber can be itself used as the gain media reducing the overall noise figure and creating a lossless transmission media. The introduction of RFA based on low cost technology will allow the consolidation of this amplification technique and its use in future optical networks.

This paper reviews the challenges, achievements, and perspectives of Raman amplification in optical communication systems. In Raman amplified systems, the signal amplification is based on stimulated Raman scattering, thus the peak of the gain is shifted by approximately 13.2 THz with respect to the pump signal frequency. The possibility of combining many pumps centered on different wavelengths brings a flat gain in an ultra wide bandwidth.

An initial physical description of the phenomenon is presented as well as the mathematical formalism used to simulate the effect on optical fibers.

The review follows with one section describing the challenging developments in this topic, such as using low cost pump lasers, in-fiber lasing, recurring to fiber Bragg grating cavities or broadband incoherent pump sources and Raman amplification applied to coarse wavelength multiplexed networks. Also, one of the major issues on Raman amplifier design, which is the determination of pump powers in order to realize a specific gain will be discussed. In terms of optimization, several solutions have been published recently, however, some of them request extremely large computation time for every interaction, what precludes it from finding an optimum solution or solve the semi-analytical rate equation under strong simplifying assumptions, which results in substantial errors. An exhaustive study of the optimization techniques will be presented.

This paper allows the reader to travel from the description of the phenomenon to the results (experimental and numerical) that emphasize the potential applications of this technology.

## 1. INTRODUCTION

The deployment of optical communication systems through long haul networks required the development of transparent optical amplifiers, for replacement of the expensive and limitative optoelectronic regeneration. The increasing distance between amplification sites saves amplification huts reducing by this way the investment and operational cost in the network management.

The first choice for transparent optical amplification pointed out to the Erbium Doped Fiber amplifiers (EDFA), which was a mature technology by the beginning of the last decade of the XX century. However, the growing demand in terms of transmission capacity has been increasing dramatically, fulfilling the entire spectral band of the EDFA, and wideband amplifiers are now required. Raman fiber amplifiers (RFA) have emerged as a key technology for the optical networks.

In lumped amplified systems (using for example EDFAs) the amplification modules are placed every 40~50 km of span. This module amplifies back to the initial power level, the transmission signal attenuated during propagation. The distance between amplifiers is determined by the span loss, by the limit imposed from the maximum admissible power allowed in the fiber without inducing nonlinear effects and by the minimum acceptable power that avois a degradation of the signal-to-noise-ratio.

The use of Raman amplification allows the confinement of the signal inside the limits imposed by the nonlinearities and of the signal-to-noise-ratio degradation resulting from higher span distances. This advantage of the distributed (Raman) over lumped amplification is illustrated in figure 1.

The distributed amplification scheme can be used to cover very long span links or to increase the distance of ultra-long haul systems.

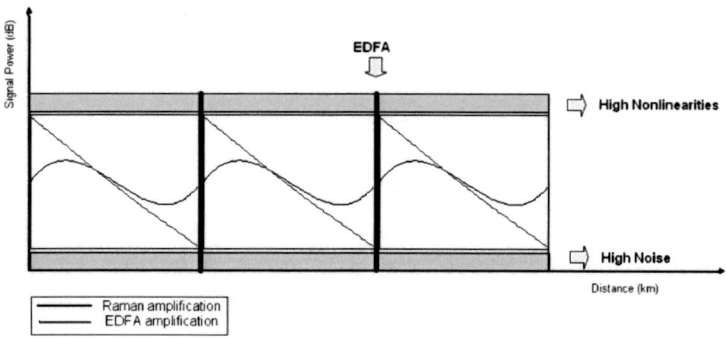

Figure 1. Distributed and lumped amplification signal evolution.

Raman fiber amplifiers are based on the power transfer from pump(s) signal(s) to information carrying signals (usually described as probes) due to stimulated Raman scattering (SRS) which occurs when there is sufficient pump power within the fiber. Since the gain peak of this amplification is obtained for signals downshifted approximately by 13.2 THz (for Silica), relative to the pump frequency, to achieve gain at any wavelength we need to select a pump whose frequency complies with this relation. In this way, it is possible to optimize the number of pumps to obtain a wide and flat gain [1-3]. However, it is necessary to bear in mind, that due to the pump-to-pump interaction, the shorter wavelength pumps demand more power to be effective [4,5].

From a telecommunications point of view, the pump wavelengths must be placed around 1450 nm because the signal wavelengths used on the so called 3$^{rd}$ transmission window are centered around 1550 nm and the maximum gain occurs for a Stokes frequency shift of 13.2 THz.

The RFA had become attractive just after the development and commercialization at a reasonable cost of a key component: the high power pump laser [6]. Typically a high power laser for Raman amplification, provides an optical power of 300 mW, launched over an optical fiber, which for a standard single mode fiber (SMF) is equivalent to a power density of 3.75 GW/m$^2$. This high power injected into the fiber, especially when multipump lasers are utilized, imposes new concerns in terms of safety.

Therefore, the use of RFAs requires the utilization of automatic power reduction or automatic laser shut down systems to prevent the hazard of high power leakage from the optical cables or service cabinets. Also, as the optical power rises, the nonlinear effects, such as the fiber fuse effect, start to become relevant. This effect has threshold intensity of 10~30 GW/m$^2$ and it is responsible by a catastrophic destruction of the fiber core. This destruction once started propagates in direction to the optical source, resulting also in the destruction of the pumping laser [7]. For operating wavelengths of 1550 nm, the fuse effect power threshold is ~1.5 W for SMF fibers, while for dispersion shift fibers (DSF) this power is reduced to ~1.2 W [8]. This effect is also responsible by the damage of the optical connectors interface [8].

In terms of implemented systems, several architectures have been proposed, based in all Raman or hybrid Raman/EDFA amplification [10]. The use of bidirectional Raman amplification has also been reported for long reach access networks. Experimental results have shown the feasibility of systems with symmetric up-and-downstream signals with bitrates up to 10 Gb/s, supported by distributed Raman amplification over 80 km of fiber [11].

Field transmission experiments have been reported with 8 × 170 Gb/s over 210 km of single mode standard fiber, achieving spectral efficiency of 0.53 bit/s/Hz [12].

As the traffic increases, wavelength division multiplexing (WDM) arises to enlarge the transmission capacity. This, in turn, requires flexible and broadband architectures which reinforces the interest in Raman amplification. Nowadays, WDM exists in two formats: Dense WDM (DWDM) working at C and L spectral windows, allocating a maximum of 150 channels spaced by 0.8 nm [13], and Coarse WDM (CWDM) working at O, E, S, C, and L spectral windows, allocating a maximum of 18 channels spaced by 20 nm [14]. The DWDM solution is extensively used in long haul systems, sending as much information as possible. CWDM is a good solution whenever less information is transmitted over short distances in a less expensive way than DWDM. As CWDM works with far apart signals, it can make use of uncooled distributed feedback (DFB) lasers [15,16] needing multiplexing components with flexible tolerances. However, as the channels in CWDM systems are far apart, optical amplification is still a matter of concern. Traditional EDFA bandwidth (20~40 nm) cannot support the full band of CWDM channels [17].

Other technical solution to amplification of signals is the semiconductor optical amplifier, which presents a low saturation power (around 13 dBm) when compared with other fiber based amplifiers, but with a signal-to-noise ratio degradation quite considerable. A good solution for the amplification of both DWDM and CWDM relies on Raman amplifiers. A wide and flat spectral gain profile is achievable thanks to the combination of several pumping lasers operating at specific powers and wavelengths. The composite amplification is determined from the mutual interactions among the pump and signal wavelengths. Gain spectra as large as 100 nm were obtained using multiple pumps. Emori *et al.* have presented an experimental Raman amplifier with a 100 nm bandwidth using a WDM laser diode unit with 12 wavelengths ranging from 1405 to 1510 nm, whose maximum total power was equal to 2.2 W [4, 18]. Therefore, a gain equal to 2 dB is obtained over a 25 km SMF link and a 6.5 dB gain using a 25 km DSF link, both with 0.5 dB of maximum ripple. Kidorf *et al.* provided a mathematical model to implement a 100 nm Raman amplifier using low power pumps with maximum power of each pump equal to 130 mW [14]. They used 8 pumps from 1416 nm to 1502 nm along 45 km of SMF, obtaining a gain around 4 dB with a maximum ripple equal to 1.1 dB.

The growing maturity of high pump module technologies is providing competitive solutions based on Raman amplification and currently many alternative techniques are being developed to overcome the ordinary one pump and dual pumping methods [19, 20]. In particular, we report here two major techniques. First, the use of low power pumping lasers provides gain comparable to the ordinary one pump Raman amplification. This technique is especially interesting for combining commercial and low cost lasers [21]. The second particular technique corresponds to an evolution of the cascaded Raman amplification. Actually, a sixth order cascade Raman amplifier was recently proposed [22]. In the cascade Raman amplification, the pump power is downshifted in frequency by using a pair of fiber Bragg gratings (FBG) placed in spectral positions multiples of 13 THz, from the pump frequency. In a particular case, the generation of the fiber pump laser is obtained by using only one passive reflector element and distributed reflectors over the long optical fiber, established by a nonlinear fiber intrinsic effect called Rayleigh backscattering.

The enlargement of the bandwidth of Raman amplifiers is also achieved using incoherent pumping instead of multi-pump schemes [23-27]. Vakhshoori *et al.* proposed a high-power incoherent semiconductor pump prototype that uses a low-power seed optical signal, coupled into a long-cavity semiconductor amplifier. It was achieved 400mW of optical power over a 35nm spectral window [27]. A 50 nm bandwidth amplifier was obtained with an on/off gain equal to 7 dB. It was also demonstrated that the use of six coherent pumps is less efficient, in terms of flatness, than the use of two incoherent pumps [24]. The signal wavelengths were comprised between 1530 nm and 1605 nm and the transmission occurs over 100 km of optical fiber. Another advantage of using incoherent pumping is the reduction of nonlinear effects, such as Brillouin scattering, four wave mixing of pump-pump, pump-signal and pump-noise [28].

RFAs have become a crucial component for the implementation of fiber optic communication systems [9]. An exponential increase on the product distance × capacity of the transmission experiments on optical communication systems was observed in the last decade. The majority of these experiments, especially since the year 2000, have employed RFA as amplification technology [9]. This survey attempts to cover the most recent aspects in the field of Raman amplification for fiber communication systems.

## 2. THEORETICAL DESCRIPTION OF RAMAN SCATTERING

In 1928 Raman scattering was discovered independently and almost simultaneously by two research groups, one working in India and lead by Sir C. V. Raman [29], and the other by G. S. Landsberg and L. I. Mandelstam working in Russia [30]. In 1930, the Nobel committee distinguished Sir C. V. Raman for his discovery of the molecular scattering of light and since then this effect has been known as the Raman effect.

Raman effect is a scattering effect of light. Light scattering occurs as a consequence of fluctuations in the optical properties of a medium. In optical fibers three types of scattering effects are relevant: Rayleigh, Brillouin and Raman scattering.

Rayleigh scattering is an elastic process, i.e., the incident and the scattered photon have the same energy, therefore the same frequency. Rayleigh scattering in fibers couples light from guided modes to unguided ones leading to optical attenuation. Indeed, in modern fibers operating in the near infrared, Rayleigh scattering is the major source of attenuation, as absorption is practically negligible. In fact, Silica lattice and electronic resonances are in the mid infrared and in the ultra-violet, respectively. Therefore in the near infrared, fibers operate, essentially in an off-resonance regime, apart from impurities, which in nowadays fibers are reduced to an extremely low level [31]. However, besides the off-resonant interaction with bound electrons, optical waves also interact with molecules inside Silica fibers, through scattering.

Raman and Brillouin scattering are both inelastic processes, i.e., the incident and scattering photons have different energies. The energy lost by the incident field is stored into the medium in the form of vibrational energy, named phonons. Indeed, the origin of both Raman and Brillouin scattering effects resides in the interaction of light with these vibrational states (phonons). In the Brillouin scattering low frequency vibrational states are involved, usually referred as acoustic phonons. In the Raman process high frequency vibrational states are presented, named as optical phonons.

Raman scattering can occur in two distinguished forms: Spontaneous Raman Scattering, and Stimulated Raman Scattering (SRS).

In the spontaneous form, Raman scattering occurs when the incident field interacts with vibrational modes, mainly excited by thermal effects, of the molecules constituting the medium. From this interaction, it can result another optical phonon, with frequency $\Omega$, and a down shifted optical photon with frequency $v_S = v_0 - \Omega$, or a up shifted photon of frequency $v_A = v_0 + \Omega$ and in this case an optical phonon is annihilated, $v_0$ is the frequency of the incident signal. As the frequency $\Omega$ is related to the normal vibrational modes of the molecules constituents of the medium, by analyzing the scattered light, information about the medium can be retrieved. This is the main idea behind Raman spectroscopy, a widely used technique for materials characterization. In amorphous materials, like Silica, $\Omega$ can assume a value belonging to a broad spectral range, starting from zero and going up to 40 THz. Experimentally both down shifted and up shifted frequencies waves have been observed and have been named as Stokes and anti-Stokes, respectively.

Stimulated Raman Scattering was discovered by E. J. Woodbury and W. K. Ng, almost accidentally in 1962, when working with a Ruby laser [32]. They observed a strong spectral line not coincident with any spectral line of the fluorescence spectrum of Ruby. To understand this process let us assume that an incident photon is scattered by an optical phonon in the medium, and in this process a down shifted photon and an optical phonon are created. We can see that we have two ways of creating phonons, the scattering process and the thermal mechanism. If the intensity of the incident light is small, the rate of phonons created by scattering is low and due to thermal equilibrium the density of phonons in the medium is unchanged, and therefore the medium maintains the same optical properties. If the intensity of light is increased above a certain threshold, the optical properties of the medium can be changed in a way that the scattering process is enhanced [33]. In this situation, the incident light stimulates the scattering process and we are in the presence of Stimulated Raman Scattering. Through this positive feedback the scattering process can be enhanced by several orders of magnitude. Due to the Bosonic nature of the photons, this process can indeed provide gain. The photon emission process by a scattering center, it can be stimulated by the presence of another photon, and this stimulated emission is the origin of the gain. The term emission is used in this context in a quite abusive way because there is no absorption to a real state, but this process can be treated considering that the scattering photon is initially absorbed to a virtual state and after re-emitted.

If we consider that the decay from the virtual states only occurs spontaneously, the Stokes power grows linearly with the pump power. In the other way, if we consider that the decays from the virtual states must be triggered by another photon, the Stokes power grows exponentially with the pump power. Off course, in reality both spontaneous and stimulated emission occurs. If the photon that triggers the stimulated emission is part of a signal we are in the presence of optical gain, which can be beneficial for optical communication systems [34]. If this photon was initially generated by spontaneous emission we are in the presence of amplified spontaneous emission noise which is usually considered as harmful, at least for telecommunications purposes. The spontaneous emission process always leads to an excess of noise in the system.

The optical gain provided by the Raman process can be completely characterized by the Raman-gain coefficient $g_R(\Omega)$, which is related with the imaginary part of the third-order nonlinear susceptibility. The characterization of the amplified spontaneous emission process requires, besides the Raman-gain coefficient $g_R(\Omega)$, another coefficient named noise spontaneous emission factor $n_{sp}(\Omega)$. However, it turns out that another source of noise must be also considered to characterize the noise in Raman amplifiers. This source of noise arises from Rayleigh scattering. Most of the Rayleigh scattered photons are lost through non-guided modes, but some of them are coupled to the counter-propagating mode. Those photons can be amplified and through another Rayleigh scattering process can appear as extra-noise at the amplifier output. This effect is usually named as double Rayleigh scattering and will be described in more detail in section 4.3.

## 3. MODELING OF RAMAN AMPLIFIERS

The implementation of RFA, using an optical fiber as gain medium, requires that the pump and information signals must be injected into the same fiber. A basic scheme for a RFA architecture is displayed in figure 2. The signal and pump waves are launched into the optical fiber (the gain medium) by a coupler, so, that stimulated Raman scattering can occur. Since the SRS effect occurs uniformly for all the orientations between pumps and signals, Raman amplifiers can work both in forward and/or backward pumping configuration.

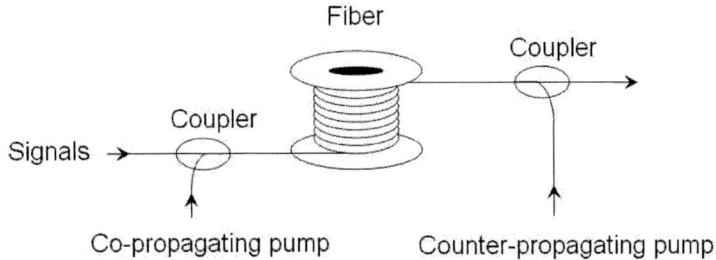

Figure 2. General scheme for a distributed Raman amplifier. For simplicity the optical isolators used to protect the pumps and signals sources, were omitted.

The model for power evolution in Raman amplifiers assuming a multipump multisignal configuration is often based on an unified treatment of channels, pumps and spectral components of the amplified spontaneous emission (ASE). The major interactions can be reasonably drawn by considering the pump-to-pump, signal-to-signal and pump-to-signal power transfer, attenuation, Rayleigh back scattering, spontaneous Raman emission and its temperature dependence. Other effects, such as noise generation due to spontaneous anti-Stokes scattering, polarization and nonlinear index are neglected, but they can reach considerable importance in certain regimes of transmission. It must be noted that signal channels and Raman pumps are treated as fields at single frequencies, so ignoring the interactions due to the spectral shape of signals and pumps.

In a general approach, the power evolution of pumps, signals and ASE (in forward and backward directions), with time along the fiber distance is given by the following set of coupled differential equation [35]. For $N_p$ pumps, $N_s$ probe signals and $N_{ASE}$ spectral components for ASE, the system is formed by $N_p+N_s+2N_{ASE}$ equations.

$$\pm \frac{\partial P_i^{\pm}(z,t)}{\partial z} \mp \frac{1}{V_i}\frac{\partial P_i^{\pm}(z,t)}{\partial t} = \left[-\alpha_i + \sum_{j=1}^{i-1} g_{ji}\left[P_j^+(z,t)+P_j^-(z,t)\right] - \sum_{j=i+1}^{m}\frac{\upsilon_i}{\upsilon_j}g_{ij}\left[P_j^+(z,t)+P_j^-(z,t)\right] - 2h\upsilon_i\sum_{j=i+1}^{m}g_{ij}\Gamma(1+\eta_{ij})\Delta\upsilon\right]P_i^{\pm}(z,t) + \gamma_i P_i^{\mp}(z,t)$$
$$+ h\upsilon_i \sum_{j=1}^{i-1} g_{ji}\Gamma\left[P_j^+(z,t)+P_j^-(z,t)\right](1+\eta_{ji})\Delta\upsilon \quad (1)$$

$V_i$ is the frequency dependent group velocity. The ± signs stand for the forward or backward propagating waves, being $\alpha_i$ and $\gamma_i$ the coefficients of attenuation and Rayleigh of the $i^{th}$ wave at frequency $\upsilon_i$. $h$ and $k_B$ are the Planck and Boltzmann constants, respectively, and $T$ is the fiber absolute temperature. The phonon occupancy factor is given by:

$$\eta_{ij} = \left[\exp\left(\frac{h(\upsilon_i - \upsilon_j)}{k_B T}\right) - 1\right]^{-1} \quad (2)$$

The frequencies $\upsilon_i$ are numbered by their decreasing value (the lower order corresponds to the higher frequency). Thus, the terms in the summation in expression 1, from $j=1$ to $j=i-1$ cause amplification since the wave $i$ is receiving power from the lower order waves (with higher frequency). For the same reason, the terms in the summation from $j=i+1$ to $j=m$ originate depletion. For mathematical convenience the gain spectrum was divided into slices of width $\Delta\upsilon$, spanning the range over which ASE spectral components are significant.

The terms that contain a product of powers describe the coupling via stimulated Raman Scattering, being its strength determined by the Raman gain coefficient of the fiber, $g_{ij}$ obtained by equation 3.

$$g_{ij} = \frac{g_R(\upsilon_i - \upsilon_j)}{\Gamma A_{eff}} \quad (3)$$

where $A_{eff}$ is the effective area of the fiber and the factor $\Gamma$ is a dimensionless quantity comprised between 1 and 2 that takes into account the polarization random effects. The achieved gain, as well as the slope of the gain spectrum, depends on the transmission fiber [36, 37]. In figure 3, two Raman gain coefficient spectra are displayed, showing the different strengths of the Raman coupling of a SMF fiber and a dispersion compensating fiber (DCF).

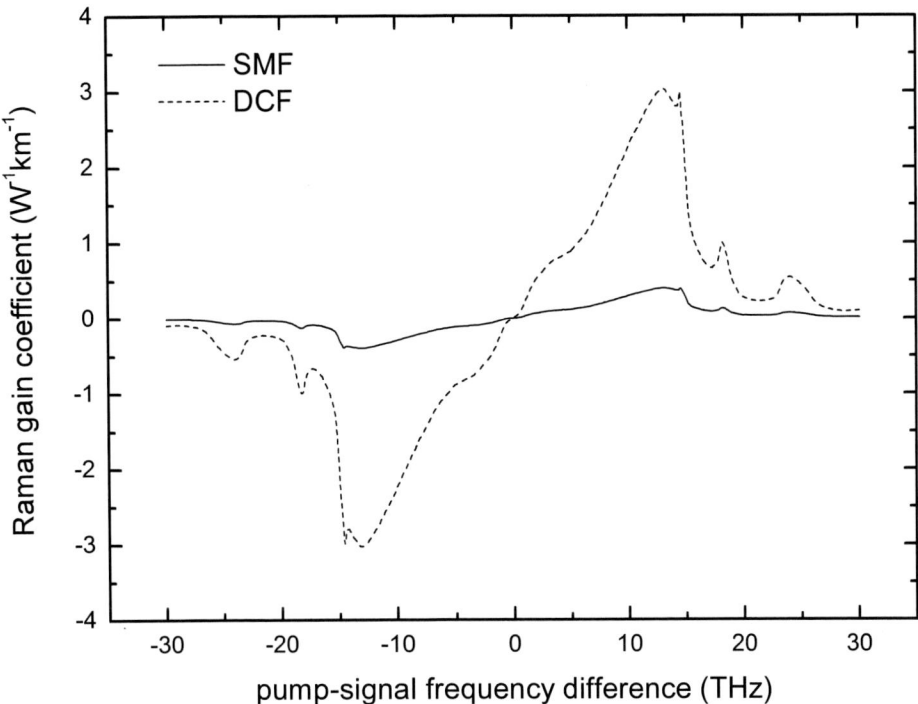

Figure 3. Raman gain coefficient spectra for two germanosilicate fibers: Single mode fiber (SMF) and dispersion compensating fiber (DCF), for a pump wavelength of 1450 nm.

As a matter of fact, the small effective area of the DCF (15 μm$^2$) is determinant for its higher Raman gain coefficient when compared to the SMF (80 μm$^2$) or when compared with DSF fibers (50 μm$^2$). Those spectra also show peaks that are broader than those presented by crystalline materials, since the amorphous nature of Silica allows a continuum of molecular vibrational frequencies.

To obtain a steady-state power distribution, the time derivative in equation 1 is settled equal to zero, and the set of equation takes the form of expression 4.

$$\pm \frac{dP_i^{\pm}}{dz} = \left[ -\alpha_i + \sum_{j=1}^{i-1} g_{ji} \left[ P_j^+ + P_j^- \right] - \sum_{j=i+1}^{m} \frac{\upsilon_i}{\upsilon_j} g_{ij} \left[ P_j^+ + P_j^- \right] - 2h\upsilon_i \sum_{j=i+1}^{m} g_{ij} \Gamma \left( 1 + \eta_{ij} \right) \Delta\upsilon \right] P_i^{\pm} + \\ + \gamma_i P_i^{\mp} + h\upsilon_i \sum_{j=1}^{i-1} g_{ji} \Gamma \left[ P_j^+ + P_j^- \right] \left( 1 + \eta_{ji} \right) \Delta\upsilon \qquad (4)$$

In spite of the simplification, the modeling is still computationally intensive, especially for the situation of backward or bidirectional pumping. In those situations, the mathematical problem that describes the power evolution of pumps and signals along the fiber is a boundary value problem (BVP) which is more difficult to solve than the initial value problem (IVP) in the forward pumping scheme. An immediate approach to the numerical solution of such problem is the shooting method [38]. There are other allowable numerical methods, such as relaxation methods, or collocation methods [39]. Generally, shooting methods are faster than relaxation ones. In shooting methods, we choose values for all the dependent variables at one boundary, solve the system of ordinary differential equation (ODE) as an IVP and verify if the obtained values on the other boundary are consistent with the stipulated values (boundary conditions) [40]. Then, the parameters are repeatedly changed using some correction scheme until this goal is attained. The selection of the correction scheme is crucial for stability and efficiency of the resulting algorithm. An other variant of the shooting method, we can guess boundary values at both ends of the domain, integrate the equation to a common midpoint and repeatedly adjust the guessed boundary values so that the solution tends to the same value at the middle point. This adjustment task is usually performed by the Newton-Raphson method.

Recently, some shooting algorithms with different correction schemes for the design of Raman fiber amplifiers have been proposed in order to improve convergence of the solutions even for larger fiber lengths [41]. This scheme is obtained by modifying the numerical method used to perform the IVP integration (fourth order Runge-Kutta, Runge-Kutta-Felhberg, etc). Other approaches to solve the equation 4 propose a shooting method to a fitting point using a correction scheme based on a modified Newton approach. Therefore, by introducing the Broydens rank-one method into the modified Newton method, the algorithm becomes more efficient and stable. This happens because the intensive numerical calculations of the Jacobi matrix are substituted by simpler algebraic calculations [41].

The use of projection methods such as collocation, gives a continuous approximation of the solution as a function of the fiber length. The basic idea is to approximate the BVP solution by a simpler function that represents an approximation.

Nevertheless, the Raman equations (equation 4) are also solvable through semi-analytical methods, using the average power analysis (APA) presented by Min *et al.* [42]. The amplifier is split into $n$ small segments, in order to avoid the position dependency of the powers of equation 4. The equations are then solved analytically in each segment, considering as input conditions the outputs provided by the solution on the previous segment. Equations 5 to 8 show how the powers are iteratively computed. The output pump/signal power at each section end is given by:

$$P_{out}^{\pm} = P_{in}^{\pm} G(z,\upsilon) \quad (5)$$

being $G(z,\upsilon)$ the section gain,

$$G(z,\upsilon) = \exp[(-\alpha(\upsilon) + A(\upsilon) - B(\upsilon))\Delta z] \quad (6)$$

The constants, $A(\upsilon)$ and $B(\upsilon)$ are obtained through:

$$A(\upsilon) = \sum_{j=1}^{i-1} g_{ij} \overline{P}_j \quad (7)$$

$$B(\upsilon) = \sum_{j=i+1}^{m} g_{ji} \overline{P}_j$$

The optical power term in each section can be substituted by its length averaged values given by:

$$\overline{P} = P_{in}^{\pm} \frac{G(\upsilon)-1}{\ln(G(\upsilon))} \quad (8)$$

For a RFA, the net gain is usually defined as the ratio between the signal powers at the end and at the beginning of the fiber link, as defined in equation 9:

$$G_{net} = \frac{P_{signals}(z=L)}{P_{signals}(z=0)} \quad (9)$$

The so-called on/off gain is another useful quantity that measures the increase in signal powers at the amplifier output when the pumps are turned on, as follows:

$$G_{on/off} = \frac{P_{signals}(z=L) \text{with pumps on}}{P_{signals}(z=L) \text{with pumps off}} \quad (10)$$

The numerical issues due to the backward pumping can be surpassed by assuming that the pump inputs are located at the same fiber end that the signal inputs. Therefore, the pump equations are integrated reversely as if they were backward by multiplying them by (−1). A guessed initial input is necessary to perform the integration, but the algorithm is able to adjust it using an optimization routine that adjust the initial input until the output at the fiber end reaches the real backward pump input. The use of the APA approach has shown a reduction of two orders of magnitude in the computation time, being the obtained results in agreement with the ones resulting from traditional numerical methods.

To demonstrate the numerical resolution of the steady-state Raman propagation equations, we assume the scheme in figure 4, where three bidirectional pumps (two backward and one forward) and four probe signals are considered. The counterpropagated pumps have power levels set equal to 0.1W, working at 1450 nm and 1460 nm, respectively. The copropagated pump is working at 1470 nm with an output power also equal to 0.1W. The forward pumping signal are then injected into 40 km of SMF fiber and combined with 4×1000 GHz spaced C band probe signals with an initial optical power equal to 1μW. The spatial evolution of pumps and probe signals are displayed in figure 4.

The implementation of equation 4 also allows the calculation of the total noise for each signal (forward and backward ASE) within the amplifier, whose spatial evolution for this system can be followed in figure 5.

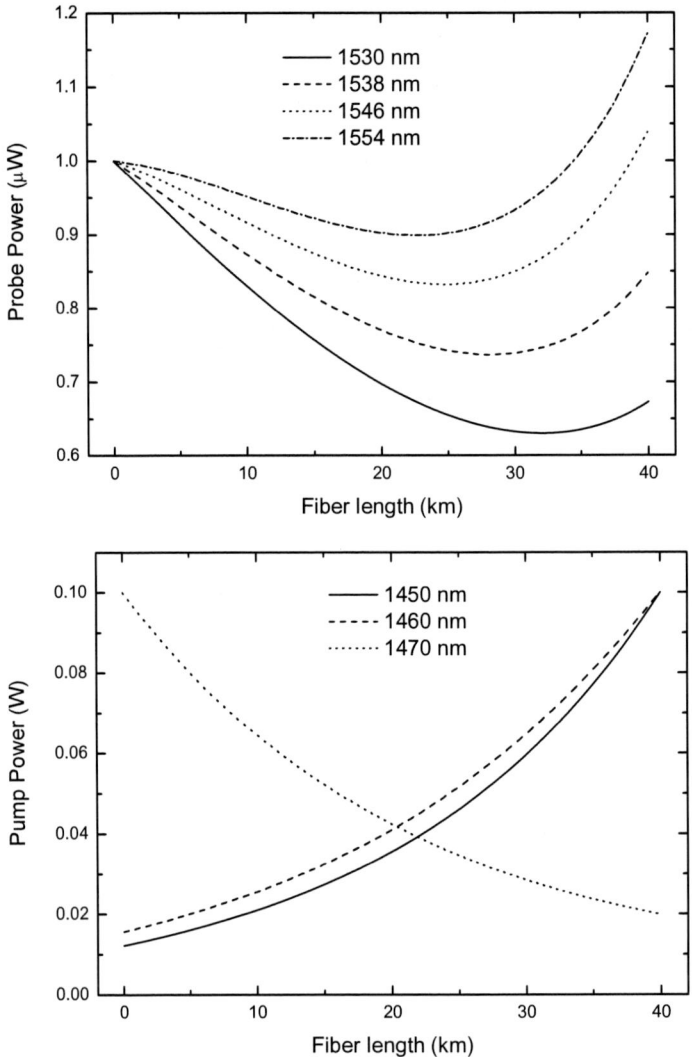

Figure 4. Spatial evolution of two counterpropagated pumps, one copropagate pump and four probe signal along a 40 km SMF fiber span amplifier. Probe signals evolution (top) and pump signals evolution (bottom).

The noise figure of an optical amplifier amounts the degradation of the signal to noise ratio (SNR) when the signals are amplified. The most important source of noise in optical amplifiers is ASE, which, for Raman amplifiers is due to spontaneous scattering. Assuming that the signals are initially as noiseless as possible, and that their degradation is due to signals spontaneous beat noise produced by ASE, the noise figure, in linear units, is given by equation 11 [36]:

$$\mathrm{NF} \approx \left( \frac{2P_{ASE}^{+}(z=L)}{h\upsilon\Delta\upsilon} + 1 \right) \frac{1}{G_{net}} \qquad (11)$$

where $h\upsilon$ is the photon energy and $P_{ASE}^+$ is the forward ASE measured over the reference bandwidth $\Delta\upsilon$. The first term corresponds to the noise from the signal spontaneous beating and the second one to shot noise.

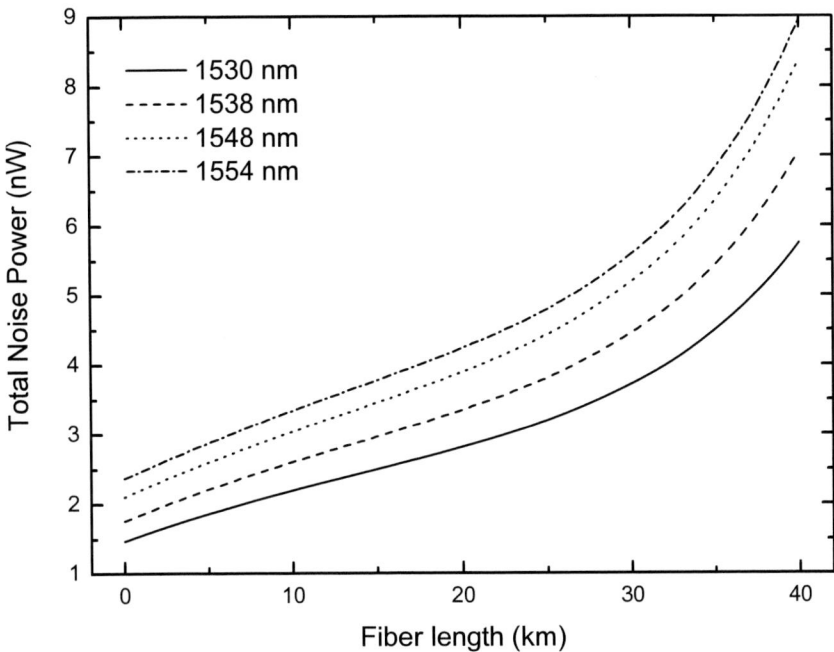

Figure 5. Spatial evolution of the total noise (forward and backward ASE) power along 40 km SMF fiber span.

Another quantity, named effective noise figure, accounts the noise that a discrete amplifier placed at the end of an unpumped fiber link would need to have the same noise performance that a distributed Raman amplifier. In decibel units, the effective noise figure is computed using:

$$NF_{eff}^{dB} = NF^{dB} - (\alpha L)^{dB} \quad (12)$$

Typically, WDM systems allocate a large number of channels spaced over wide bandwidths. Considering the previous system but doubling the pump powers and using 64 probe signals (instead of 4), we obtained the spectra of the gain and noise figure which are plotted in figure 6.

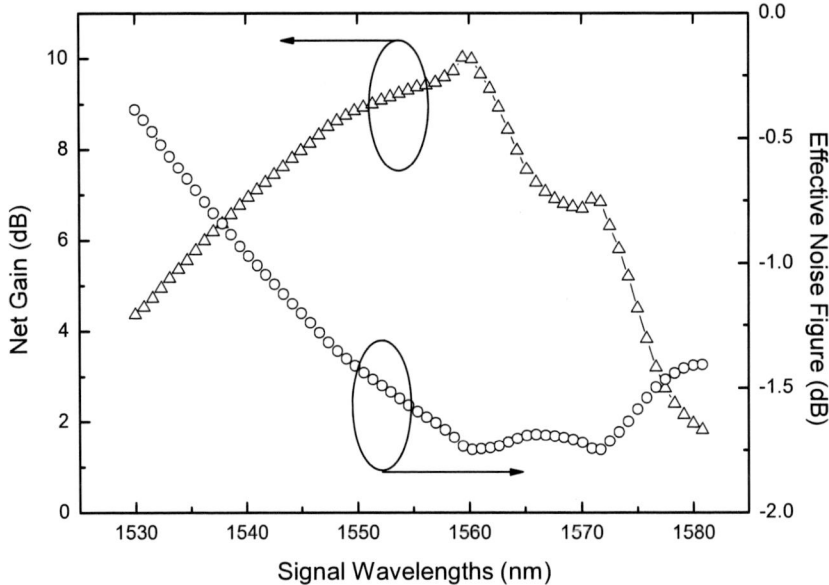

Figure 6. Net gain and effective noise figure spectra for a system with two counterpropagated pumps, one copropagate pump and 64×100 GHz probe signals along 40 km SMF fiber span.

As depicted in this section and despite some remaining numerical issues, the modeling of a multipump Raman amplifier anticipates many valuable applications for WDM systems, namely the broadband gain. It is important to notice that gain spectra as wide as 100 nm are achievable and that the gain value can be kept quite constant by an appropriate tailoring of the amplifier architecture. This procedure involves solely the proper dimensioning of the pump power levels and operating wavelengths, as discussed more extensively in section 4.

Another interesting feature of RFA is the noise performance. The ASE noise in RFA is intrinsically low (as suggested by the negative effective noise figure presented above). The reason relies in the fast relaxation of the optical phonons, the absorption of signal photon to the upper virtual state is extremely small. The inversion of population is almost complete.

## 4. CHALLENGES IN RAMAN AMPLIFICATION

### 4.1. Gain Profile Optimization

One of the most impressive features of Raman fiber amplifiers is assuredly the possibility to achieve gain at any wavelength, by selecting the appropriate pump wavelength. Therefore, it is possible to operate in spectral regions outside the Erbium doped fiber amplifiers bands over a wide bandwidth (encompassing the S, C and the L spectral transmission bands). Nevertheless, some studies have been reporting that despite the Raman gain dependence is essentially due to the pump-signal frequency difference; there is also some weaker dependence on the pump absolute frequency [43]. However, since a deeper study of this topic is beyond the scope of this work, we will not consider it in the gain optimization.

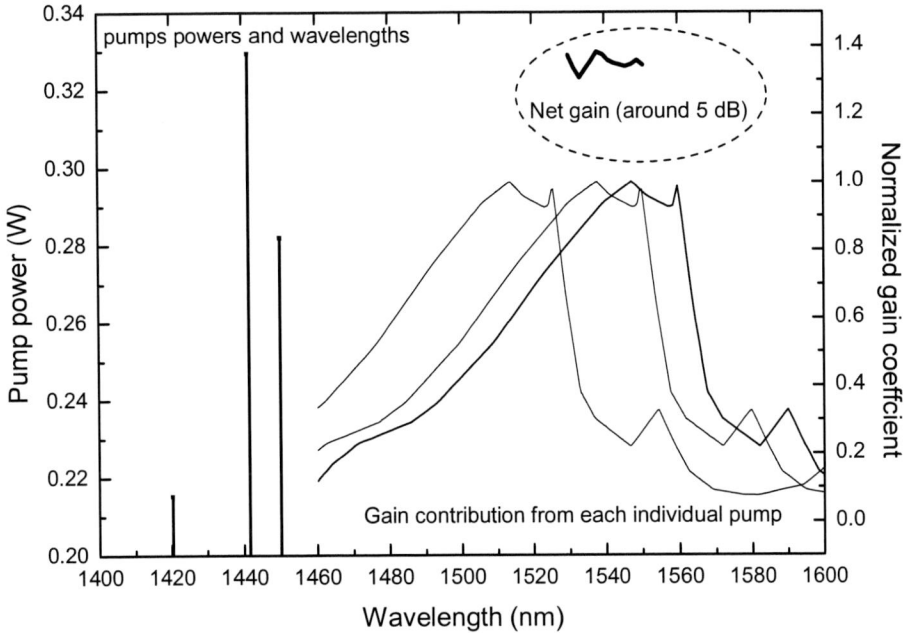

Figure 7. Numerical simulation of broadband Raman amplifier gain. Bars show backward input pump powers and wavelengths. Ticker line show 14×400 GHz probe signals optimized net gain and thin lines the gain contribution of each individual pump. The simulation was carried out through 25 km of SMF fiber.

A flat spectral gain profile is achievable with the combination of several pumping lasers operating at specific powers and wavelengths. The Raman gain created by pumps at different frequencies is slightly shifted from each other to partly overlap and form a composite gain. When the pump powers and frequencies are properly chosen, this wide gain can also be considerably flat. Another important feature to take into account when designing a flat gain scheme, is the strong Raman interaction between the pumps, since the higher frequency pump is responsible for the amplification of the lower frequency signals, more pumping power is needed, as some will also be transferred to the lower frequency pumps. This interaction between pumps also affects the noise properties of the amplifier. However, some novel pumping schemes have been recently proposed in order to prevent those unwanted effects: copumping, time dependent Raman pumping, higher order pumping and broad-band pumping [44].

Typically, laser diodes with output powers in the 100-200 mW range can be used in a multipump scheme. This scheme is normally composed of a set of laser diodes operating in the 14XX nm region, whose spectral width is narrowed and stabilized by a FBG. Optical couplers are used to combine and depolarized them, in order to suppress the polarization dependent gain. The multipumping allows bandwidth upgradeability by the addition of new laser diodes. Theoretically, the larger the number of pumps the better the gain ripples.

Nevertheless, there are economic issues that prohibit the use of an arbitrary number of pumps. For this reason, we have to find a balance between the system performance and the cost of amplification.

Optimization of the gain spectrum has been widely performed making use of several global search methods, such as neural networks [45], simulated annealing [46] and genetic algorithm (GA) [47]. During the search process, the pump powers and frequencies are directly substituted into the system of propagation equations to calculate de gain profile. Depending on the speed of the numerical method used to integrate the system of equations, the amount of numerical computations involved can be considerably large and the optimization inevitably time consuming. Those solutions are not suitable for practical applications where the real optimal solution must be provided in a short time.

However, some alternatives can be found by replacing the usual intensive numerical integrations with simpler algebraic calculations using the APA method while integrating the Raman propagation equations.

Another simple but important issue when using a global optimization method relies in a proper dimensioning of the search domain. Using the APA method, all the inputs are located at the same fiber end, even for the counter pump situations. Therefore, the pump power inputs are chosen by presuming a typical propagation profile. By this way, it is advisable to try lower power values for the higher frequency pumps and higher power values for the lower frequency pumps (the opposite happens at other fiber end). Regarding to the optimization of the pump frequencies, it is advisable to divide our spectral range into the number of pumps and then shift those values by 13 THz.

A second approach to speed up the search of the optimal pump configuration uses the genetic algorithm (GA) method only to search the pump frequencies and a quadratic programming method to solve the power integral [48]. The search domain of the GA method is by this way reduced to a half, enabling faster convergence.

Another approach combines GA with the Nelder-Mead search. This so called hybrid GA can be useful in certain situations for the purpose of saving some function evaluations and consequently to perform the optimization in the least time possible [49]. The hybrid GA follows the routine depicted in Figure 8. Firstly, the initial population, as well as the other GA operators are dimensioned: selection, crossover and mutation. The selection, together with the crossover, is responsible for the bulk of GA processing power. The mutation is an operator that plays a secondary role in the GA. Since, the genetic operators can be performed by different methodologies, it is important to choose the ones that are more adequate to the problem we are dealing with, in order to improve the GA search procedure [50]. It must be noted that if the search space is not large, it can be searched exhaustively and the best possible solution will be probably found. The maximum number of allowed generations is also an important feature because, when carefully chosen, it can save a large number of function evaluations. The Nelder-Mead method uses a simplex in a n-dimensional space, characterized by the $n+1$ distinct vectors that are its vertices. At each step of the search, a new point in or near the current simplex is generated. The function value at the new point is compared with the function values at the vertices of the simplex and one of the vertices is replaced by the new point, giving a new simplex. This step is repeated until the diameter of the simplex is less than the specified tolerance

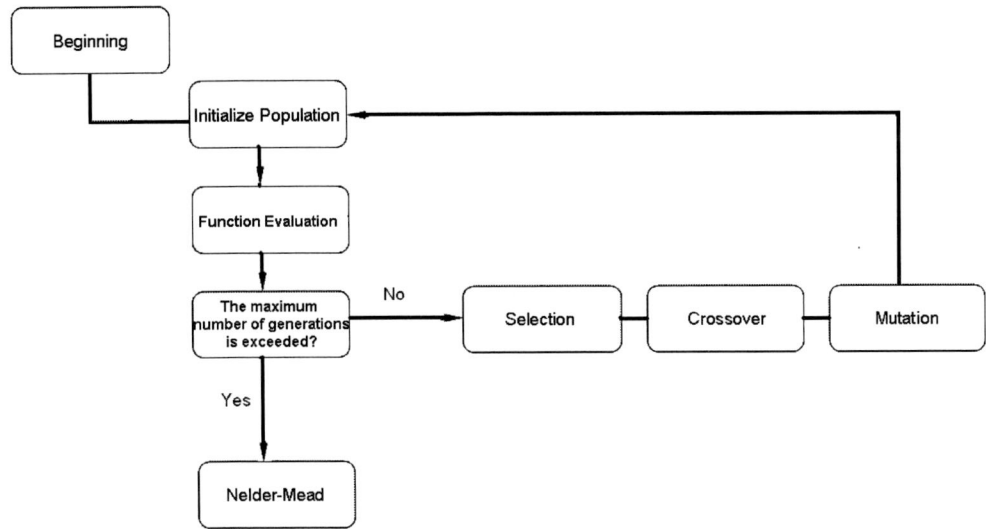

Figure 8. Scheme of the hybrid GA implementation.

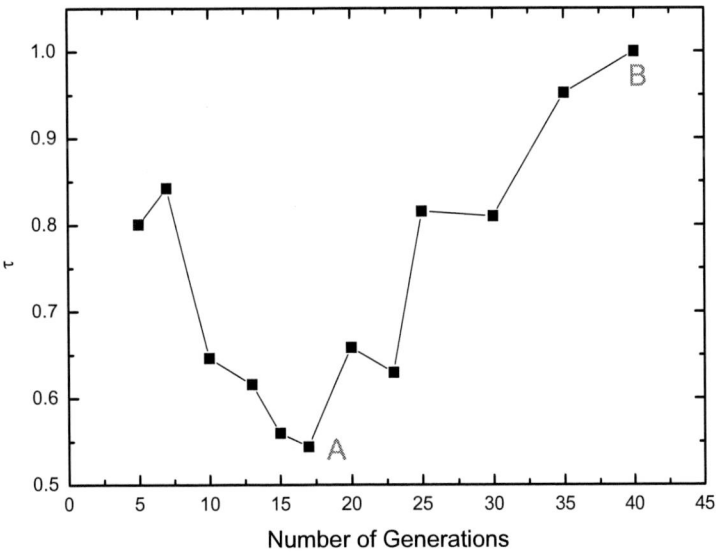

Figure 9. Total simulation time for a hybrid GA against the number of generations for a population size of 50 individuals. The line is a visual guide.

By determining properly the right moment to switch from one method to another, it is possible to reduce the simulation time to a half when compared to simple GA. This result can be observed in figure 9. The normalized total simulation time ($\tau$) against the number of generations is plotted. Here, the reference is the slowest simulation, the one that use 40 generations, identified by the letter $B$. The best situation (tagged as $A$) tooks a simulation time equal to about a half of the time of the worse situation and it was attained with 17 generations. An heuristic explanation relies on the intrinsic nature of the GA. We verify that for small number of generations (bellow 15) the GA time is small but the system reaches a worse fittest solution. Thus, the Nelder-Mead method needs more time to reach a desirable solution. When the number of generations increases the GA reaches a best solution but the needed computation time increase accordingly.

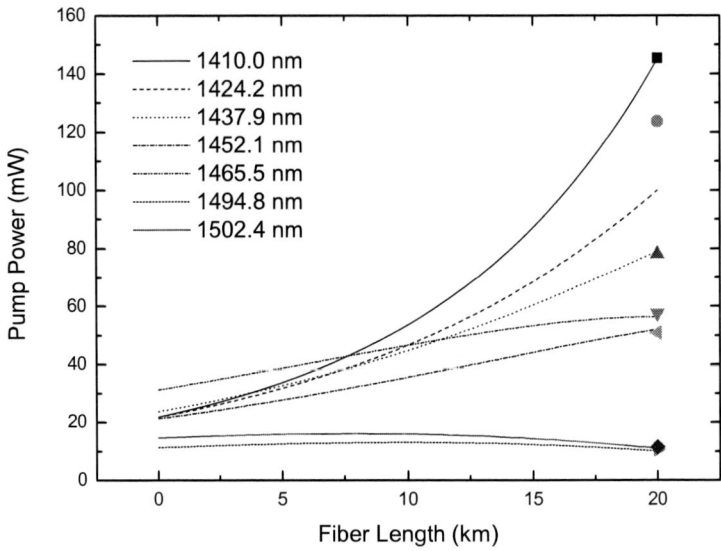

Figure 10. Power evolution of optimized pumps along 20 km of SMF (lines). The geometric shapes stand for the used experimental values.

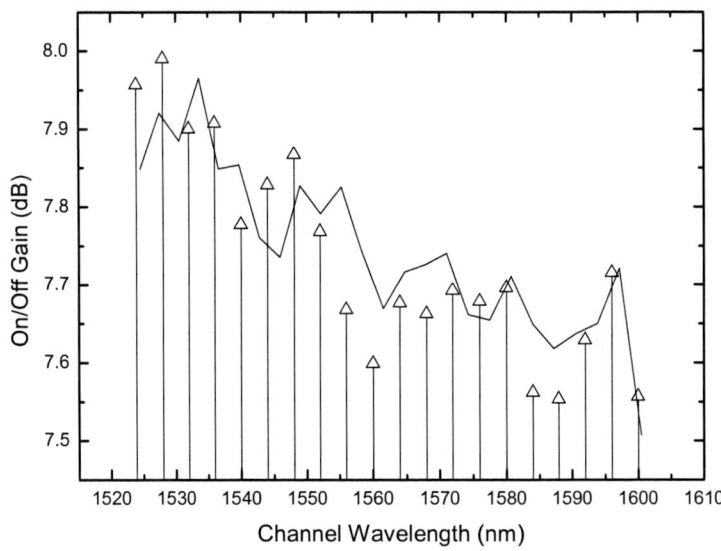

Figure 11. Experimental (arrows) and simulation (line) on/off spectral gain for the 20 probe signals and 7 counter propagated pumps, over 20 km of SMF fiber.

In order to enlighten the conclusions provided by the hybrid GA algorithm, a laboratorial implementation was carried out to test the optimization results. A Raman amplified system with 20 km of SMF fiber, 20 probe signals and 7 backward pumps was implemented. Since the pump wavelengths are already settled, only the optimization of the power levels is needed. The simulation used the stochastic uniform method for selection, the scattered crossover method and the uniform mutation. A population of 50 individuals and a number of generations equal to 35 were considered. The spatial evolution of the pumps signals optimized values are displayed in figure 10 jointly with the pump signal experimental values. In figure 11, the optimized and experimental on/off gain spectra are presented.

This is a good agreement between the optimization modeling and the experiment. The maximum ripple attained by the optimization is 0.41 dB being the experimental maximum ripple equal to 0.23 dB. The mean square deviation between simulation and experimental results is equal to 0.0036. Indeed, a flat gain over a wide bandwidth (~80 nm) was attained, using seven pumps with a total input power equal to 453 mW.

## 4.2. Raman Amplification Using Multiple Low Power Lasers

One of the main issues in Raman amplification is related to the stability of the high power lasers, the costs and the need for efficient cooling. To go around these problems, the usual solution is the use of several pump signals, what results in added advantages, like high, flat and wide-gain bandwidth [51-53].

The technology evolution allowed that high power pumps are nowadays commercially available, although some problems still limited [54]. The pressure on optical components prices, lead to the creation of CWDM standards [55]. This is reflected specially on price dropping of uncooled lasers with relatively high powers (>10mW). The price to pay is wavelength wondering, however, neither for CWDM nor for Raman, wavelength stability is not a stringent requirement, allowing simple control. With this technology the possibility of achieving Raman gain by combining multiple of these low power lasers was successfully implemented [21].

Teixeira et el proposed the use of an array of low cost lasers to achieve wideband Raman amplification, providing both experimental and simulation results [21]. In this work a counterpropagating topology was implemented, using 40 C band lasers with 20 mW output power spaced by 0.8 nm (1533 nm – 1557 nm). These lasers were combined using a multiplexer, bringing up a total power of more than 200mW (23 dBm). This power is enough to generate SRS. Several fibers were tested: True Wave and dispersion compensating. To characterize the gain profile, an array of 40 L-band 0.8nm spaced probe lasers (1565 nm - 1605 nm) with a total optical power of 1 mW was used. Figure 12 (a) shows the implemented setup. Figure 12 (b) presents the simulation results for the implemented system to four different pumping configurations. The first curve corresponds to the traditional approach, where one high power pump (23.6 dBm) at a single wavelength (located at 1530 nm) is used. In the second case, three lasers spaced by 0.8 nm starting at 1530 nm having total power of 23.6 dBm were multiplexed. The results for the two above pumping configurations are approximately equal, having only a wavelength shift of 0.8 nm as expected due to the average pumps wavelength difference. Similar simulation was experienced considering 40 lasers, each with 7.6 dBm (after the multiplexer), resulting in a total power of 23.6 dBm. In this case, the gain curve appears even smothered and the peak shifted by ~16 nm; the peak gain is similar, however a small enhancement on the 3 dB gain bandwidth was obtained and the gain profile smothered. In order to explore the advantages of the methodology (gain flatness), the power distribution for the pumps was optimized to reach an equalized gain, while maintaining the same total pump comb. The peak gain was decreased at the expense of an increased flattened profile.

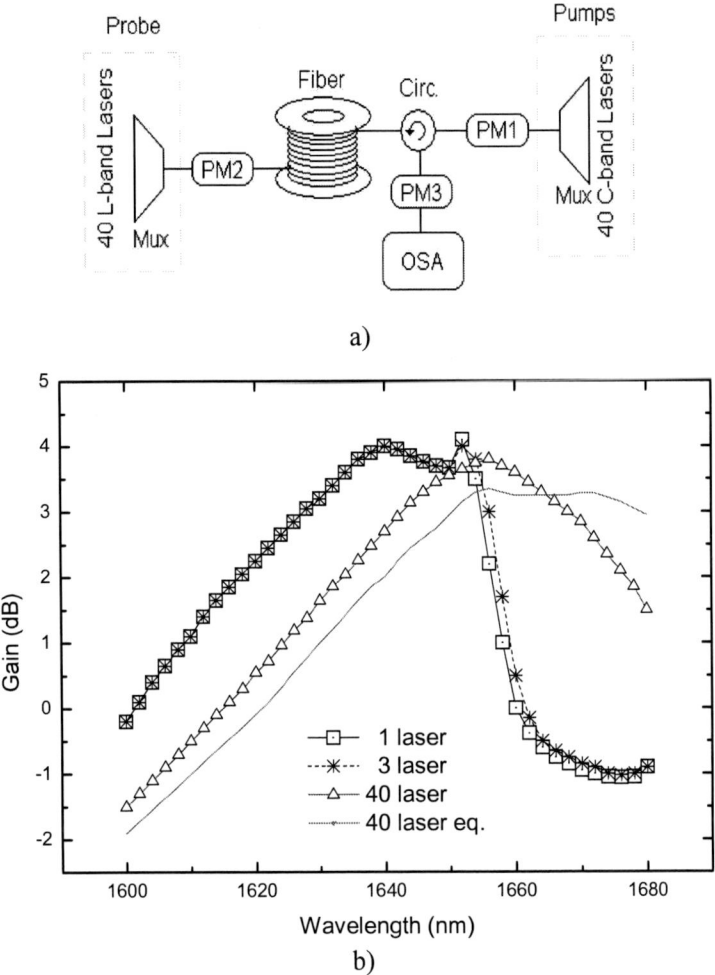

Figure 12. a) Implement setup for the simulation and experimental systems, b) simulated Raman gain profiles for several sets of pumping configurations, with 23.6 dBm of total power. PM demotes an optical power meter, OSA denotes an optical spectrum analyzer and MUX is an optical multiplexer.

Due to limitations on available probe and pump signals, the experimental implemented system only can scope part of the spectral bands used in simulation. The gain is only measurable when the pump powers go above 10 dBm. A maximum of 3 dB net gain was achieved in the L band for full pump power, 23.6 dBm, as displayed in figure 13 b). Also, in the same figure, a minimum of 2dB gain over more than 30nm, with 1dB ripple, was achieved without any power distribution optimization. The results have demonstrated the effectiveness of the technique to achieve Raman amplification.

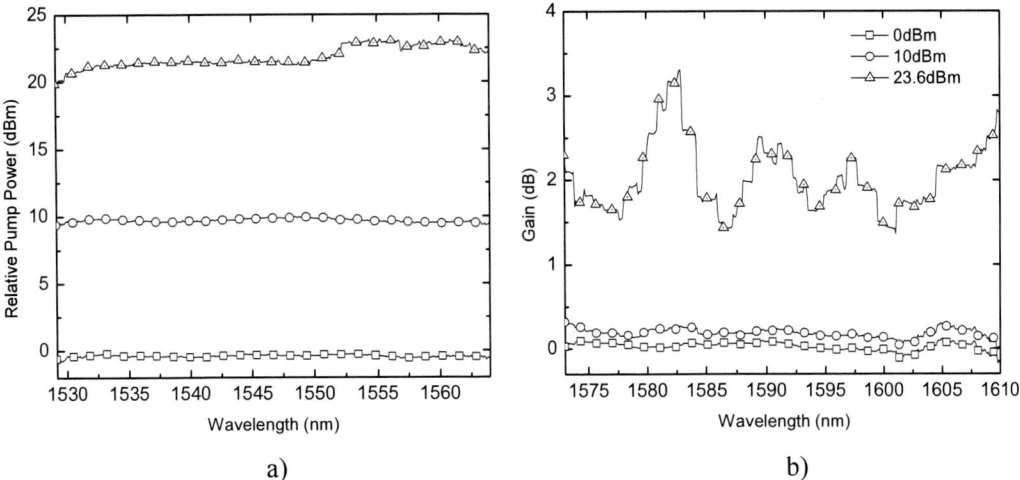

Figure 13. a) Pump to pump Raman effect; b) experimental Raman gain achieved for several values of the pump power, with probes at 0 dBm.

In figure 13 (a) it is illustrated the pump to pump effect which is commonly occurring in dense WDM systems. This effect starts to be noticeable above 10 dBm of total power and is evident for 7.6 dBm per channel. This phenomenon can be harmful due to uneven distribution of power during transmission, however, if correctly considered can be used to obtain beneficial extra gain in the system, if a pre equalization is also implemented .

## 4.3. Raman Amplification Using Rayleigh Backscattering

Raman amplification pumping can also be achieved by recurring to the traditional methods of shifted gain [19, 20]. In these methods, several FBG reflector pairs are used to generate resonant cavities in the maximum of the Raman gain spectrum. Thus, with a Ytterbium laser operating in the vicinity of 1090 nm, where it exhibits its maximum efficiency, it is possible to generate pumps in the E band, as demonstrated by Papernyi *et al*, where a set of 6 FBG reflector pairs were used to generate pumping in the E-band [22]. The latter amplifies the C band, where the probe signals transmission usually occurs.

The main penalties of traditional Raman amplification are associated with intrinsic nonlinear phenomena such as nonlinear refraction and Rayleigh backscattering, since it is required to use high powers and long fiber spans. This last effect occurs when a fraction of scattered light is backreflected towards the launch end of the optical waveguide. This reflection is called single Rayleigh backscattering (SRB). Part of this scattered light is also backreflected in the forward direction and it is called double Rayleigh backscattering (DRB), as shown in figure 14 [56]. SRB and DRB can be controlled by actuating properly on the fiber drawing process or by a correct power design [57]. The Rayleigh backscattering has been studied, modeled and characterized by many authors [56-59]. It is known that the process results from multiple reflections of light inside the fiber and therefore spontaneous and unstable lasing can occur [60]. However, this phenomenon has been observed as an impairment to signal transmission [61, 62].

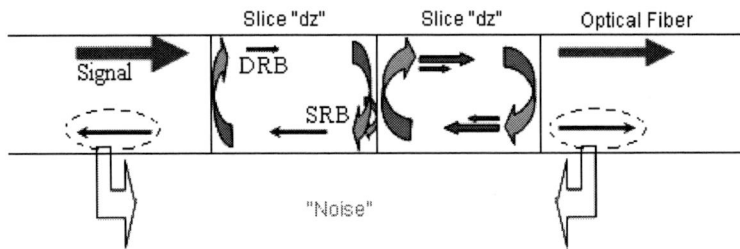

Figure 14. Simple Rayleigh backscattering (SRB) and double Rayleigh backscattering (DRB) representations over an infinitesimal length of fiber.

Recently, a method that, up to some extent, allows the control of this phenomenon was reported [56,63]. With the possibility of controlling the SRB and DBR effect, novel applications can be drafted. One suggestion is the use of this effect to generate distributed resonant cavities, which will degenerate in lasing if enough gain is achieved. These are achieved with the help of only one end FBG set [63]. This is advantageous when compared to the previously described methods to obtain cascaded Raman amplification, since it needs only one FBG set, minimizing the need for identical FBG to be used and tuned at different sites which can be not colocated.

In order to demonstrate the application of this technique to control SRB and DRB, the experimental system reported in figure 15 was implemented. A Raman pump in the E-band, at 1428 nm, was coupled to the transmission fiber, with controllable power up to 1.5 W. A circulator was used to protect the laser from back reflections and, simultaneously, to allow the measurements of the back reflected power spectrum. Two different scenarios were observed: the FBGs are absent between the fiber and the coupler; and the setup was complete as described in figure 15. These two scenarios target to show the controlling effect achieved by the FBGs.

A set of three FBG with wavelengths centered at: 1520 nm, 1531.6 nm, 1535.6 nm all having 95% reflectivity, were placed after the pump and act as reflective elements.

In a first setup, a 14 km DCF fiber with dispersion parameter equal to -1393 ps/nm and Raman coefficient of $3.05 \times 10^{-3}$ $m^{-1}W^{-1}$ was used as transmission medium.

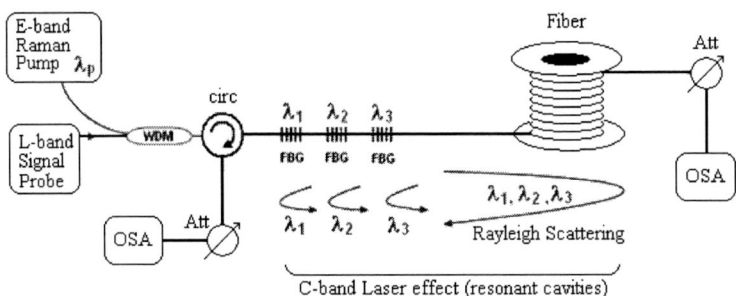

Figure 15. Experimental setup for the double shifted Raman experiments; WDM denotes a band coupler and Att denotes an optical attenuator.

Considering the first scenario, where no FBGs were present, the common Raman effect in the C band was observed, figure 16 a) for a pump power of 350 mW. When the power of the pump was increased to 600 mW, Rayleigh backscattering spontaneous lasing effect is observed, as displayed in figure 16 b). This effect presents random behavior, both in wavelength and power, being the spectrum time dependent.

In a second scenario the FBGs were present, control of the random process generated by the SRB and DRB was achieved and the lasing was stabilized in the FBGs wavelengths. In this situation a virtual cavity was established, formed by the FBG and the Rayleigh backscattered light. To generate more than one laser in the C-band a set of cascaded wavelength mismatched FBGs were used. These gratings are responsible for a multipeak frequency dependent reflection back into the fiber of the amplified spontaneous emission and DRB light from the fiber. This, in conjunction with the FBG, create resonant cavities, which generate stable wavelength constant lasing actions, from now on called as FBG-DRB lasing.

Due to the different reflectivities of the FBGs and the Raman gain profile, different lasing powers for each configuration occur in the C-band. Whenever the power of the generated lasers in the C-band is high, cascaded Raman effects will occur that generate gain in the far L and U-band. The FBG-DRB lasing and consequent stabilization process with the simultaneous L-U band spontaneous emission is reported in figure 16 c), where a pump power of 1.2 W was used [63].

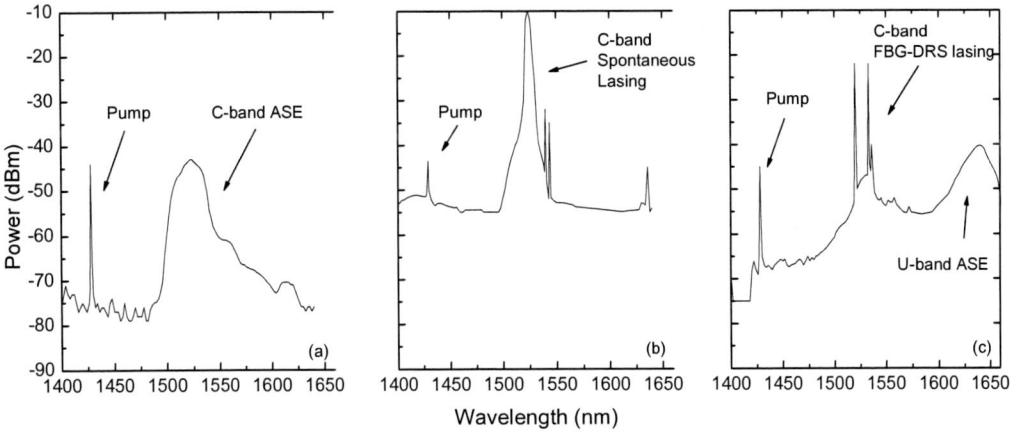

Figure 16. Transmission spectra for 14 km DCF fiber: a) Spontaneous ASE for a pump of 300mW; b) spontaneous lasing for a pump of 600mW; c) C-Band FBG-DRB lasing and far L and U-band Raman generated ASE for a 1.2 W pump.

The results show a 38 nm flattened ASE bandwidth in the U-band, generated by the FBG-DRB. By introducing a copropagating probe at 1625 nm, a gain of 10 dB was measured for an E-band pump power of 1 W.

In a second setup, different optical fibers were tested in order to compare the pump power laser threshold. A 14 km long DCF fiber, a 50 km long DSF fiber and a 50 km long non zero dispersion shift fiber (NZDSF) were used [60]. Figure 17 shows the different lasing thresholds and curve shapes resulting from the intrinsic differences between the optical fibers. From figure 17, it can be observed that this process is more efficient in the DCF fibers, where the threshold power is 350 mW, while for the NZDSF fiber is 650mW and 1W for the DSF fiber.

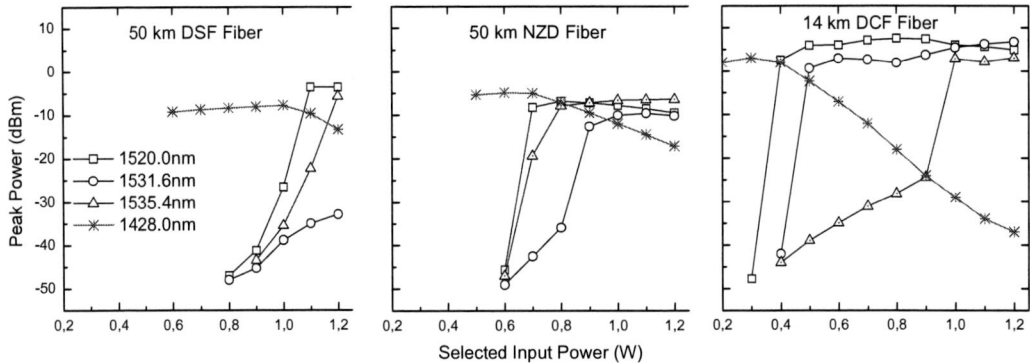

Figure 17. Depletion of the E-band pump and peak power of the C band lasers as a function of the pump power for several fiber types; from left to right: DSF, NZD and DCF.

Usually, the simulation of Raman amplification as convergence and stability problems, especially for high pump powers, has reported in previous sections. The simulation of the lasing effect with high pump power has similar difficulties. To avoid such problems the solving method for the differential equation system is simplified to an analytical method based on the transfer matrix (APA) as proposed in section 4.1. Inside these fiber slices, the parameters are considered to have small variations and the solution of the equation system is obtained by stabilization after multiple passes along the length of the fiber [64]. The initial solution uses an analytical approach that was based in the undepleted case. The approach to the pump depletion is included in the attenuation of the pump.

In each fiber slice, the Rayleigh backscattering is calculated at the boundary and this backscattering power is added to the signals in the same direction and wavelength, that also suffer amplification and depletion.

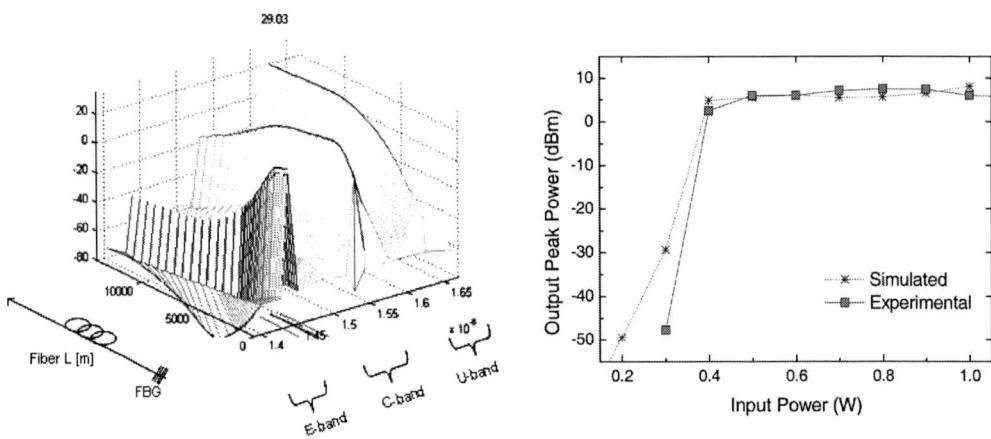

Figure 18. (a) Optical power density spectra for 14 km DCF fiber from E-band to U-band; (b) Output power evolution of the lasing effect of the FBG-DRB at 1520 nm.

Figure 18 (a) presents the simulation results of such algorithm for 29.03 dBm of pump power. The evolution of the power densities from E-band to U-band spectra is shown for a long fiber span. The E-band pump signal suffers depletion in the long propagation fiber. This pump works as a seed of the C-Band FBG-DRB lasing, which generate the L-U-Band Raman gain.

Since the process of lasing is not stable, the simulation process presents a slow stabilization, but, the boundary powers over the FBG are quickly stabilized. Figure 18 (b) presents a comparison of the threshold laser power obtained by experimental and semi-analytical methods. The output peak power of the FBG-DRB signal at 1520 nm is related with the input pump power.

As observed for the DCF fiber, stable multiple laser actions were achieved for moderate pump powers (350 mW) for both simulation and experiment.

## 4.4. Amplification with Incoherent Pumps

A technique to increase the bandwidth and decrease the spectral ripple of RFA is available with incoherent pump lasers. A Raman amplifier with incoherent pumps can be modeled as a multipump Raman amplifier. In such case, the spectrum of the incoherent pump is well approximated by a large number of pumps of infinitesimal spectral width and whose power sum equals the integral power of the incoherent pump. Therefore, the theoretical model used for incoherent pump schemes is based on the model, previously presented, for coherent multipump configurations.

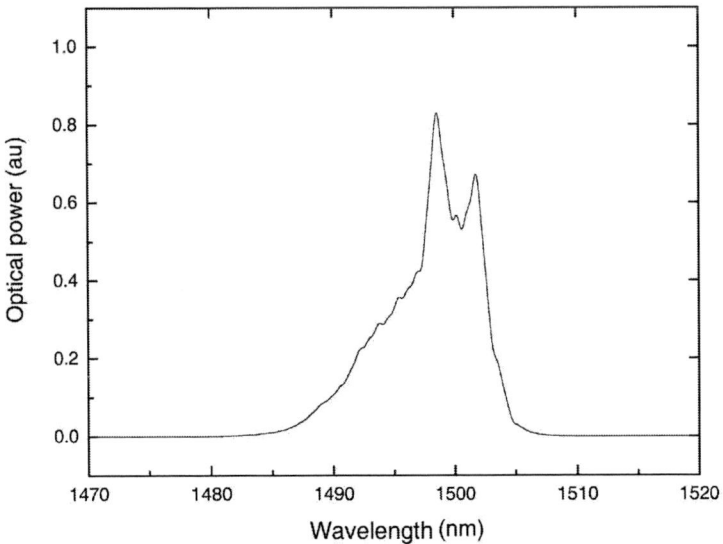

Figure 19. Pump spectrum for the incoherent pump.

An incoherent pump spectrum, as displayed in figure 19, with 10 nm FWHM, can be approximated by 100 pumps of infinitesimal spectral width, having an aggregate power equal to the integral power of the incoherent pump. The incoherent pump here considered was obtained from a high power FBG (Fiber Bragg Grating) laser, from which the stabilization grating was removed [65].

To evaluate the advantages of this technique, the Raman on/off gain and the noise figure were measured for coherent and incoherent pumping over 40 km of SMF fiber. The probe signal combo consists of 13 channels, with 1 mW power, spaced by 100 GHz over the 1546-1556 nm spectral region. Both co-propagating and counter-propagating architectures were considered. The coherent pumping source was a high power FBG laser with a wavelength of 1490 nm. In both cases the pump power was 290 mW.

The results of Raman on/off gain and effective noise figure are shown in figure 20. The relatively low on/off gain is due to the fact that the pump wavelengths have not been optimized for this signal band.

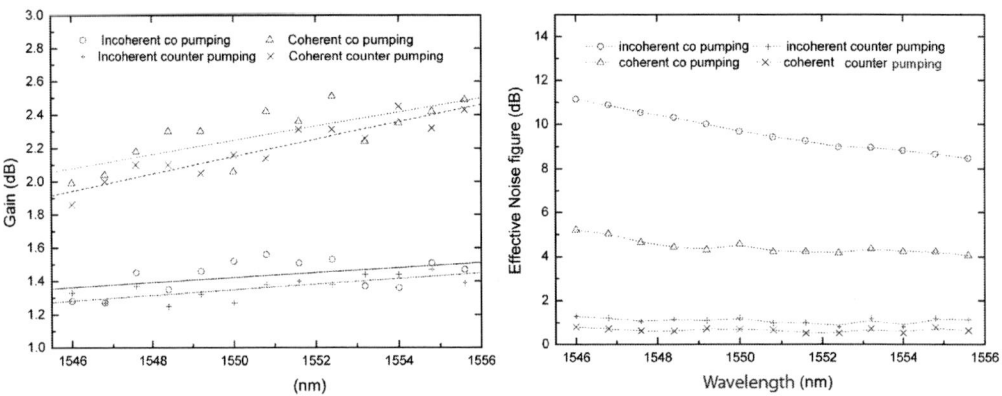

Figure 20. Raman gain and Effective Noise Figure. Lines are simulated results and points represents to experimental data.

The incoherent pumping gain slopes are 0.015±0.008 dB/nm and 0.017±0.004 dB/nm for co and counter propagation configurations, respectively. For coherent pumping, the gain slopes are 0.042±0.01 (co-propagation) and 0.052±0.005 dB/nm (counter-propagation). Such results show that the incoherent pumping configuration presents a flatter gain.

The noise figure is approximately the same for coherent and incoherent pumping in the counter-propagating configuration. However, in the co-propagating case, the noise figure is considerably lower for coherent pumping.

In agreement with previous works [23-26], these results indicate that the incoherent pumping technique can be used to decrease the spectral ripple of the Raman gain.

## 4.5. Raman in CWDM Systems

Another important challenge is the deployment of RFA for access networks, namely for CWDM networks. Since CWDM systems require large bandwidths to guarantee the transmission of a reasonable number of channels, spaced by 20 nm, wide band Raman amplifiers are well suited for this purpose.

The Raman amplifier bandwidth can be enlarged by using multiple pumps. Optimization of the number of pumps and their wavelengths enables the large needed gain spectra and that could be placed in any range of wavelengths used in optical communications.

The design of an amplifier that fits more than two CWDM channels can be achieved, with the following procedure. The number of channels to be transmitted is determined in order to define the required bandwidth. The optical fiber characteristics impose a minimum to the required gain, and finally the number of pumps as well as their characteristics are decided. The scheme of figure 21 illustrates the important issues to be considered to design a multi-pumped Raman amplifier for a CWDM system.

Since the number of CWDM channels and the length of the link as well as its losses are defined, the minimum required gain to compensate the transmission losses and the minimum bandwidth to transmit all the required channels may be determined using the rectangle shown in figure 21. The gain has to be high enough to compensate the losses caused by the optical fiber and the bandwidth should be large enough to support all the transmitted channels. The purpose is to obtain a spectrum that encloses this rectangle.

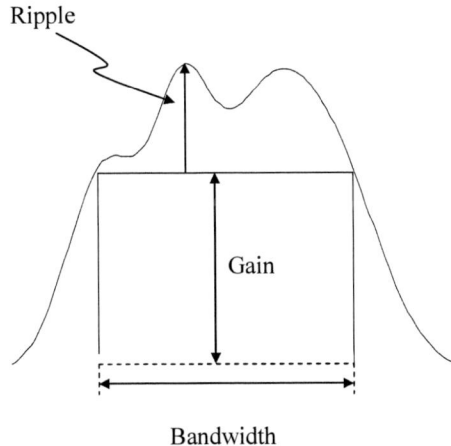

Figure 21. Design concerns for a multi-pumped Raman amplifier for CWDM systems.

Another point is to guarantee that the maximum deviation between the values of the designed and the needed gain as smallest as possible. The curved line in figure 21 represents the obtained spectrum after optimization of pumps characteristics. The ripple represents the maximum deviation cited above.

Another concern in designing the spectrum is to make it flat, with all the channels at the same level, in order to avoid reception constrains.

The multipumped Raman amplifier can be designed using a set of coupled nonlinear equations as equation 4. Solving the coupled equations for one signal and one pump, may be simplified when pump depletion is ignored. This approximation is valid because the pump power is higher than the signal power, $P_p \gg P_s$ [66]. However, whenever multiple pumps are used this simplification cannot be used due to the interaction between pumps which enhances the effect of depletion due to the higher powers involved.

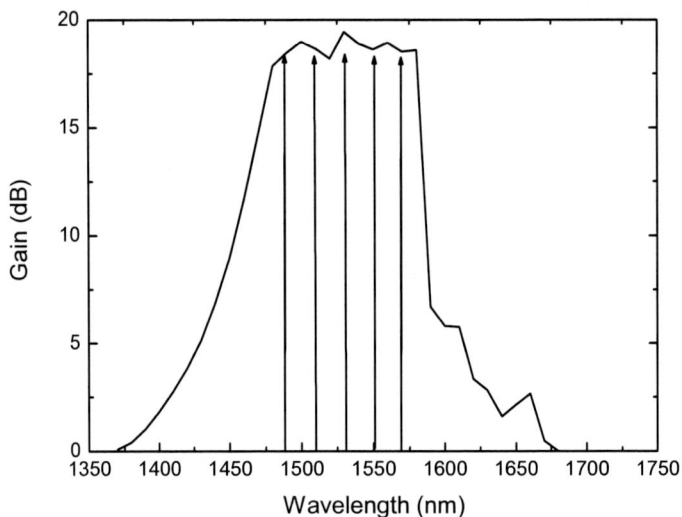

Figure 22. Example of a multi-pumped Raman amplifier applied to CWDM systems.

Figure 22 shows an example of an optimized Raman amplifier spectrum applied to CWDM systems. It was designed to transmit five probe channels at 1490 nm, 1510 nm, 1530 nm, 1550 nm and 1570 nm. The transmission link is based on 80 km SMF fiber, with 0.23 dB/km losses, which implies a 18.4 dB gain with a minimum spectral bandwidth of 80 nm.

The graph in figure 22 is the result of a forward pumping configuration. The number of pumps used in this example was six. The continuous line represents the gain spectrum obtained with the six pumps the arrows represent the transmitted probe channels.

The bandwidth is 100 nm, 20 nm larger than the minimum required bandwidth, in order to guarantee that all the signals are amplified. The gain is around 18.4 dB with a maximum deviation between the designed and needed gain equal to 1 dB, and a maximum gain deviation for each channel being 0.9 dB.

The six pumps used are centered at 1380 nm, 1393 nm, 1405 nm 1428 nm, 1444 nm, and 1468 nm with powers of 450 mW, 200 mW, 330 mW, 160 mW, 45 mW, and 55 mW, respectively. The pump of lower wavelength needs the highest power due to the interactions between pumps: The lower wavelength pump loses energy to the higher wavelengths, causing its depletion.

This optimization scheme was verified experimentally, with 3 probe channels CDWM system, pumped with 3 pump signals at 1470 nm, 1490 nm and 1510 nm. For the optimization the hybrid GA algorithm, previously presented, was used. This pump allocation problem is less exigent, in terms of ripple, than for a DWDM system, since the probe signal are far apart.

The implemented scenario consists of a 40 km SMF fiber, with a counterpropagating pump scheme and a 7 dB gain target. The optimized pump powers were 128.1 mW, 65.0 mW and 146.9 mW, respectively. The maximum gain excursion was 0.002 dB and 0.12 dB for the simulation and experimental systems, respectively.

Experimentally, we can observe that the Raman amplification improves the eye opening penalty of a signal transmitted along a fiber link allowing a good reception at the end of the transmission path. To illustrate this behavior, the eye diagram of a signal after a 40 km link is shown in figure 23.

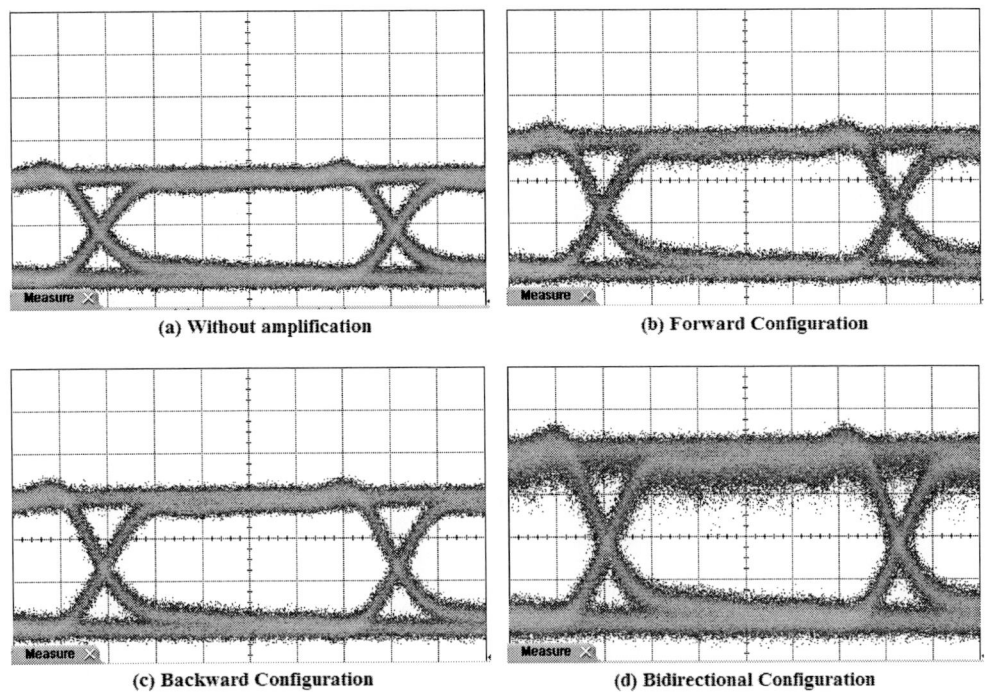

Figure 23. Comparison between eye diagrams with and without Raman amplification.

Figure 23 illustrates a real case where there is a signal centered at 1567 nm and two pumps centered at 1508.8 nm, one in forward configuration and another in backward configuration. The powers of the pumps are chosen to be 100 mW each.

The eye openings are given in Volt and the gain was obtained using the on/off definition (equation 10). Using the relation between power and voltage, $P=V^2/R$, the on/off gain becomes $G_{Voltage} = 10\log_{10}(V_{with\ pump}/V_{without\ pump})$.

It is notorious that an increase of the eye opening obtained when both pumps are turned on. The scale is the same to all the graphs in figure 23 to allow comparisons of the eye opening amplitude. The eye opening to the bidirectional configuration is higher due to the higher pump power, while the forward and backward systems have 100 mW, the bidirectional system uses 200 mW. The respective gains of the eye openings are 1.94 dB, 2.35 dB, and 2.98 dB to the forward, backward and bidirectional systems, respectively. The results show a higher gain for the counter propagating situation.

## 5. CONCLUSION

Raman fiber amplifiers are a technological key component that fulfill the challenging strict requirements of the beginning of this century, enabling applications not feasible with conventional EDFAs.

In this contribution, we have discussed the origin of Raman scattering and the critical properties for system design, such as pumping allocation, cascade pump and broadband amplification for multiple CDWM networks. It was also presented solutions that provide that gain, such as the use of low power pumps or incoherent pumps.

These issues are, in the authors point of view, the relevant questions and challenges associated with Raman amplification on communication systems.

## ACKNOWLEDGMENTS

This work was supported by the POSC program, financed by the European Union FEDER fund and by the Portuguese scientific program. The authors also greatly acknowledge the ARPA (POSI/EEA-CPS/55781/ 2004) and TECLAR (POCI/A072/2005) projects and to FCT and ALBAN scholarship program.

## REFERENCES

[1] Hu, J.; Marks, B. S.; Menyuk, C. R. *J Lightwave Technol.* 2004, 22, 1519-1522.
[2] Giltrelli, M.; Santagiustina, M. *IEEE Photon Technol Lett.* 2004, 16, 2454-2456.
[3] Cui, S.; Liu, J.; Ma, X. *IEEE Photon Technol Lett.* 2004, 16, 2451-2453.
[4] Kidorf, H.; Rottwitt, K.; Nissov, M.; Ma, M.; Rabarijaona, R. *IEEE Photon Technol Lett.* 1999, 11, 530-532.
[5] Bromage, J. *J Lightwave Technol.* 2004, 22, 79-93
[6] Namiki, S.; Seo, K.; Tsukiji, N.; Shikii, S. *Proceedings IEEE* 2006, 94, 1024-1035.
[7] Shuto, Y.; Yanagi, S.; Asakawa, S.; Kobayashi, M.; Nagase, R. *IEEE Journal Quantum Electronics* 2004, 40, 1113-1121.
[8] Seo, K.; Nishimura, N.; Shiino, M.; Yuguchi, R.; Sasaki, H. *Furukawa Review* 2003, 24, 17-22.
[9] Bromage, J. *J Lightwave Technol.* 2004, 22, 79-93.
[10] Chen D. Z., Wellbrock G., Peterson D. L., Park S. Y., Thoen E., Burton C., Zyskind J., Penticost S. J., Mamyshev P., *ECOC* 2006, Cannes.
[11] Monroy I. T., Kjaer R., Palsdottir B., Koonen A: M. J., Jeppesen P., *ECOC* 2006, Cannes.
[12] Scheiders, M.; Vorbeck, S.; Leppla, R.; Lach, E.; Schimidt, M.; Papernyi, S.; Sanapi, K, *Optical fiber conference* 2005, Anaheim, Post-deadline paper.
[13] ITU -T G.694.1, 2002.
[14] ITU -T G.694.2, 2002.
[15] Eichenbaum, B. R.; Das S. K. *National Fiber Optic Engineers Conference* 2001, Baltimore, USA, 1444-1448.

[16] Wang, D.; He, C.; Li, Y. *FibreSystems Europe/Lightwave Europe*. 2005, 13-15
[17] Desurvire E. *Erbium-Doped Fiber Amplifiers - Principles and Applications*; John Wiley & Sons, New York, NY: 1994.
[18] Emori, Y.; Tanaka, K.; Namiki, S. *Electron Lett.* 1999, 35, 1355-1356.
[19] D. I. Chang, H. K. Lee and K. H. Kim, *Electron. Lett.* 1999, 35, 1951-1952.
[20] S.B.Papernyi,V.I.Karpov,andW.R.L.Clements, *Optical Fiber Communication (OFC)* 2002, Anaheim, USA, PostdeadlinePaperFB4-1.
[21] Teixeira, A.; Andre, P.; Stevan, S.; Silveira, T.; Tzanakaki, A.; Tomkos, I. *International Conference on Internet and Web Applications and Services/Advanced International Conference AICT-ICIW* 2006, on 19-25 Feb. 2006 Page(s):85 - 85 ,Digital Object Identifier 10.1109/AICT-ICIW.2006.158
[22] Papernyi, S.B., Ivanov, V.B. ,Koyano, Y., Yamamoto, H., "Sixth order cascaded Raman Amplification .- Ivanov", *Optical Fiber Communication Conference*, 2005. Technical Digest. OFC/NFOEC
[23] Han, B.; Zhang, X. P.; Zhang, G. D.; Lu, Z. G.; Yang, G. X. *Opt Express*. 2005, 13, 6023-6032.
[24] Zhang, T.; Zhang, X.; Zhang, G. *IEEE Photon Technol Lett*. 2005, 17, 1175-1177.
[25] Wen, S. F. *Opt Express*. 2006, 14, 3752-3762.
[26] Wen, S. F. *Opt Express*. 2006, 15, 45-55.
[27] Vakhshoori, D.; Azimi, M.; Chen, P.; Han, B.; Jiang, M.; Knopp, K. J.; Lu, C. C.; Shen, Y.; Vander Rhodes, G.; Vote, S.; Wang, P. D.; Zhu, X. *OFC*. 2003, 3, PD47-P1-3.
[28] X. Zhou, M. Birk and S. Woodward, *IEEE Photon. Technol. Lett* 2002, vol 14, 1686-1688.
[29] C. V. Raman, K. S. Krishnan, "A New Type of Secondary Radiation", *Nature*, Vol. 121, pp. 501-502, March, 1928.
[30] G. S. Landsberg, L. I. Mandelstam, "Eine neue Erscheinung bei der Lichtzerstreuung in Krystallen", *Naturwissenschaften*, Vol. 16, pp. 557-558, July, 1928.
[31] P. S. André, A. N. Pinto, *Chromatic Dispersion Fluctuations in Optical Fibers Due to Temperature and Its Effects in High-Speed Optical Communication Systems, Optics Communications*, Vol. 246, Issues 4-6, 15 February 2005, pp. 303-311, 2005;
[32] E. J. Woodbury, W. K. Ng, "Ruby Laser Operation in the Near IR", *Procedings of IRE* (correspondence), Vol. 50, No. 11, pp. 2367, November, 1962.
[33] R. W. Boyd, Nonlinear Optics, second edition, San Diego, Academic Press, 2003;
[34] C. Headley, G. P. Agrawal, *Raman Amplification in Fiber Optical Communication Systems*, San Diego, Academic Press, 2005.
[35] Karásek M., Kanka J., Honzátko P., Peterka P., *Int. J. Numerical Modelling: Electronic Network, Devices and Fields* , 2004, vol. 17, n°2, 165-176.
[36] Bromage J., *J. Lightwave Technol*, 2004, vol. 22, n°1, 79-93.
[37] André, P., Teixeira, A., Kalinowsky, H., Pinto, J. L. *Optica Aplicatta* 2003, 33 559 – 573.
[38] Liu X., Lee B., *Opt. Express*, 2003, vol. 11, n°.12, 1452-1461.
[39] Neto B.; Stevan S; Teixeira A. T; André P. S.; *ICT* 2006, Funchal.
[40] Han Q., Ning J., Chen Z., Shang L., Fan G., *J. Opt A Pure Appl. Opt.,* 2005, 7, 386-390.
[41] Han Q., Ning J., Zhang H., Chen Z., *J. Lightwave Technol*, 2006, 24, 1946-1952.
[42] Min B., Lee W. J., Park N., *IEEE Photon. Technol Lett.*, 2002, 12, 1486-1488.
[43] Newbury N. R., , *J. Lightwave Technol*, 2003, 21, 3364-3373.

[44] Mollenauer L. F., Grant A. R., Mamyshev P. V., *Opt. Lett.*, 2001, 592-594.
[45] Xiao P. C., Zeng Q. J., Huang J., Liu J. M, *IEEE Photon. Technol Lett.* 2003, 15, 206-208.
[46] Yan M., Chen J., Jiang W, Li J., Chen J., Li X., *IEEE Photon. Technol Lett.* 2001, 13, 948-950.
[47] Zhou X., Lu C., Shum P., Cheng T. H., *IEEE Photon. Technol Lett.* 2001, 13, 945-947.
[48] Cui S, Liu L, Ma X, *IEEE Photon. Technol Lett.* 2004, 16, 2451-2453.
[49] Neto, B., Junior, S., Teixeira, A., André, P. *European Conf. on Networks and Optical Communications - NOC* 2006, Berlin, Germany.
[50] Goldberg D. E., *Genetic Algorithms in Search, Optimisation and Machine Learning*, Massachusetts: Addison-Wesley co, 1989, pp 28-56
[51] V. E. Perlin and H. G. Winful *J. Lightwave Technol.* 2002, 20, 250–254.
[52] X. Liu, J. Chen, C. Lu, and X. Zhou, *Opt. Express* 2004, 12, 6053-6066.
[53] X. M. Liu, et al., *J. Lightwave Technol.* 2003, 21, 3446-3455.
[54] M. Islam, *IEEE Journal of Selected Topics in Quantum Electronics* 2002, 8.
[55] T. Miyamoto, R. Lindsay, *Lightwave Magazine* 2003, 1.
[56] Kobyakov, A., Gray S. and Vasilyev M. *Electronics Letters* 2003, 39, 732 – 733.
[57] Tsujkawa K., Tajima K., Ohashi, M. *J. Lightwave Technol.* 2000, 18, 1528-1532.
[58] Essambre R.; Winzer P.; Bromage J.; Kim C. H., *Photonics Tech. Letters* 2002, 14, 914 – 916.
[59] Park, J., Kim N. Y., Choi W.; Lee, H., Park N. *Photonics Tech. Letters* 2004, 16, 1459 – 1461.
[60] Teixeira A., Stevan Jr., S. Silveira T.; Nogueira R.; Tosi Beleffi G. M., Forin D., Curti F. *European Conf. on Networks and Optical Communications - NOC* 2005,London, UK.
[61] P. B. Hansen, L. Eskildsen, A. J. Stentz, T. A. Strasser, J. Judkins, J. J. DeMarco, R. Pedrazzani, and D. J. DiGiovanni, *IEEE Photon. Technol. Lett.* 1998, 10, 159-161.
[62] Faralli, S. Di Pasquale, F., *IEEE Photonics Technology Letters* 2003, 15, 804- 806.
[63] S. Stevan Jr., A. Teixeira, T. Silveira, P. André, G. M. Tosi Beleffi, A. Reale and A. Pohl, Double shifted Raman amplification by means of spontaneous Rayleigh Backsattering lasing control, *ITC* 2006.
[64] S. Stevan Jr. , A. Teixeira, P. André, G. M. Tosi Beleffi, A. Pohl, simulation of Raman amplification and Rayleigh Scattering laser using the transference matrix method, *MTPT* 2006, Leiria, Portugal.
[65] André, P., A. N. Pinto, Teixeira, A.T., Neto, B., Junior, S., Spertti, D., Rocha, F., Bernardo, M., Fujiwara, M., Rocha, A., Facão, M. *ICTON* 2007, Rome, Italy.
[66] Agrawal, G. P. *Nonlinear Fiber Optics*, 3rd ed. San Diego: Academic Press, 2001.

*Chapter 11*

# FIBER BRAGG GRATINGS IN HIGH BIREFRINGENCE OPTICAL FIBERS

*Rogério N. Nogueira, Ilda Abe and Hypolito J. Kalinowski*
Instituto de Telecomunicacoes, polo de Aveiro,
Aveiro, Portugal

## ABSTRACT

Fiber Bragg gratings (FBG) are a key element in optical communication devices and in fiber sensors. This is mainly due to its intrinsic characteristics, which include low insertion loss, passive operation and immunity to electromagnetic interferences. Basically a FBG is a periodic modulation of the core refractive index formed by exposure of a photosensitive fiber to a spatial pattern of ultraviolet light in the region of 244–248 nm. The lengths of FBGs are normally within the region of 1–20 mm. Usually a FBG operates as a narrow reflection filter, where the central wavelength is directly proportional to the periodicity of the spatial modulation and to the effective refractive index of the fiber. The production technology of these devices is now in a mature state, which enables the design of gratings with custom-made transfer functions, crucial for all-optical processing. Recently, some work has been done in the application of FBG written in highly birefringent fibers (HiBi). Due to the birefringence, the effective refractive index of the fiber will be different for the two transversal modes of propagation. Therefore, the reflection spectrum of a FBG will be different for each polarization. This unique property can be used for advanced optical processing or advanced fiber sensing.

The chapter will describe in detail this unique device. The chapter will also analyze the device and demonstrate different applications that take advantage of its properties, like multiparameter sensors, devices for optical communications or in the optimization of certain architectures in optics communications systems.

## 1. INTRODUCTION

The development of the fiber optical technology was an important step in the revolution of global communications and in information technology. One of these developments happened in the 70's with the first optical fibers with low attenuation [1], a feature that enabled long- distance communication with high bandwidth. The intrinsic optical bandwidth of the optical fibers has also allowed the propagation of different simultaneous channels, allowing the transmission of data at Tbit/s rates [2]. In these systems, in addition to transmission and amplification, it is often necessary to do all-optical processing to the signal. This is due to the inherent advantages of the optical processing, relative to the optic-electric-optic processing, like the higher flexibility to operate at different bit rates and modulation formats and also at the higher bandwidth. The evolution of the fiber optical technology has also enabled the development of devices for all optical processing. In this way, the insertion loss is reduced and the processing quality improved. One of the factors contributing to all-fiber optical processing devices was the discovery of the photosensitivity in optical fibers. It was documented for the first time in 1978 by Hill et al. [3] and led to the development of fiber Bragg gratings (FBG).

A FBG is, generally speaking, a periodic perturbation, along the longitudinal axis, of the refractive index in the fiber core. The production of the refractive index perturbation is done optically in a photosensitive fiber. With the current techniques, it is possible to produce fiber Bragg gratings with different optical properties, which can be designed according to the desired optical processing. In addition to the high flexibility in the production of gratings with custom amplitude and phase responses, the compatibility with common transmission fiber also reduces the insertion loss and decreases the production costs.

The application in optical sensors is also a large potential market for FBG. Their intrinsic low immunity to electromagnetic interference, high dynamic range, passive operation, resistance to corrosion and the possibility of multiplexing hundreds of sensors have made FBG a quite interesting sensor for different applications including medicine, civil, aeronautics or biomechanics. Their properties enable the measurement of temperature and also deformation with extremely high resolution. Nevertheless, it can also be used to measure other parameters using indirect measurements [4-7]. The high potential of these devices has also induced the creation of several companies dedicated to the production and installation of fiber sensors.

There are already good references for the study of FBGs [8,9]. The purpose of this chapter is not to study in detail these devices, but to describe a special case when a fiber Bragg grating is written in high birefringence fibers (HiBi FBG). These special gratings have unique polarization properties that give them exclusive capabilities for optical communications. This is due to the possibility of applying a different optical processing for different polarization components of the signal being transmitted.

HiBi FBGs are also quite interesting for multiparameter sensors, due to their response to temperature variations and deformation. Sensors capable of measuring simultaneously several physical parameters have increased in importance in today's technological world. In particular, there are various applications of such sensors in civil, mechanical, biomedical or aeronautical engineering, where measurements of different parameters are required [10]. Engineering structures are an example of an application area for the multiparameters sensors, where strain sensing can lead to better understanding about their lifetime and failure. Such knowledge can be critical for some applications like smart skins for airplanes and aeronautical vehicles.

## 2. FIBER BRAGG GRATINGS

A FBG is an optical device produced within the core of a standard optical fiber (figure 1). Basically, it is a periodic modulation of the core refractive index formed by exposure of a photosensitive fiber to a spatial pattern of ultraviolet light. The length of a FBG is dependent on its application, but it generally varies between a few millimeters to a few centimeters.

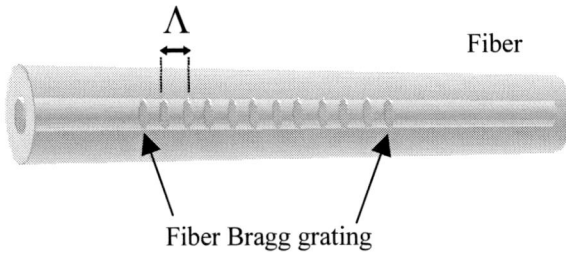

Figure 1. Scheme of a Fiber Bragg grating written in an optical fiber.

The periodic modulation of the refraction index generates a resonant condition at the Bragg's wavelength ($\lambda_B$) which is given by the Bragg's condition:

$$\lambda_B = 2n_{eff}\Lambda \quad (1)$$

where $n_{eff}$ is the effective refraction index of the fiber and $\Lambda$ is the modulation period. Therefore, when a FBG is illuminated by a broadband source, a spectral band centered at $\lambda_B$ will be reflected back. The reflection function can be determined using the coupled mode theory [11-14], since it is difficult to determine analytically. The exception is the uniform FBG, where it is possible to calculate the reflectivity in an analytical way. Considering a uniform periodic modulation of the refractive index, with amplitude $\Delta n$, the reflection coefficient of the grating can be given by

$$\rho(\lambda) = \frac{-\kappa \sinh(\varphi L)}{\delta \sinh(\varphi L) + i\varphi \cosh(\varphi L)} \quad (2)$$

where $L$ is the length of the FBG, the propagation constant mismatch, $\delta$, is given by

$$\delta = \frac{2\pi n_{eff}}{\lambda} - \frac{\pi}{\Lambda}, \qquad (3)$$

$\varphi = \sqrt{\kappa^2 - \delta^2}$, and $\kappa$ is the coupling constant given by

$$\kappa = \frac{\pi \Delta n}{\lambda} \eta \qquad (4)$$

where $\eta$ is the overlap integral and can be approximated as $\eta \approx 1$ for single mode fibers with step index.

The reflectivity is given by

$$R = |\rho|^2 = \frac{\sinh^2(\varphi L)}{\cosh^2(\varphi L) - \frac{\delta^2}{\kappa^2}} \qquad (5)$$

and the phase by

$$\phi_R = \arctan\left[\frac{\mathrm{Im}(\rho)}{\mathrm{Re}(\rho)}\right] \qquad (6)$$

Figure 2 shows the calculated reflectivity and the phase of a uniform FBG with $L = 5$ mm and $\Delta n = 2 \times 10^{-4}$ as given by the above equations.

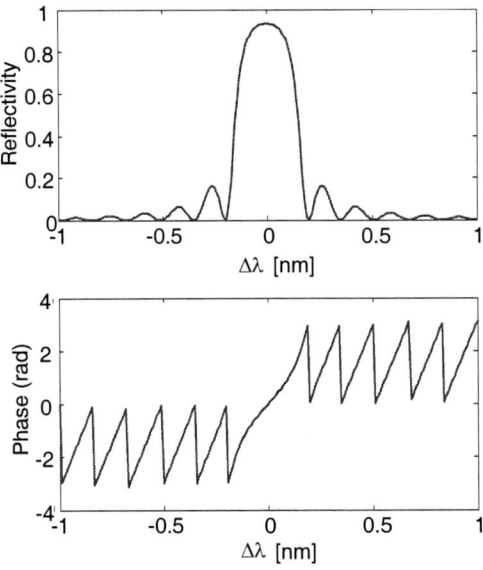

Figure 2. Reflectivity and phase of a uniform FBG. Parameters: L=mm and $\Delta n = 2 \times 10^{-4}$.

If the period changes linearly with the length of the grating, the FBG is said to have a linear chirp. Figure 3 shows the simulation of the reflectivity and group delay of a linear chirped FBG. The simulation method is based on the coupled mode theory.

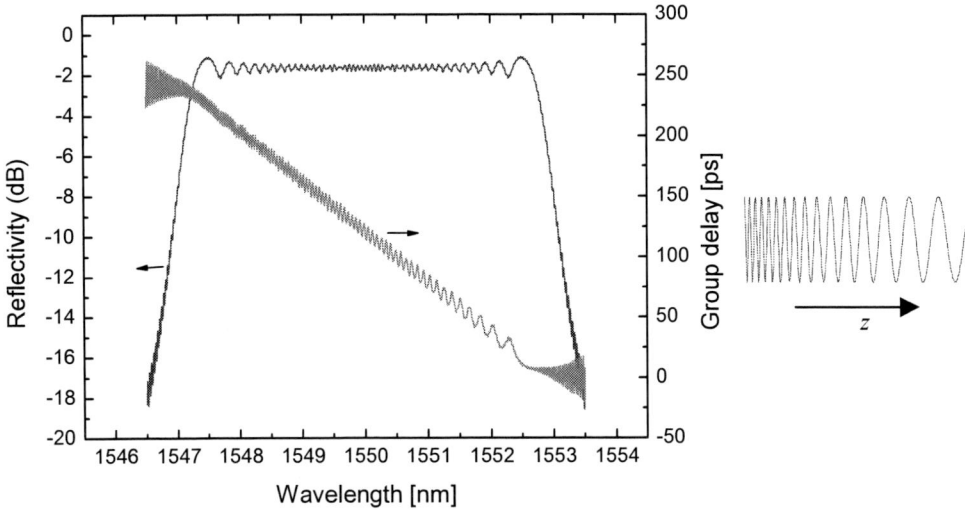

Figure 3. Reflectivity and group delay of a linear chirped FBG.

## 3. HIGH BIREFRINGENCE FIBERS

In an ideal monomode fiber, with a perfect cylindrical core, and with uniform diameter, the fundamental propagation mode is a degenerated combination of two orthogonal propagation modes. However, in real fibers, that degeneration does not exist. In fact, small variations of diameter in the fiber's core generate a birefringence in the optical fiber. The birefringence can also be a result of an anisotropic stress in the fiber. The local birefringence, $B$, in each position of the fiber, is defined as

$$B = |\bar{n}_x - \bar{n}_y| = C_f (\sigma_x - \sigma_y) \quad (7)$$

where $\bar{n}_x$ and $\bar{n}_y$ are the mean refractive index of the orthogonal polarization modes, $\sigma_x$ and $\sigma_y$ are the main stress on the polarization axes and $C_f$ is the photoelastic constant of the fiber. In monomode silica fibers $C_f$ is around $3.08 \times 10^{-6}$ mm$^2$/N for wavelengths near 1500 nm, while B is typically $B \approx 10^{-7}$. Due to this small birefringence value, the two polarization components of the light propagating in the fiber have a propagation velocity very similar. Therefore, small environmental perturbations will lead to an energy coupling between one polarization to another. As a result, a linearly polarized light will rapidly evolve to a random polarization. This situation can be avoided with high birefringence fibers. In these fibers, the core has an anisotropic stress, which is generated due to the geometric properties of the fiber.

Due to the photoelastic effect, the stress induces a birefringence in the core. Typical values are $B \approx 10^{-4}$ [15]. Due to the high birefringence, the propagation constant is different for the two orthogonal propagation modes, which means that the coupling between both transversal propagation modes is far lower as compared to standard fibers. Therefore, the higher the birefringence, the easier will be for a linearly polarized light, propagating in one of the orthogonal modes, to maintain its state of polarization. Due to this feature, HiBi fibers are also known as polarization maintaining fibers. Figure 4 shows the main structure of the most common HiBi fibers.

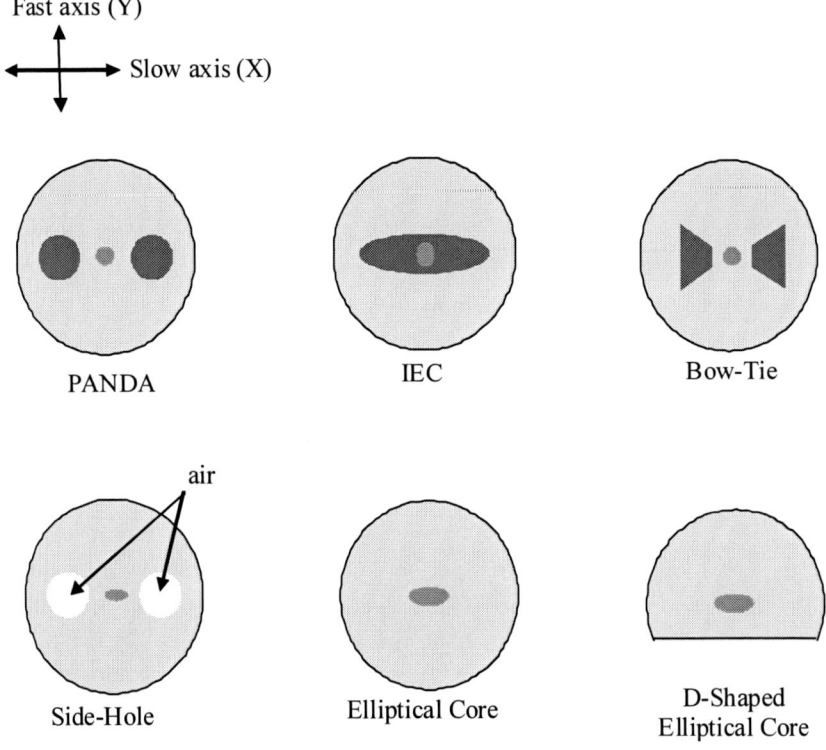

Figure 4. Schematic of the transversal section of some of the most known HiBi fibers.

The PANDA (Polarization-maintaining AND Attenuation-reducing), IEC (Internal Elliptical Cladding) and Bow tie fibers have anisotropic glass structures around the core, with a Poisson coefficient different from the rest of the fiber. These structures create the anisotropic stress in the core, which produces the birefringence. The Side-Hole, the Elliptical Core and the D-Shaped Elliptical Core fibers have an elliptical core to generate the birefringence, aided by two air structures, in the case of the Side-Hole or by the shape of the cladding, in the case of the D-Shaped elliptical core. The main axes of the HiBi fibers are designated as fast axis (Y) for the lower refraction index and slow axis (X) for the higher refraction index.

## Coherence Length

If a linearly polarized light propagates in a monomode fiber, with a polarization angle of 45°, relatively to the main axes of the fiber, both orthogonal polarization modes will be excited with equal power. If the fiber has a constant birefringence, the mismatch, $\Phi_{HB}(z)$, between the orthogonal polarization components will change as a function of the propagation distance on the fiber, $z$, and it's given by

$$\Phi_{HB}(z) = (\beta_x - \beta_y)z \quad (8)$$

where $\beta_x$ and $\beta_y$ are the propagation constants in the X and Y axes respectively. The mismatch will change periodically with the fiber, leading to a change in the state of polarization from linear to elliptical and back again to linear (figure 5).

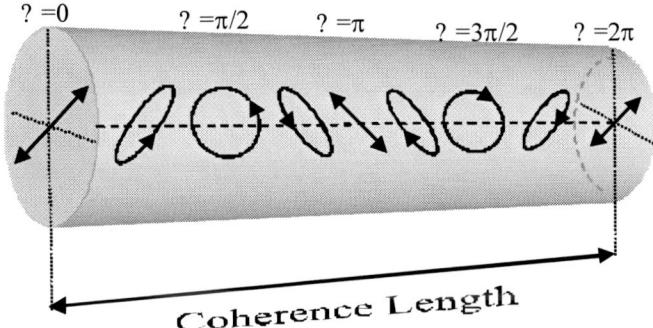

Figure 5. Evolution of the state of polarization in a birefringence fiber.

The spatial periodicity of the evolution of the state of polarization is designated as coherence length ($L_B$). It is determined by the birefringence of the fiber and can be expressed as

$$L_B = \lambda / B \quad (9)$$

where $\lambda$ is the operating wavelength. Typical coherence lengths for HiBi fibers are in the millimeter scale [16].

## 4. FIBER BRAGG GRATINGS WRITTEN IN HIBI FIBERS

HiBi fibers can have two linear polarization modes with refractive indexes $n_x$ and $n_y$ for the slow and fast modes respectively. When a FBG is written in one of these fibers, the periodic modulation will be the same for the two orthogonal polarization modes; however since the effective refraction index is different for the two polarizations, the Bragg wavelength will also be different for each mode. Consequently, expression (1) can be rewritten for the two orthogonal modes:

$$\lambda_i = 2n_i \Lambda \quad , i = X,Y \quad (10)$$

where $\lambda_i$ are the Bragg wavelengths for each polarization mode.

The wavelength difference between the two reflection peaks, $\Delta\lambda_{HB}$, can be calculated by

$$\Delta\lambda_{HB} = \lambda_x - \lambda_y = 2n_x\Lambda - 2n_y\Lambda \quad (11)$$

The reflectivity of a HiBi FBG will be given by the linear sum of the reflectivity of the two polarization components, i.e. $R(\lambda)=R_x(\lambda)+R_y(\lambda)$. $R_x$ and $R_y$ are the reflectivity for each polarization given by

$$R_i = |\rho|^2 = \frac{\sinh^2(\varphi_i L)}{\cosh^2(\varphi_i L) - \frac{\delta_i^2}{\kappa^2}} \quad , i=x,y \quad (12)$$

where

$$\delta_i = \frac{2\pi n_i}{\lambda} - \frac{\pi}{\Lambda}, \; i=x,y \quad (13)$$

and

$$\varphi_i = \sqrt{\kappa^2 - \delta_i^2} \;, i=x,y \quad (14)$$

Figure 6 shows a simulation, using the previous model, for the reflectivity of a HiBi FBG with birefringence of $B = 3.2 \times 10^{-4}$.

If the HiBi FBG is illuminated with light having the two orthogonal components, the reflection spectrum will have those two peaks at orthogonal polarizations. This feature can be very important in some applications, namely in optical communications, as it will be confirmed further in this chapter.

The production of HiBi FBGs uses the same techniques as the ones used in regular FBGs. The only difference will be in the utilization of photosensitive HiBi fiber. Generally it is used a hydrogenated HiBi fiber.

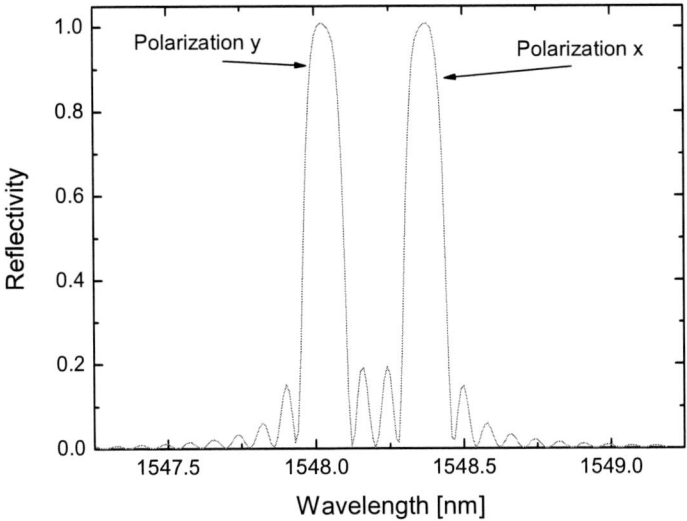

Figure 6. Reflectivity of a simulated HiBi FBG. Simulation parameters: $B=3.2 \times 10^{-4}$, $\Lambda=535$ nm, $L=10$ mm.

Table 1 shows the dimensions of the anisotropic glass structures around the core of some HiBi fibers obtained through the photographs of the transverse section. The table also displays the main characteristics of HiBi fibers obtained from the manufacturers data sheet.

**Table 1. Characteristics of different HiBi fibers. The structures of the HiBi fibers were obtained by microphotography.**

| Fiber type | Commercial provider | Wavelength (nm) | Core diameter (μm) | Cladding diameter (μm) | Intrinsic stress-applying region |
|---|---|---|---|---|---|
| IEC (FS-PM-6621) | 3M | 1300 | 8 | 125 | Ellipse: Major axis: 75 μm; Minor axis: 30 μm |
| Bow tie (F-SPPC-15) | Newport | 1550 | 8 | 125 | From core center to extremity of bow tie lobe: 18.4 μm |
| Bow tie (HB-1500G) | Fibercore | 1550 | 8 | 80 | From core center to extremity of bow tie lobe: 16.5 μm |
| PANDA (SM-13-P-7) | Fujikura | 1300 | 8 | 125 | From core center to opposite extremity of side cylinder: 41 μm; Diameter of side cylinder: 32 μm |

The reflection spectra for gratings written in the above fibers are shown on figure 7, where the plots of the best-fitted bands are also presented [19]. All the gratings were produced with the phase mask technique. The estimated length of the grating is 10 mm.

From these spectra it can be seen the effect of the intrinsic birefringence of the HiBi fibers. The IEC fiber has the higher birefringence, corresponding to larger spectral splitting between both polarizations bands, while the bow tie fiber presents the lowest birefringence.

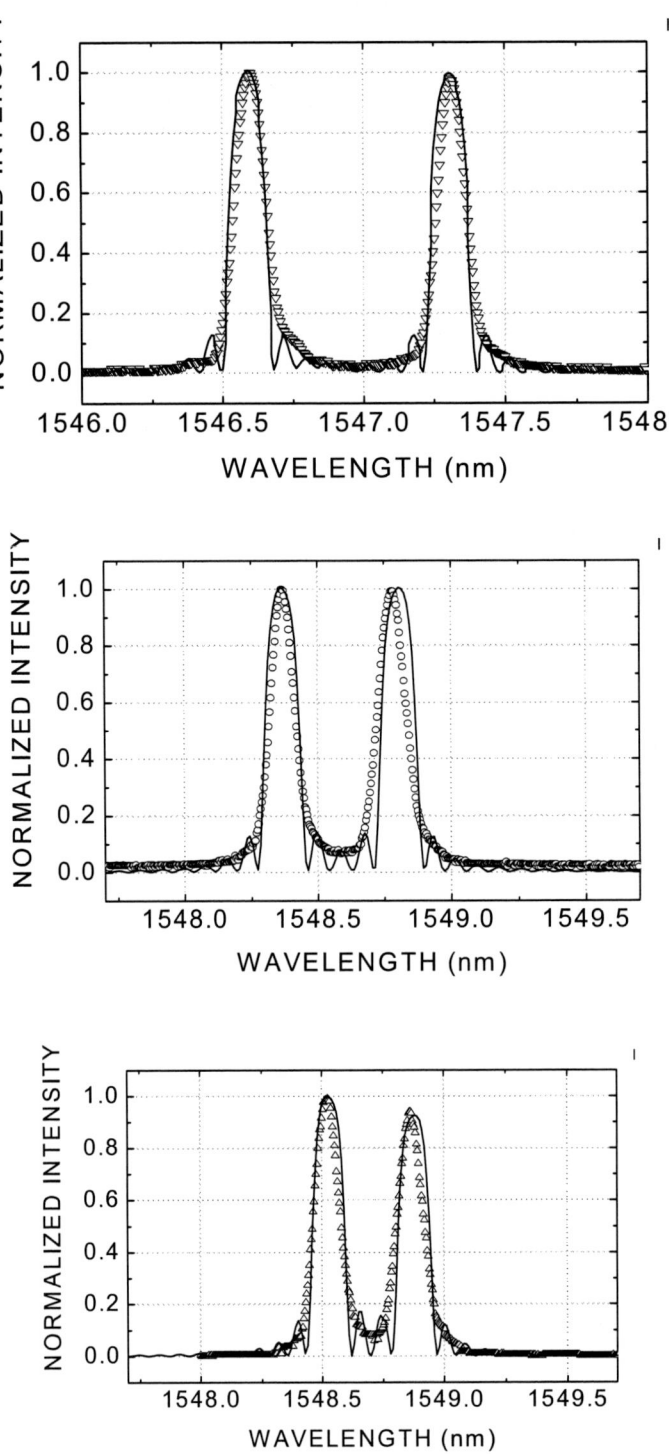

Figure 7. Reflection spectra of Bragg gratings written in different HiBi fibers: IEC ($\nabla$), Panda (O) and bow tie ($\Delta$). The continuous line represents the simulated best fit.

Table 2 shows the best-fit parameters obtained in the simulation process. From the fit it is also possible to obtain the values of birefringence of the HiBi fibers.

**Table 2. Parameters of FBGs written in HiBi fibers obtained for the best fit for the experimental data.**

| HiBi Fiber | Bands | $\lambda$ (nm) | $n_{eff}$ | kL | $\Lambda$ (nm) | B |
|---|---|---|---|---|---|---|
| IEC | $\lambda_Y$ | 1546.57 | 1.44539 | 1.7212 | 535 | 6.7 x $10^{-4}$ @ 1550 nm |
| (FS-PM-6621) | $\lambda_X$ | 1547.29 | 1.44606 | 1.7196 | | |
| PANDA | $\lambda_Y$ | 1548.39 | 1.44709 | 1.7172 | 535 | 4.1 x $10^{-4}$ @ 1550 nm |
| (15P8) | $\lambda_X$ | 1548.82 | 1.44750 | 1.7162 | | |
| Bow tie | $\lambda_Y$ | 1548.61 | 1.44730 | 1.7167 | 535 | 3.2 x $10^{-4}$ @ 1550 nm |
| (SPPC-15) | $\lambda_X$ | 1548.95 | 1.44762 | 1.7159 | | |

The Bragg wavelength peaks of the optical spectrum for both polarizations can change with temperature and strain. Therefore, considering a HiBi FBG under a temperature variation of $\Delta T$ and under a strain aligned with the main axes of the fiber $\Delta\varepsilon_X$, $\Delta\varepsilon_Y$ and $\Delta\varepsilon_Z$, the resultant wavelength shift, $\Delta\lambda_x$ and $\Delta\lambda_y$ of both wavelength peaks, $\lambda_x$ and $\lambda_y$, can be expressed as

$$\frac{\Delta\lambda_X}{\lambda_X} = \Delta\varepsilon_Z - \frac{n_X^2}{2}[p_{11}\Delta\varepsilon_X + p_{12}(\Delta\varepsilon_Z + \Delta\varepsilon_Y)] + \left[\alpha + \frac{(\partial n/\partial T)}{n_X}\right]\Delta T \quad (15)$$

$$\frac{\Delta\lambda_Y}{\lambda_Y} = \Delta\varepsilon_Z - \frac{n_Y^2}{2}[p_{11}\Delta\varepsilon_Y + p_{12}(\Delta\varepsilon_Z + \Delta\varepsilon_X)] + \left[\alpha + \frac{(\partial n/\partial T)}{n_Y}\right]\Delta T \quad (16)$$

where $p_{11}$ and $p_{12}$ are the components of the photoelastic tensor and $\alpha$ is the thermal expansion coefficient of the fiber, $\alpha = 0.55 \times 10^{-6}$ K$^{-1}$ [17]. For a fiber based on germanium and silica, $p_{11}=0.113$, $p_{12}=0.252$ and the thermo-optic coefficient is $\frac{(\partial n/\partial T)}{n} = 8.6 \times 10^{-6}$ [18].

Figure 8 shows schematically the effect on the reflection spectrum of a HiBi FBG when it is under temperature variations, under transversal strain or longitudinal strain. The effect of temperature variations or longitudinal strain in the reflection spectrum is equivalent to a translation in the wavelength. On the other hand, when under a transversal strain, the peak separation will change. This difference can be used in multiparameter sensors as it will be discussed further in this chapter.

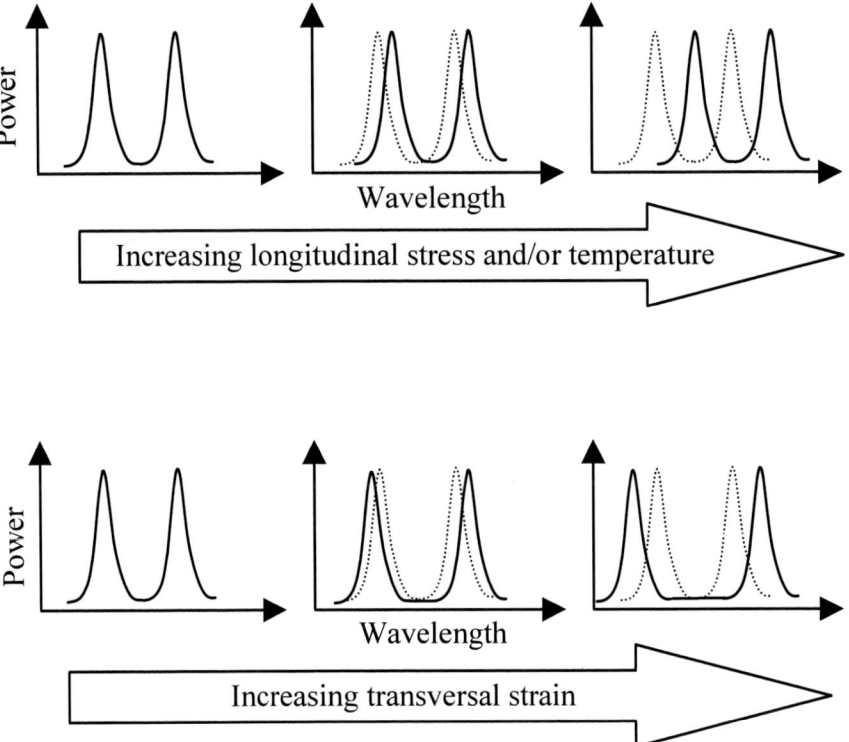

Figure 8. Evolution of the reflection spectrum of a HiBi FBG when under a longitudinal stress, temperature variation or transversal strain.

## 4.1. Characterization of Bragg Gratins Written in High-Birefringence Fiber Optics

### *4.1.1. Transverse Strain*

The sensitivity of HiBi FBG to transversal strain can be characterized using a mechanical set-up, like the one shown in figure 9. The transversal load is applied using a micro scratch mechanical system. The system uses an arm to apply a load with a precision of 0.1 N. A grating written in HiBi fiber was placed between two plates having a length of 13 mm. The apparatus arm applies the load to the upper plate. The transverse loads were made for several orientations of the birefringence axis with respect to the direction of the applied load through two fiber rotators. Figure 9 also shows the optical system used to analyze the FBG reflection spectrum. Optical spectra were recorded using an amplified spontaneous emission (ASE) of an erbium doped fiber amplifier as light source, an optical circulator and conventional optical spectrum analyzer (OSA).

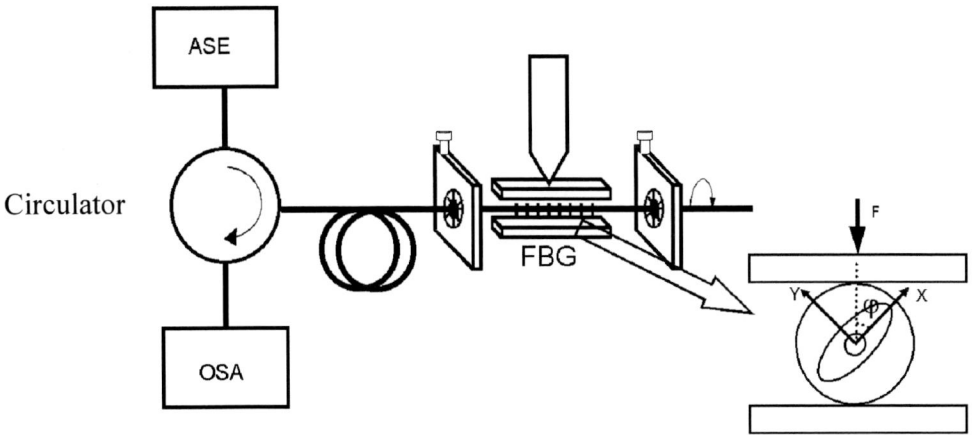

Figure 9. Set-up for the characterization of HiBi FBGs under a transversal load. The detail shows the transverse section of a HiBi fiber oriented along the angle φ.when subjected to applied force F. ASE: Broadband optical source (amplified spontaneous emission); OSA: optical spectrum amplifier.

Figure 10 (a), (b) and (c) shows an example of the reflection spectra of a FBG written in a IEC HiBi fiber as a function of an applied load of 0°, 45° and 90°, respectively.

The results show that, if a load is applied to one of the main axes, fast or slow, it leads to a change in the wavelength of the spectral band associated with the orthogonal axis, while the band associated with the correspondent axis will show a smaller variation. For the applied load angle of 45° both polarization bands present similar evolution. The figure also shows, for different applied load angles ($\varphi$), the evolution of the peak wavelength of each reflection band with the transverse strain applied to the sample. The strain calibration points in the spectra deformed areas were obtained by identifying and measuring local maximum, minimum and inflexion points. The band split that occurs in some of the spectra is due to a phase shift induced by the applied load. The resulting complex structure is known to be responsible for spectral changes of FBG subject to mechanical stress [9].

Figure 11 shows the wavelength sensitivity curves obtained for both polarization bands. The graph also displays the periodic evolution of the bands as a function of the applied load angle.

Identifying and measuring the reflection peaks as a function of the applied load can be used to obtain the calibration line for each polarization band. The respective slopes can be evaluated and, from them, the dependence of the Bragg wavelength position with the strain can be obtained. Table 3 shows some measurements obtained with FBGs written in IEC and PANDA fibers.

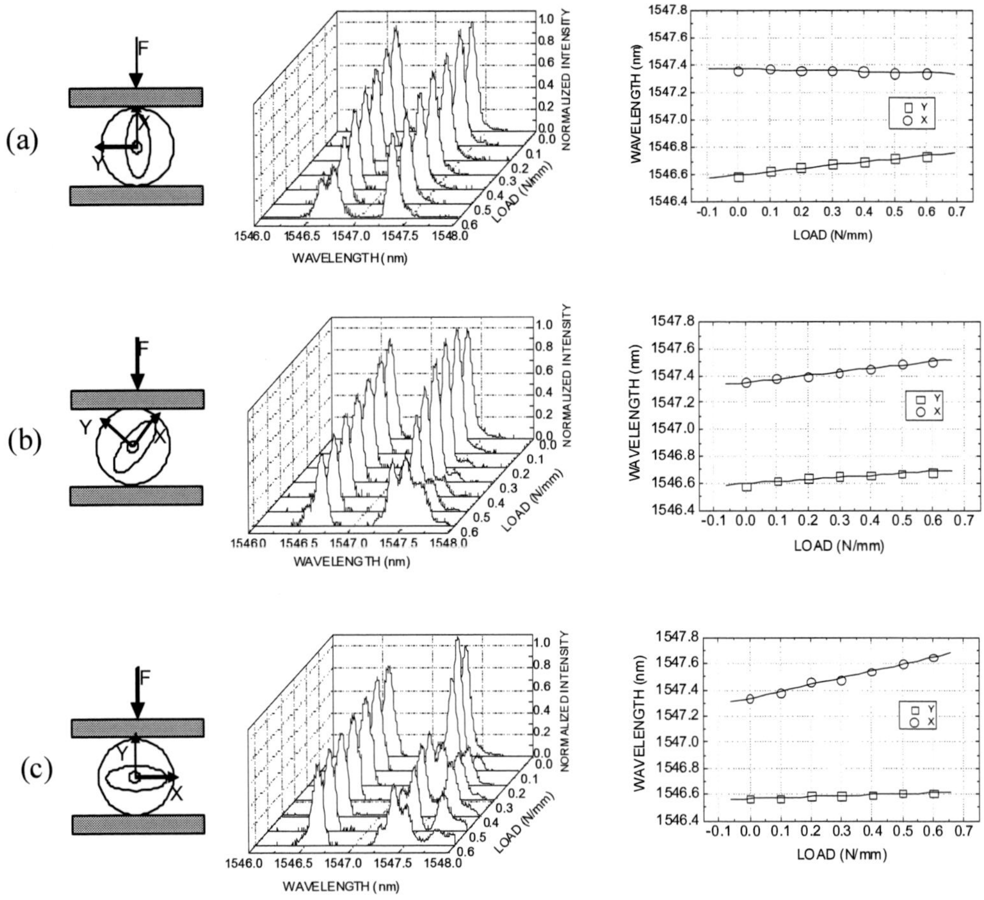

Figure 10. Left: Changes in the spectral response of a FBG written in an IEC HiBi fiber when subjected to an applied load oriented along the angle φ: (a) 0° (X-axis); (b) 45° and (c) 90° (Y-axis). Right: Peak position of each band as a function of the applied load. The lines represent the linear best fit for the experimental data.

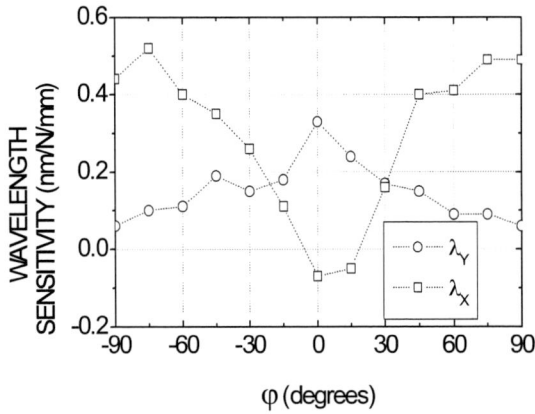

Figure 11. Curves of peak sensitivities of the FBG in IEC HiBi fiber as a function of the applied load angle.

**Table 3.** Slopes and strain sensitivities of FBGs written in IEC and PANDA HiBi fibers as a function of the direction of applied load (module values). Both fibers have a diameter of 125 μm.

| HiBi Fiber | Angle of applied load | X- polarization band | | Y- polarization band | |
|---|---|---|---|---|---|
| | | Slope (nm/N/mm) | Strain sensitivity (pm/με) | Slope (nm/N/mm) | Strain sensitivity (pm/με) |
| IEC | $\varphi = 90°$ | 0.51 | 7.02 | 0.07 | 1.02 |
| | $\varphi = 0°$ | 0.02 | 0.29 | 0.11 | 1.55 |
| PANDA | $\varphi = 90°$ | 0.46 | 3.78 | 0.02 | 0.24 |
| | $\varphi = 0°$ | 0.01 | 0.11 | 0.13 | 2.80 |

### 4.1.2. Longitudinal Strain

The Bragg wavelength dependence with the longitudinal strain can be measured by gluing one extremity of the fiber in a holder, while the other is glued to a translation stage, which applies a known deformation using a calibrated micrometer.

Figure 12 shows the reflection spectra of a FBG and peak position of each band, written in an IEC fiber as a function of longitudinal strain.

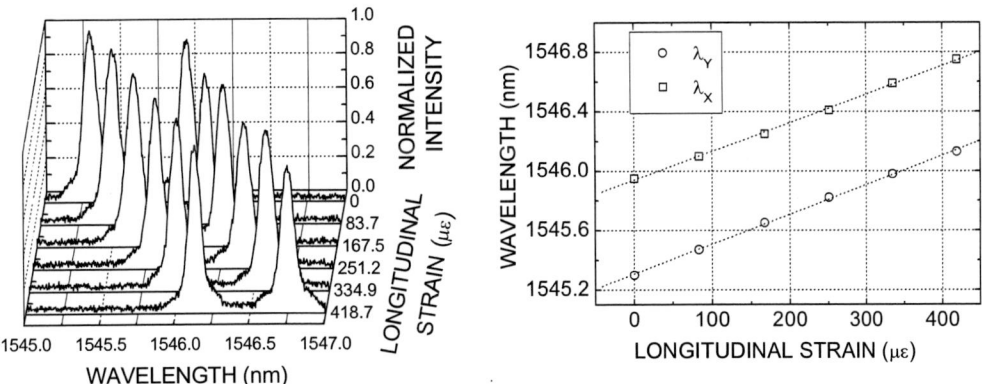

Figure 12. Left: Changes in the spectral response of a FBG written in IEC HiBi fiber when subjected to a longitudinal strain. Right: Peak position of each band as a function of the longitudinal strain. The lines represent the linear best fit for the experimental data.

Both bands show the same behavior, which is an increase of peak wavelengths as the strain increases. The slopes and the Bragg wavelength sensitivity to longitudinal strain are given in table 4. The obtained ratios between strain and applied load were 758 με/N (X-axis) and 755 με/N (Y-axis).

**Table 4. Slopes and longitudinal strain sensitivity of a FBG written in an IEC HiBi fiber.**

| Bands | Slope (nm/N) | Longitudinal strain sensitivity (pm/µε) |
|---|---|---|
| X – polarization | 1.44 | 1.9 |
| Y - polarization | 1.51 | 2.0 |

### 4.1.3. Temperature

The temperature dependence of the reflection bands of HiBi FBGs can be characterized using a cooling/heating system. Figure 13 shows the evolution of the reflection bands and peak position of each band of a Bragg grating written in an IEC fiber as a function of temperature.

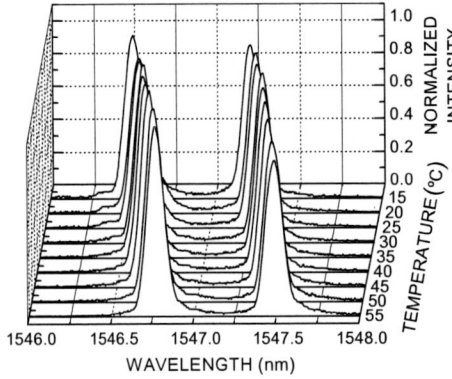
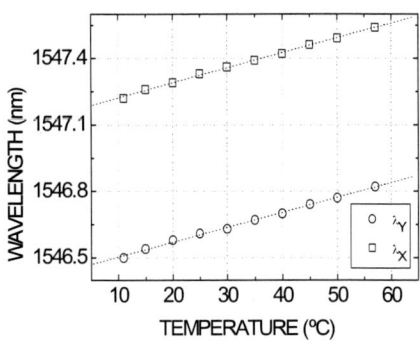

Figure 13. Left: Changes in the spectral response of a FBG written in an IEC HiBi fiber when subjected to different temperatures. Right: Peak position of each polarization band as a function of the temperature. The lines represent the linear best fit for the experimental data.

Table 5 shows the temperature sensitivity values for a FBG written in IEC, PANDA and bow tie HiBi fibers.

**Table 5. Slopes of temperature for FBGs written in HiBi fibers.**

| HiBi fiber | Slope (pm/°C) | |
|---|---|---|
| | X – polarization band | Y – polarization band |
| IEC 125 µm | 6.76 | 6.71 |
| PANDA 125 µm | 3.28 | 3.40 |
| Bow Tie 125 µm | 10.93 | 11.12 |
| Bow Tie 80 µm | 8.02 | 8.46 |

The results show that there are quite large variations between the sensitiveness to temperature for different HiBi fibers. The values changed between ~ 3 pm/°C for the PANDA fiber to ~ 11 pm/°C for the IEC fiber. There are also differences in the coefficients between polarization bands of the same fiber. For example, in the PANDA fiber this difference is 0.24 pm/°C. These results can be used for simultaneous measurements of temperature and longitudinal strain with only one FBG in a HiBi fiber. This approach, along with others, will be described in the next section.

## 5. APPLICATION IN MULTIPARAMETER SENSORS

FBG sensors are generally based on a unique grating written in a standard fiber optic. The wavelength shift in the reflection spectrum may be used to measure a single component of strain or temperature variation, but not both simultaneously. An adequate measurement of both temperature and strain requires a suitable sensor with a differential sensitivity between parameters. HiBi FBGs can be used as sensors to simultaneously measure one component of transverse strain, temperature and/or longitudinal strain. As it was shown previously in this chapter there are differences in the calibration coefficients of both polarization bands, which can be used to simultaneous measure the temperature and longitudinal strain with only one HiBi FBG. Since the variations of temperature or longitudinal strain causes both bands to shift, and the variation of strain causes asymmetric spectral response in the polarization bands depending of the direction of the applied load, allows the FBG in the HiBi fiber to measure simultaneously transverse strain and temperature or transverse strain and longitudinal strain.

Several types of optical sensors using FBG written in HiBi fibers, which simultaneously measure longitudinal strain and temperature have been proposed and demonstrated [20-24]. Some of the methods include the recording by a CCD camera of the $LP_{01}$ and $LP_{11}$ spatial modes [22], using a HiBi FBG partially exposed to chemical etching [20] or by using a quasi-rectangular HiBi fiber to increase the birefringence [21]. In those works, only the longitudinal strain component was measured in simultaneous with temperature. But, there are many applications where it is desirable to determine the transverse strain components in addition to longitudinal strain. Several techniques based in HiBi FBG have already been reported for transverse strain sensing [19, 25-29]. However, when a transverse strain is applied to a HiBi FBG, depending of the fiber orientation relatively to the applied load, the separation of the two Bragg wavelengths can be quite low, so it becomes impossible to resolve the two peaks. To overcome this problem, it can be used an interrogation system capable of detecting independently and simultaneously the two orthogonally polarized signals reflected from the HiBi FBG [26].

There are many applications where it is necessary an ultra small sensor to measure simultaneously components of transverse strain, longitudinal strain and temperature. The use of two superimposed Bragg gratings in HiBi fiber have been described in the literature like potential sensors for monitoring four parameters, two components of transverse strain, longitudinal strain and temperature. [30-34].

### 5.1. Simultaneous Measurement of Transverse Strain and Temperature

The change in the Bragg wavelength of a HiBi FBG, for each polarization, due to a temperature change $\Delta T$ and a transversal strain $\Delta \varepsilon$, is given by

$$\Delta \lambda_X = \frac{\partial \lambda_X}{\partial T} \Delta T + \frac{\partial \lambda_X}{\partial \varepsilon} \Delta \varepsilon \qquad (17)$$

$$\Delta\lambda_Y = \frac{\partial \lambda_Y}{\partial T}\Delta T + \frac{\partial \lambda_Y}{\partial \varepsilon}\Delta \varepsilon \qquad (18)$$

were $\partial\lambda_X/\partial T$ and $\partial\lambda_Y/\partial T$ are the temperature coefficients and $\partial\lambda_X/\partial\varepsilon$ and $\partial\lambda_Y/\partial\varepsilon$ are the transverse deformation coefficients.

Expressions (17) and (18) can be rearranged and written in matrix form in order to calculate the transverse strain and temperature, given the measured wavelength shifts for each polarization band:

$$\begin{bmatrix} \Delta T \\ \Delta \varepsilon \end{bmatrix} = K^{-1} \begin{bmatrix} \Delta\lambda_X \\ \Delta\lambda_Y \end{bmatrix} \qquad (19)$$

where K is a matrix given by

$$K = \begin{bmatrix} \frac{\partial \lambda_X}{\partial T}, \frac{\partial \lambda_X}{\partial \varepsilon} \\ \frac{\partial \lambda_Y}{\partial T}, \frac{\partial \lambda_Y}{\partial \varepsilon} \end{bmatrix} \qquad (20)$$

A simultaneous measurement of transverse strain and temperature can be obtained by determining the coefficients of K, which are determined with previous characterization.

Two examples of simultaneous measurement of these parameters are shown in the table 6 for an IEC fiber and table 7 for a PANDA fiber. The results were obtained using the values of the Bragg wavelength changes for both polarizations bands.

**Table 6. Simultaneous measurements of temperature and transverse strain using a FBG written in a IEC HiBi fiber. The set values were determined by the experimental system equipment. Direction of applied load: 0°. [19].**

| Set values | 12 °C | 23 °C | 31 °C | 46 °C |
|---|---|---|---|---|
| 61 με | 11.8 °C | 24.6 °C | 31.5 °C | 45.5 °C |
|  | 65 με | 71 με | 70 με | 75 με |
| 76 με | 12.0 °C | 25.4 °C | 32.6 °C | 46.3 °C |
|  | 69 με | 79 με | 73 με | 81 με |
| 91 με | 12.7 °C | 26.8 °C | 34.0 °C | 48.0 °C |
|  | 76 με | 83 με | 78 με | 79 με |

Table 7. Simultaneous measurements of temperature and transverse strain using a FBG in written in a PANDA HiBi fiber. The set values were determined by the experimental system equipment. Direction of applied load: 90°.

| Set values | 7 °C  | 21 °C | 22 °C | 40 °C | 53 °C |
|---|---|---|---|---|---|
| 11 με | 7 °C  | 20 °C | 23 °C | 39 °C | 51 °C |
|       | 12 με | 12 με | 9 με  | 11 με | 10 με |
| 21 με | 7 °C  | 19 °C | 22 °C | 38 °C | 51 °C |
|       | 21 με | 20 με | 17 με | 16 με | 18 με |
| 31 με | 6 °C  | 18 °C | 22 °C | 38 °C | 53 °C |
|       | 32 με | 31 με | 31 με | 22 με | 24 με |
| 41 με | 5 °C  | 18 °C | 21 °C | 37 °C | 55 °C |
|       | 39 με | 44 με | 44 με | 37 με | 39 με |
| 51 με | 5 °C  | 18 °C | 21 °C | 37 °C | 55 °C |
|       | 43 με | 42 με | 48 με | 41 με | 43 με |

## 5.2. Simultaneous Measurement of Transverse Strain and Longitudinal Strain

For the measurement of the longitudinal ($\Delta\varepsilon_Z$) and transverse ($\Delta\varepsilon_X$ or $\Delta\varepsilon_Y$) strain, the equations can also be written in matrix form, given the measured wavelength shifts for each polarization band:

$$\begin{bmatrix} \Delta\varepsilon_Z \\ \Delta\varepsilon_{X,Y} \end{bmatrix} = K^{-1} \begin{bmatrix} \Delta\lambda_X \\ \Delta\lambda_Y \end{bmatrix} \quad (21)$$

where K is now given by:

$$K = \begin{bmatrix} \dfrac{\partial \lambda_X}{\partial \varepsilon_Z}, \dfrac{\partial \lambda_X}{\partial \varepsilon_{X,Y}} \\ \dfrac{\partial \lambda_Y}{\partial \varepsilon_Z}, \dfrac{\partial \lambda_Y}{\partial \varepsilon_{X,Y}} \end{bmatrix} \quad (22)$$

Table 8 shows an example of simultaneous measurements of longitudinal and transversal strain obtained using the wavelength changes of both polarizations bands.

**Table 8.** Simultaneous measurements of longitudinal and transverse strain using an FBG written in a IEC HiBi fiber. The set values were determined by the experimental system equipment. Direction of applied load: 90°.

| Set values | 0 με | 9 με | 14 με |
|---|---|---|---|
| **83 με** | 1 με | 8 με | 13 με |
|  | 64 με | 66 με | 68 με |
| **167 με** | 1 με | 8 με | 13 με |
|  | 160 με | 161 με | 154 με |
| **251 με** | 0 με | 7 με | 12 με |
|  | 240 με | 246 με | 244 με |
| **335 με** | 1 με | 8 με | 13 με |
|  | 320 με | 316 με | 313 με |

## 5.3. Simultaneous Measurement of Transverse Strain, Longitudinal Strain and Temperature

Two superimposed Bragg gratings can be written in high birefringence fiber optics to measure simultaneously temperature, transverse and longitudinal strain.

This section demonstrates the use of a pair of Bragg gratings written in high birefringence fiber optics to measure, simultaneously, three physical parameters [31]. The Bragg gratings are superimposed in the same position of the fiber optic, in order to behave as a single sensor with reduced dimension.

### 5.3.1. Superimposed Bragg Gratings

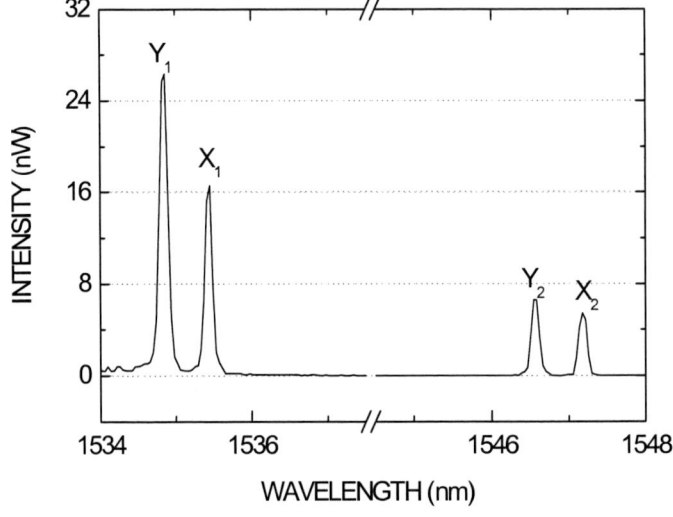

Figure 14. Optical reflection spectrum of two superimposed Bragg gratings written in HiBi IEC fiber [31].

Figure 14 shows an optical reflection spectrum of two gratings recorded at the same fiber position. The two FBG were written with different periods in an IEC HiBi optical fiber with 125 μm diameter. The figure shows the polarization bands (Y-polarization and X-polarization) of each pair. Their relative intensity is not the same as the optical source was not flat along the full wavelength range.

The superimposed HiBi FBGs were characterized by longitudinal, transversal strain and temperature. The measurements of transversal load were made with the fiber oriented with the fast or slow birefringence axis in the direction of the applied load.

Figure 15 shows the dependence of the peak position of each reflection band against the transversal strain applied to the sample (load applied along the Y-axis direction). The best-fitted lines are not parallel; their slopes are different depending on the polarization band. This asymmetric behavior can be used to distinguish the effects of longitudinal and transversal strain acting upon the grating pair.

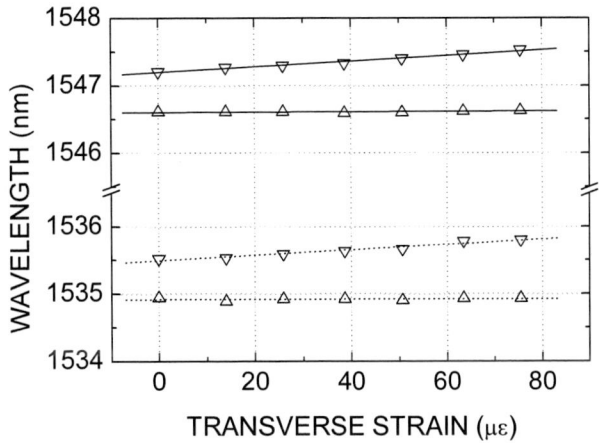

Figure 15. Dependence of the peak wavelength on transverse strain for the reflection bands [X (▽) and Y (△)] of the two superimposed FBGs written in an IEC HiBi fiber. Direction of applied load: 90°. The lines represent the linear best fit to the experimental data [31].

The behavior of the reflection bands, when the sensor is under longitudinal strain, is the same for both gratings. The temperature dependence of the reflection bands of the both FBGs has also approximately the same behavior, which is an increase in the wavelength with an increase of temperature.

**Table 9. Slopes of temperature, longitudinal and transverse strain for the two superimposed FBGs in an IEC HiBi fiber. Direction of applied transverse load: Y-axis [31].**

| Slopes | Polarization bands | | | |
|---|---|---|---|---|
| | $Y_1$ | $X_1$ | $Y_2$ | $X_2$ |
| $\partial\lambda/\partial T$ (pm/°C) | 8.4 | 7.8 | 7.8 | 7.5 |
| $\partial\lambda/\partial\varepsilon_Y$ (pm/με) | 0.08 | 4.02 | 0.19 | 4.11 |
| $\partial\lambda/\partial\varepsilon_Z$ (pm/με) | 1.3 | 1.39 | 1.39 | 1.36 |

The corresponding slopes of temperature, longitudinal and transversal strain for both polarization bands, for the best-fitted lines of superposing FBGs in IEC HiBi fiber, are given in table 9.

### 5.3.2. Simultaneous Measurements

The change in the Bragg wavelength of the reflection spectrum of the both FBGs, due to a temperature change $\Delta T$, a transverse strain ($\Delta \varepsilon_X$ or $\Delta \varepsilon_Y$) and longitudinal strain $\Delta \varepsilon_Z$, for each polarization, is given by

$$\Delta \lambda_{X1} = \frac{\partial \lambda_{X1}}{\partial T} \Delta T + \frac{\partial \lambda_{X1}}{\partial \varepsilon_{X,Y}} \Delta \varepsilon_{X,Y} + \frac{\partial \lambda_{X1}}{\partial \varepsilon_Z} \Delta \varepsilon_Z \qquad (23)$$

$$\Delta \lambda_{Y1} = \frac{\partial \lambda_{Y1}}{\partial T} \Delta T + \frac{\partial \lambda_{Y1}}{\partial \varepsilon_{X,Y}} \Delta \varepsilon_{X,Y} + \frac{\partial \lambda_{Y1}}{\partial \varepsilon_Z} \Delta \varepsilon_Z \qquad (24)$$

$$\Delta \lambda_{X2} = \frac{\partial \lambda_{X2}}{\partial T} \Delta T + \frac{\partial \lambda_{X2}}{\partial \varepsilon_{X,Y}} \Delta \varepsilon_{X,Y} + \frac{\partial \lambda_{X2}}{\partial \varepsilon_Z} \Delta \varepsilon_Z \qquad (25)$$

$$\Delta \lambda_{Y2} = \frac{\partial \lambda_{Y2}}{\partial T} \Delta T + \frac{\partial \lambda_{Y2}}{\partial \varepsilon_{X,Y}} \Delta \varepsilon_{X,Y} + \frac{\partial \lambda_{Y2}}{\partial \varepsilon_Z} \Delta \varepsilon_Z \qquad (26)$$

where $\partial \lambda_{X1}/\partial T$, $\partial \lambda_{X2}/\partial T$, $\partial \lambda_{Y1}/\partial T$ and $\partial \lambda_{Y2}/\partial T$ are the temperature coefficients, $\partial \lambda_{X1}/\partial \varepsilon_{X,Y}$, $\partial \lambda_{X2}/\partial \varepsilon_{X,Y}$, $\partial \lambda_{Y1}/\partial \varepsilon_{X,Y}$ and $\partial \lambda_{Y2}/\partial \varepsilon_{X,Y}$ are the transversal deformation coefficients, and $\partial \lambda_{X1}/\partial \varepsilon_Z$, $\partial \lambda_{X2}/\partial \varepsilon_Z$, $\partial \lambda_{Y1}/\partial \varepsilon_Z$ and $\partial \lambda_{Y2}/\partial \varepsilon_Z$ are the longitudinal deformation coefficients.

Equations (23) to (26) can be rearranged and written in matrix form, in order to calculate the transverse, longitudinal strain and temperature, given the measured wavelength shifts for each polarization band. In this way, the calculation of the three parameters being measured can be made using the following (the choice of reflection bands was arbitrary):

$$\begin{bmatrix} \Delta T \\ \Delta \varepsilon_{X,Y} \\ \Delta \varepsilon_Z \end{bmatrix} = K^{-1} \begin{bmatrix} \Delta \lambda_{Y1} \\ \Delta \lambda_{X1} \\ \Delta \lambda_{Y2} \end{bmatrix} \qquad (27)$$

where K is assembled from the several sensitivities for temperature and deformation:

$$K = \begin{bmatrix} \dfrac{\partial \lambda_{Y1}}{\partial T} & \dfrac{\partial \lambda_{Y1}}{\partial \varepsilon_{X,Y}} & \dfrac{\partial \lambda_{Y1}}{\partial \varepsilon_Z} \\ \dfrac{\partial \lambda_{X1}}{\partial T} & \dfrac{\partial \lambda_{X1}}{\partial \varepsilon_{X,Y}} & \dfrac{\partial \lambda_{X1}}{\partial \varepsilon_Z} \\ \dfrac{\partial \lambda_{Y2}}{\partial T} & \dfrac{\partial \lambda_{Y2}}{\partial \varepsilon_{X,Y}} & \dfrac{\partial \lambda_{Y2}}{\partial \varepsilon_Z} \end{bmatrix} \qquad (28)$$

After a previous characterization, in order to obtain K, the temperature, longitudinal and transversal strain components can be simultaneously measured. Some of the obtained results with the grating pair described above are given in table 10.

**Table 10. Simultaneous measurements of temperature, transverse and longitudinal strain using two superimposed FBGs in IEC HiBi fiber. The set values were determined by the experimental system equipment [31].**

| Set values | 167 µε | | 251 µε | |
|---|---|---|---|---|
| | 15 °C | 45 °C | 15 °C | 45 °C |
| 12 µε | 12 °C | 42 °C | 12 °C | 37 °C |
| | 13 µε | 10 µε | 16 µε | 10 µε |
| | 117 µε | 99 µε | 228 µε | 177 µε |
| 22 µε | 16 °C | 33 °C | 16 °C | 43 °C |
| | 18 µε | 16 µε | 16 µε | 24 µε |
| | 141 µε | 132 µε | 252 µε | 187 µε |
| 32 µε | 16 °C | 36 °C | 18 °C | 43 °C |
| | 32 µε | 29 µε | 21 µε | 32 µε |
| | 116 µε | 139 µε | 236 µε | 178 µε |

## 5.3. Bragg Gratings in Reduced Diameter High Birefringence Fiber Optics

Bragg gratings written in reduced diameter high birefringence fiber optics can also be used for multiparameter sensing. Changes in the stress profile of HiBi fibers due to reduced diameter can modify the response of a FBG sensor system to strain or temperature optimizing the simultaneous measurement of those parameters. Chemical etching can be a good tool to reduce the fiber diameter. The changes in the birefringence properties of HiBi fibers as a function of fiber diameter can be analyzed using fiber samples chemically etched in hydrofluoric acid (HF), while the optical spectra of pre-recorded gratings are measured [34]. The diameter of the fibers during the etching can be measured by having several samples of the fiber in the acid. The samples are removed successively from the acid, rinsed in distilled water, dried, and then measured under a microscope with a calibrated scale.

The evolution of the diameter, as a result of etching, for an IEC fiber is presented in figure 16. HF acid was diluted to 20 % (parts per volume) in order to reduce the velocity of chemical etching and to increase the sampling points along the process. Figure 17 shows the changes in the transversal section of the IEC fiber, with 125 µm of diameter (left) and after etching (right), with 86 µm of diameter. The internal elliptical cladding can be observed in these photographs. The major axis of the ellipse has approximately 75 µm. The etched IEC fiber shows a higher asymmetry on the borders close to the axes along the major axis of the internal elliptical cladding.

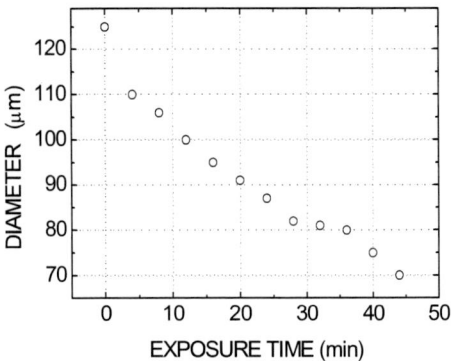

Figure 16. Diameter of an IEC HiBi fiber as a function of the exposure time. HF concentration: 20% [34].

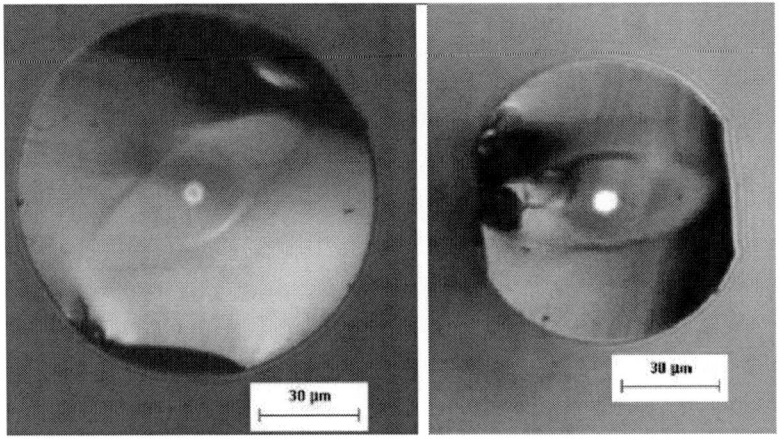

Figure 17. Microphotographs of the transverse section of an IEC HiBi fiber. Left: standard HiBi fiber with 125 μm of diameter. Right: etched HiBi fiber with 86 μm of diameter [34].

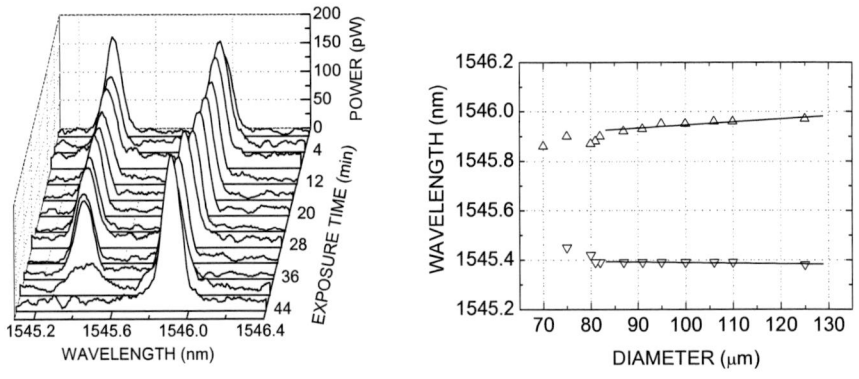

Figure 18. Left: evolution of the reflection bands of a FBG written in an IEC HiBi fiber as a function of the etching time. Right: peak position of the polarized bands (Y-polarized ($\nabla$) and X-polarized ($\Delta$))as a function of the fiber diameter. The lines represent the linear best fit for the experimental data. HF concentration: 20 % [34].

Figure 18 (left) illustrates the optical reflection spectra of the FBG in IEC fiber, obtained as a function of HF exposure time. After 36 minutes of exposition time, the optical spectrum had a single band, which means that, the fiber birefringence was almost zero. That is a consequence of the stress release due to the etching.

Figure 18 (right) shows the changes in the peak position of the reflected polarized bands as a function of the IEC fiber diameter. The different slopes for the X and Y polarized bands can be related to asymmetric changes of the internal stress applied by the internal elliptical cladding.

The evolution of the birefringence, as a function of the diameter, can be seen in figure 19.

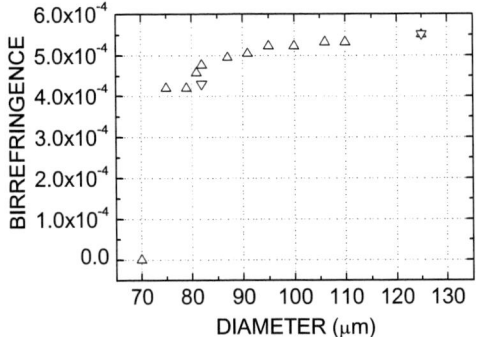

Figure 19. Calculated birefringence of the IEC HiBi fiber as a function of diameter. HF concentration: 40 % ($\nabla$) and 20 % ($\Delta$) [34].

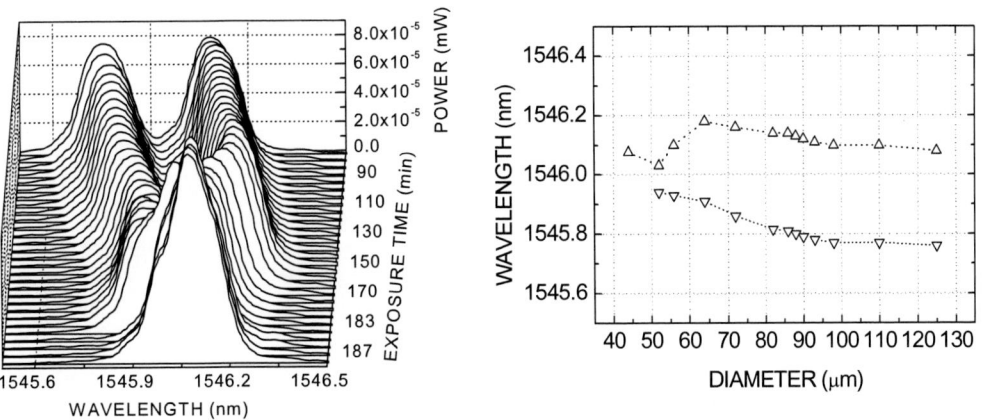

Figure 20. Left: evolution of the polarized bands of a FBG written in a bow tie HiBi fiber as a function of the etching time. Right: peak position of polarized bands (Y-polarized ($\nabla$) and X-polarized ($\Delta$)) as a function of the fiber diameter. The lines represent the linear best fit for the experimental data. HF concentration: 20 % [34].

A similar characterization can be made to other types of HiBi fibers. For example, figure 20 (left) shows the effect of chemical etching in the optical spectrum of a Bragg grating written in a bow tie fiber. The etching rate is lower and it is possible to observe that the two polarization bands collapse. Initially both bands show a trend to longer wavelengths on their peak position, as the diameter changes from 100 μm to 65 μm (figure 20 (left)). Further etching now causes the X polarized band to shift sharply to shorter wavelengths, until both bands collapse when the diameter reaches approximately 40 μm. This value agrees with the intrinsic stress-applying region dimensions, where the distance between the boundaries of the two internal side-lobes is approximately 37 μm.

Figure 21 shows the birefringence for a bow tie fiber as a function of the diameter. The results show that IEC and bow tie fibers have vanishing birefringence for diameters that are close to the value of the maximum dimension of the stress-applying region.

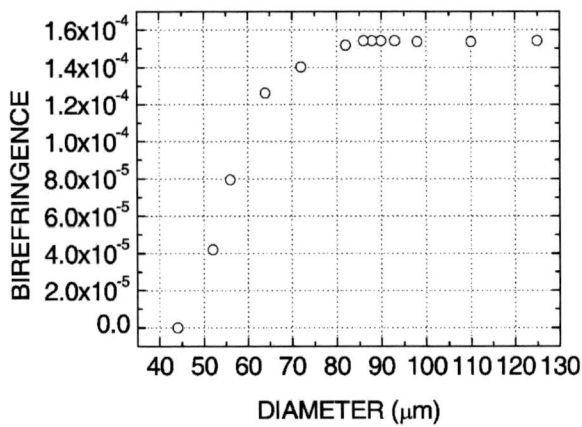

Figure 21. Measured birefringence of bow tie HiBi fiber as a function of diameter. HF concentration: 20 % [34].

### 5.4.1. Reduced Diameter for the Simultaneous Measure of Transverse Strain and Temperature

A FBG in an etched HiBi fiber can be applied as a sensor to simultaneously measure the transverse strain and temperature. Once again, a previous calibration of the different sensitivities must be made. The temperature and transverse strain coefficients for an etched IEC fiber is shown in table 11. It also displays the coefficients for a non-etched bow tie fiber with a similar diameter.

Table 11. Slopes of temperature and transverse strain of a FBG written in etched IEC and non-etched bow tie HiBi fibers [34].

| Fiber (diameter) | Temperature | | Transversal strain | |
|---|---|---|---|---|
| | $\partial\lambda_x/\partial T$ (pm/°C) | $\partial\lambda_y/\partial T$ (pm/°C) | $\partial\lambda_x/\partial\varepsilon$ (pm/με) | $\partial\lambda_y/\partial\varepsilon$ (pm/με) |
| Etched IEC (82 μm) | 7.00 | 6.90 | 0.7 (X-axis) 3.4 (Y-axis) | 2.23 (X-axis) 0.1 (Y-axis) |
| Bow tie (80 μm) | 8.02 | 8.46 | 0.02 (X-axis) 1.2 (Y-axis) | 1.16 (X-axis) 0.3 (Y-axis) |

The results of simultaneous transversal strain and temperature measurements obtained with matrix K and the values of $\Delta\lambda_X$ and $\Delta\lambda_Y$ of the reflection spectra are displayed in table 12.

**Table 12. Simultaneous measurements of temperature and transverse strain using etched FBGs in IEC HiBi fiber (diameter of 82 μm). The set values are determined by the experimental system equipment. Direction of applied load: 90° [34].**

| Set values | 16 °C | 26 °C | 36 °C | 46 °C | 56 °C |
|---|---|---|---|---|---|
| **33 με** | 15 °C | 28 °C | 33 °C | 44 °C | 56 °C |
|  | 37 με | 37 με | 39 με | 42 με | 42 με |
| **48 με** | 17 °C | 29 °C | 36 °C | 47 °C | 56 °C |
|  | 54 με | 57 με | 56 με | 54 με | 59 με |
| **64 με** | 17 °C | 28 °C | 36 °C | 46 °C | 57 °C |
|  | 49 με | 48 με | 48 με | 54 με | 53 με |
| **79 με** | 17 °C | 29 °C | 36 °C | 48 °C | 58 °C |
|  | 66 με | 80 με | 71 με | 65 με | 73 με |
| **94 με** | 17 °C | 29 °C | 36 °C | 48 °C | 58 °C |
|  | 80 με | 100 με | 94 με | 91 με | 85 με |

The errors obtained using a FBG in normal and reduced diameter HiBi fibers as a sensor are of comparable magnitude, but the dynamic range for strain measurements with the later ones is almost doubled as compared to the former sensors. This fact is important for technological applications where FBG can be tailored to attend a specific measurement range.

## 6. APPLICATIONS TO OPTICAL COMMUNICATIONS

All optical processing devices are becoming a key element in the next generation of optical communication systems, since they play a critical role in pulse formatting, spectral shaping and optimized all-optical routing and switching. These devices don't have the typical bottleneck associated to the optical-electrical-optical conversion and the majority is transparent to modulation format and bit-rate. FBGs are quite interesting for these applications, due to their low insertion loss and due to the avoidance of the decoupling of the signal outside the fiber. Moreover, the production technology is now in a mature state, which enables the design of gratings with custom made transfer functions, crucial for all-optical processing. Some advanced processing can be made if the transfer function is different for the two transversal modes of propagation in the fiber. This can be achieved by a HiBi FBG. One of the devices that take full advantage of the optical processing capabilities of the HiBi FBG is the orthogonal pumps source [35-37], which can be used in all optical wavelength converters [38, 39]. A tunable PMD compensator can also be developed based on the polarization processing properties of these special gratings [40, 41]. Also, a tunable microwave-photonic notch filter that makes use of a time delay element based on tunable HiBi chirped FBG has been demonstrated [42, 43] In addition, the interference due to laser coherence, typical in those micro-wave photonic filters was also reduced due to the polarization properties of the HiBi FBGs.

The following sections describe some example application of HiBi FBG in optics communications.

### 6.1. Optical Delay Line for PMD Compensation

In a linearly chirped grating, written in a HiBi fiber, each position of the grating will reflect two wavelengths at orthogonal polarizations (figure 22). This means that the group delay of these gratings is a combination of two linear functions, one for each polarization, with the same slope ($D_{FBG}$) and shifted by $\Delta\lambda_{HB}$:

$$\tau_y(\lambda) = D_{FBG}\lambda + b$$
$$\tau_x(\lambda) = D_{FBG}(\lambda - \Delta\lambda_{HB}) + b \quad (29)$$

where $b$ in (29) is a constant.

Therefore, the relative group delay induced by a linearly chirped FBG written in a HiBi fiber ($\Delta\tau = \tau_x - \tau_y$) is calculated using the following expression

$$\Delta\tau = -D_{FBG}\Delta\lambda$$
$$\approx -2D_{FBG}B\Lambda \quad (30)$$

Expression (30) shows that the dynamic tuning of the induced PMD can be made by adjusting the birefringence of the fiber, which can be done by applying a transversal stress in the fiber, as shown before in this chapter.

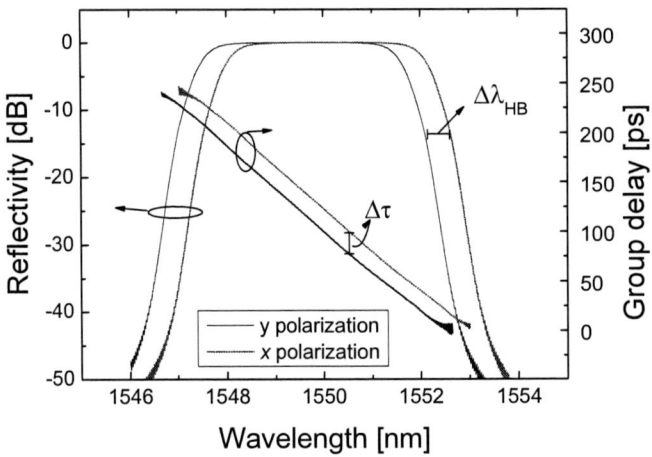

Figure 22. Reflectivity and group delay of a linearly chirped HiBi FBG for both transversal propagation modes [47].

### 6.1.1. Compensation Using a Linear Chirp

As can also be observed in expression (30), it is also possible to tune the PMD by adjusting the dispersion of the grating. That can be done using different methods [44-46]. One of them is by using thermal gradients to induce a linear chirp to a uniform FBG. Let us consider a uniform HiBi FBG put in a thermal contact with metal substrate. By applying different temperatures to the substrate, different linear temperature gradients will be generated. This gradient will induce a linear chirp to the FBG, due to thermo-optic and photoelastic effects. By changing the temperature gradient, the dispersion will also change, inducing a tunable differential delay line [47]. Figure 23 shows the experimental results of the evolution of $\Delta\tau$ as a function of the applied temperature gradient to a 24 mm uniform HiBi FBG.

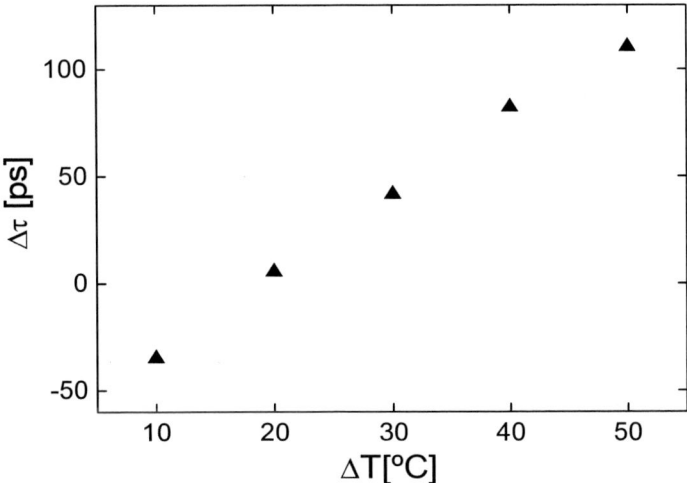

Figure 23. Relative group delay as a function of the applied linear gradient to a uniform HiBi FBG with 24 mm length.

Therefore, the presented device can be included in a PMD compensator as a tunable optical relative group delay line.

### 6.1.2. Compensation Using a Nonlinear Chirp

Let us now suppose that we have a HiBi FBG with a quadratic chirp. The group delay is now composed by two parabolic functions (one for each polarization) shifted by $\Delta\lambda_{HB}$. If the grating is tuned by temperature or longitudinal stress, the relative induced delay between the orthogonal polarizations, for a specific wavelength will change [40]. Figure 24 shows a simulation of a quadratic chirped FBG, with a length of 25 mm, written in a HiBi fiber with birefringence $B = 5 \times 10^{-4}$. For a tuning of 4.5 nm in the central wavelength, the relative group delay at 1550 nm changed from 41.6 ps to 12.1 ps. In this way, with this method, it is possible to do small corrections in the relative group delay.

The advantages of this method are its tuning simplicity and the flexibility in the operation range. However, the technique needs a FBG with a nonlinear chirp, which is quite complex to produce. It is generally produced with a custom made phase mask with a nonlinear chirp.

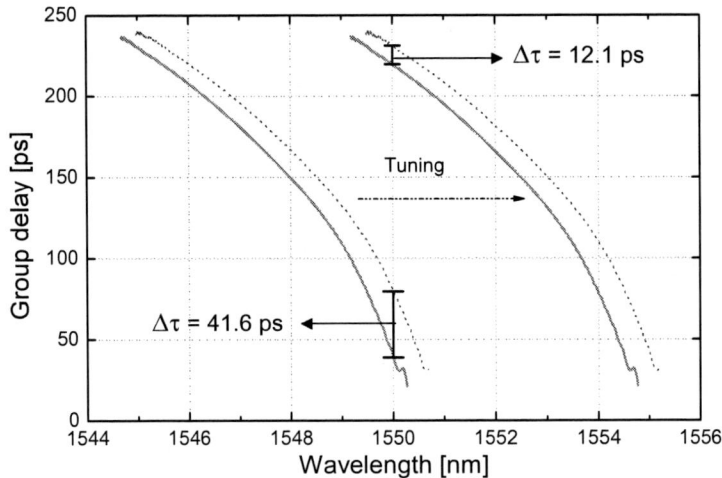

Figure 24. Simulation of the group delay of a HiBi FBG with quadratic chirp. Line: Y polarization; dots: X polarization [47].

## 6.2. Tunable Multiwavelength Linear Polarized Fiber Lasers

Fiber lasers have different applications in sensors and telecommunications due to their reduced linewidth, power and spectral profile. Like other lasers, fiber lasers need two components: a gain medium and a resonant cavity. For a fiber laser operating around 1550 nm, it is generally based on an optical pump with 980 or 1480 nm of wavelength, an erbium-doped fiber and an optical filter. The gain is obtained from the amplified spontaneous emission due to the optical pump.

Generally, fiber optical lasers based on an optical ring with erbium-doped fiber don't enable the generation of more than one laser line [48,49]. This is a consequence of the fact that erbium is a medium with homogeneous gain at room temperature, resulting in strong mode competition, which induces laser instability. A method was proposed to reduce the homogeneity of the fiber by cooling the fiber to 77 K [50, 51]. However, by obvious reasons, it is not very practical. Other methods used special fibers like the elliptical core fibers [52] or the twincore fibers [53].

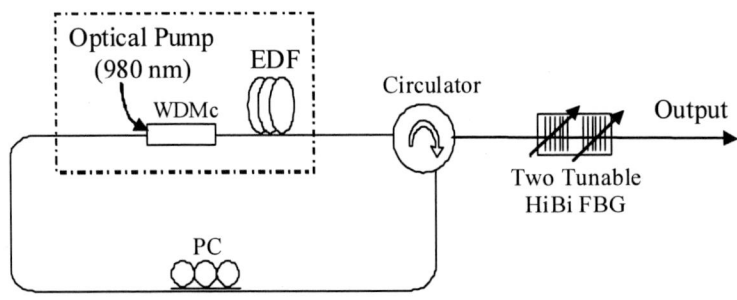

Figure 25. Diagram of a multiwavelength fiber laser based on HiBi FBGs. EDF: Erbium doped fiber; PC: Polarization controller; WDMc: WDM optical coupler;

Another way to reduce the homogeneity of the fiber is to use different laser lines operating at different longitudinal modes. For this kind of implementation, HiBi FBGs can have an important role, since they will reflect two wavelengths at orthogonal polarizations. An implementation method for a tunable laser with up to four laser lines is depicted in figure 25.

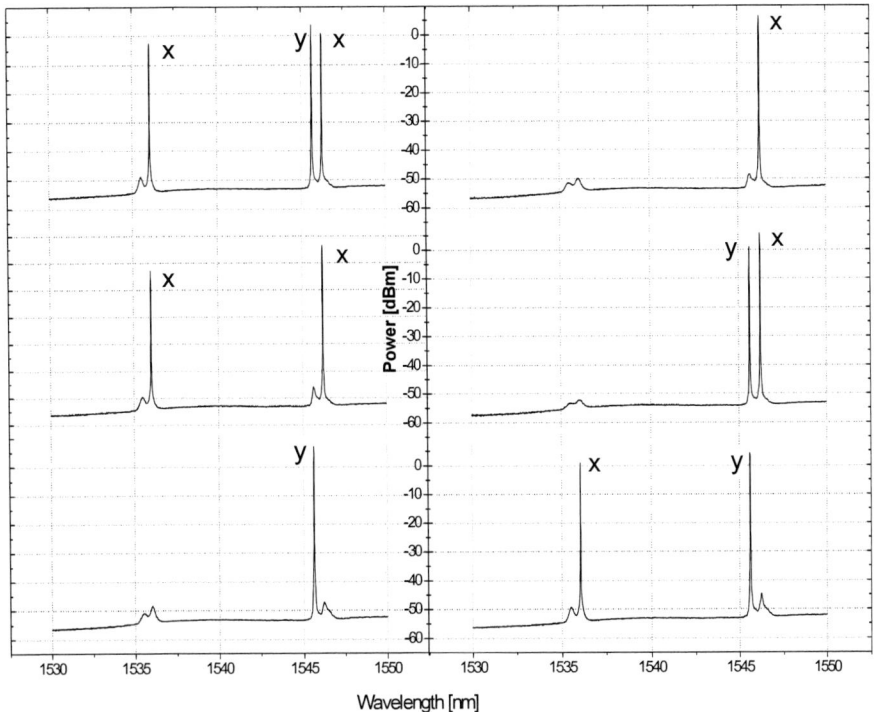

Figure 26. Optical spectra at the output of the fiber laser with different operation modes. The operating laser lines are at a linear polarization (x or y).

The two tunable HiBi FBGs enable the selection of 4 different wavelengths. By tuning the polarization controller (PC) inside the optical cavity, it is possible to select the appropriate laser lines. Figure 26 shows some of the possibilities that can be achieved with just two HiBi FBGs.

One of the advantages of this technique is its ability to generate two laser lines at orthogonal polarizations (see last spectrum of figure 26). Therefore, it can be used as two orthogonal pumps in a polarization insensitive wavelength converter [38].

## 6.3. Optical Networks Architectures Using HiBi FBG for Performance Improvement

### 6.3.1. Optical Code Division Multiple Access

Metro optical code division multiple access (OCDMA) networks can benefit from the polarization multiplexing, since two users using codes in the same time-wavelength chip can be given orthogonal polarizations to operate, therefore reducing interference. One of the implementation techniques is the "polarization assisted OCDMA with HiBi FBG" [54]. The technique uses the polarization properties of the HiBi FBG along with a special code generation scheme to improve the performance of OCDMA based networks. The coders are based on HiBi FBGs. To implement the suggested polarization assisted OCDMA, each HiBi FBG will reflect a pair of wavelengths $\lambda_i\lambda_j$, which are consecutive and cross polarized. In the case of the proposed method, a set of three of these HiBi FBGs, spaced by the fiber length needed for achieving the corresponding time chip spacing, results in two subsequent codes. Here, $\lambda_i$ corresponds to a X polarized reflection and $\lambda_j$ to a Y polarized one. This allows two consecutive user spreading sequences to share the same encoder. An implementation example is depicted in figure 27.

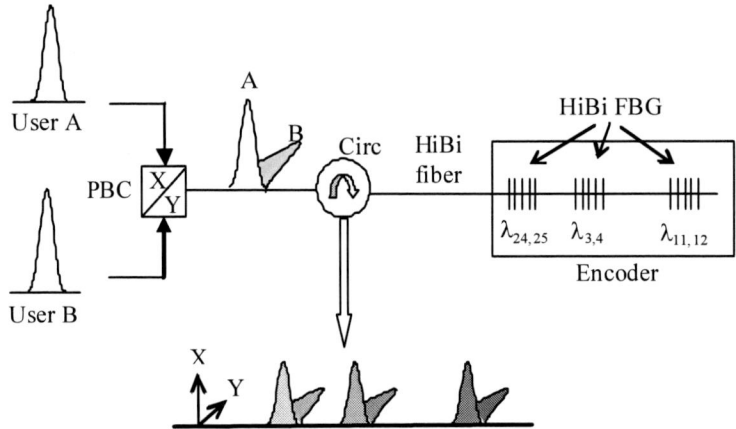

Figure 27. Schematic of the proposed encoder implementation showing the use of the polarization to encode simultaneously two users with different wavelengths at orthogonal polarizations. Legend: PBC: polarization beam combiner; Circ: optical circulator [54] (© 2006 IEEE).

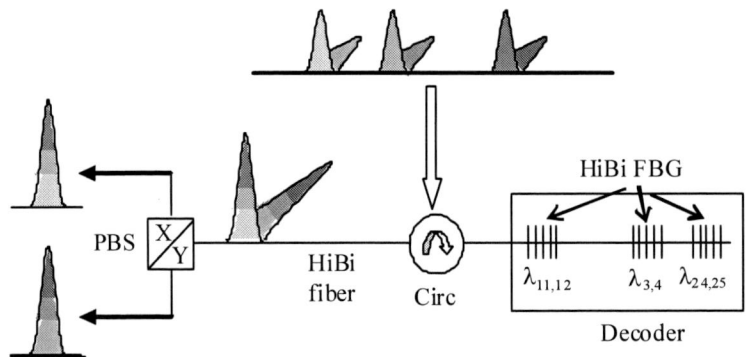

Figure 28. Schematic of the proposed implementation for the decoder based on HiBi FBG. Legend: PBS: polarization beam splitter; Circ: optical circulator [54] (© 2006 IEEE).

Each bit of information from users A and B is a wavelength comb which includes at least the wavelengths of the correspondent code (or a standard modulated broadband source can be used). Both bit sequence signals are multiplexed using a polarization beam combiner, thereby ensuring that they enter the encoder at the correct orthogonal polarizations. The encoder is based on three HiBi FBG reflecting the wavelengths $\lambda_3$, $\lambda_{11}$ and $\lambda_{24}$ for X polarization and $\lambda_4$, $\lambda_{12}$ and $\lambda_{25}$ for Y polarization. To achieve networking operation many such encoders need to operate simultaneously, and due to the properties of the technique, the number of encoders needed reduces to almost half.

The decoder can be imprinted with standard FBG, since each user has its own code. However, for reduced user interference and to reduce the number of decoders needed, it can be based on HiBi FBG like the one exemplified in figure 28.

The decoding process is similar to the encoding, where the HiBi FBG correlates two codes simultaneously. Afterwards a polarization beam splitter is used to separate both users. If no polarization maintaining fiber is used in the transmission link between the encoder and the decoder, the former must be preceded by a polarization rotator to ensure correct polarization coupling to the receiver. The polarization rotator can be automatically controlled by the receiver of one of the users using simple electronics. Even if no alignment of the polarization is made between non adjacent users, on average, only half the power will induce interference since the decoder will process only one of the two available polarizations. In the same way, the sensitivity to heterodyne crosstalk is also reduced since the power of the adjacent user, generated by the same encoder, is orthogonally polarized. In opposition to other coding/decoding techniques, like the ones based on arrayed waveguide gratings and optical delay lines, the proposed coder/decoders are quite compact, simple to use and have low insertion losses. On the other hand, since the gratings can have a length down to 1-2 mm and still have a high reflectivity, the time slots can be as low as a few picoseconds which can be considered enough for the majority of applications.

*6.3.2. Radio over Fiber*

In radio over fiber systems (RoF), using the same central station to transmit to different local stations, one can use frequency interleaving to improve the bandwidth efficiency, exploiting the unused band between the carrier and data when high modulation frequencies are used with single side band (SSB) format. However, frequency interleaving also increases the bit error rate (BER), due to the interference of adjacent carriers. This drawback can be minimized if polarization multiplexing is used, i.e., the carriers and data are at orthogonal polarizations (figure 29).

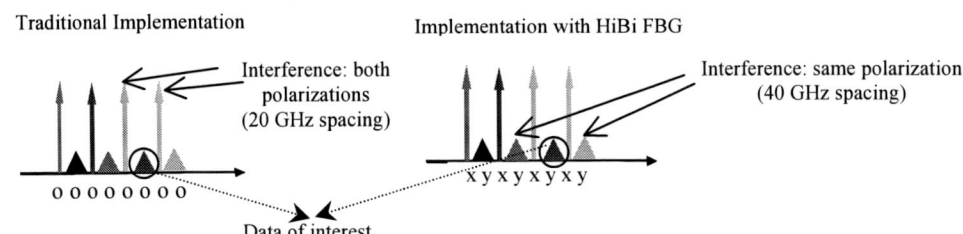

Figure 29. Diagram of the concept of interleaving using polarization multiplexing between carriers and data.

The implementation of this concept can be made using a HiBi FBG filter at the transmission which creates the SSB format and, at the same time, selects one polarization for the carrier and the orthogonal one for the data. At the local station another HiBi FBG removes the selected channel with reduced interference, since the interference will only be made by the data of the adjacent channels, which are at higher wavelength spacing and with lower power, relatively to the adjacent carriers. This technique has an impact on the overall performance of the system since the bandwidth efficiency can be improved without increasing the BER [55].

## 7. CONCLUSION

This chapter described some of the characteristics and functionalities associated with HiBi FBG. Their anisotropic behavior, relative to stress and/or strain, make them well suited for multiparameter sensors, including temperature, transversal strain and longitudinal strain. Moreover, their polarization processing capabilities also give them an interesting potential for different applications in optics communications. These applications include the development of new devices, like multiwavelength fiber lasers or in the optimization of certain architectures, like OCDMA. Some of the applications for sensing and optical communications were described but many more are yet to come.

## REFERENCES

[1] Kapron, F. P.; Keck, D. B. *Appl. Phys. Lett.* 1970, vol.17, 423-425.
[2] Yamada, Y.; Nakagawa, S. I.; Kurosawa, Y.; Kawazawa, T.; Taga, H.; Goto, K. *Elect. Lett.* 2002, vol. 38, 328-330.
[3] Hill, K. O.; Fufii Y.; Johnson D. C.; Kawasaki, B. S. *Appl. Phys. Lett.* 1978, vol. 32, 647-649.
[4] Mora, J.; Díez, A.; Cruz, J. L.; Andrés, M. V. *IEEE Photon. Technol. Lett.* 2000, vol. 12, 1680-1682.
[5] Guan, B.; Tam, H.; Liu, S. *IEEE Photon. Technol. Lett.* 2004, vol. 16, 224-226.
[6] Tjin, S. C.; Suresh, R.; Ngo, N. Q. J. *Lightwave Technol.* 2004, vol. 22, 1728–1733.
[7] Liu, Y.; Chiang, K. S.; Chu, P. L. *IEEE Photon. Technol. Lett.* 2005, vol. 17, 450-452.

[8] Kashyap, R. *Fiber Bragg gratings*, Publisher: Academic Press, San Diego, CA, 1999.
[9] Othonos, A.; Kalli, K. *Fiber Bragg gratings*, Publisher: Artech House, Norwood, NA, 1999.
[10] Higuera, J. M. L. *Handbook of Optical Fiber Sensing Technology*; Publisher: John Wiley & Sons, New York, NY, 2002.
[11] Erdogan, T. J. *Lightwave. Technol.* 1997, vol. 15, 1277-1294.
[12] Lam, D. K. W.; Garside, B. K. *Appl. Optics.* 1981, vol. 20, 440-456.
[13] Yariv, A. J. *Quantum Electron.* 1973, vol. 9, 919-933.
[14] Yariv, A.; Nakamura, M. J. *Quantum Electron.* 1977, vol. 13, 233-253.
[15] Dyott, R. B. *Elliptical Fiber Waveguides*; Publisher: Artech House, Boston, MA, 1995.
[16] Rashleigh, S. C. J. *Lightwave. Technol.* 1983, vol. LT-1, 312-330.
[17] Kersey, A. D.; Davis, M. A.; Patrick, H. J.; LeBlanc, M.; Koo, K. P.; Askins, C. G.; Putnam, M. A.; Friebele, E. J. J. *Lightwave Technol.* 1997, vol. 15, 1442-1463.
[18] Othonos, A. *Rev. Sci. Instrum.* 1997., vol. 68, 4309-4341,
[19] Abe, I.; Kalinowski, H. J.; Nogueira, R. N.; Pinto, J. L.; Frazão, O. *IEE P-Circ. Dev. Syst.* 2003, vol. 12, 495-500.
[20] Frazão, O.; Pereira, D. A.; Santos, J. L.; Araújo, F. M.; Ferreira, L. A. *Proc. SPIE* 2005, vol.5855, 765-758.
[21] Chen, G; Liu, L; Jia, H; Yu, J; Xu, L.; Wang, W. *IEEE Photon. Technol. Lett.* 2004, vol. 16, 221-223.
[22] Urbanczyk, W.; Chmielewska, E.; Bock, W. *Meas. Sci. Technol.* 2001, vol. 12, 800-804.
[23] Ferreira, L. A.; Araújo, F. M.; Santos, J. L.; Farahi, F. *Opt. Eng.* 2000, vol. 39, 2226-2234.
[24] Sudo, M.; Nakai, M.; Himeno, K.; Suzaki, S.; Wada, A.; Yamauchi, R. *Proc. 12 th Int. Conf. Optical Fiber Sensors* 1997, vol. 16, 170-173.
[25] Chehura, E.; Ye, C. C.; Staines, E. S.; James, S. W.; Tatam, R. P. *Smart Mater. Struct.* 2004, vol.13, 888-895.
[26] Ye, C.C.; Staines, S. E.; James, S. W.; Tatam, R. P. *Meas. Sci. technol.* 2002, vol. 13, 1446-1449.
[27] Bosia, F.; Giaccari, P.; Facchini, M.; Botsis, J.; Limberger, H.; Salathé, R. *Proc. SPIE* 2002, vol. 4694, 175-186.
[28] Lawrence, C. M.; Nelson, D. V.; Udd, E.; Bennett, T. *Exp. Mech.* 1999, vol. 39,202-209.
[29] Lawrence, C. M.; Nelson, D. V.; Udd, E. *Proc. SPIE* 1997, vol. 3042, 218-228.
[30] Caucheteur, C.; Ottevaere, H.; Nasilowski, T.; Chah, K.; Statkiewicz, G.; Urbanczyk, W.; Berghmans, F.; Thienpont, H.; Mégret, P. *Proc. SPIE* 2005, vol. 5952, 1-10.
[31] Abe, I.; Kalinowski, H. J.; Frazão, O.; Santos, J. L.; Nogueira, R. N.; Pinto, J. L. *Meas. Sci. Technol.* 2004, vol. 15, 1453-1457.
[32] Udd, E.; Schulz, W. L.; Seim, J. *Proc. SPIE* 1999, vol. 3538, 206-214.
[33] Udd, E.; Nelson, D., Lawrence, C.; Ferguson, B. *Proc. SPIE* 1996, vol. 2718,104-107.
[34] Abe I.; Frazão O.; Schiller M. W.; Nogueira R. N.; Kalinowski H. J.; Pinto, J. L. *Meas. Sci. Technol.* 2006, vol.17, 1477–1484.
[35] Nogueira, R. N.; Teixeira, A. L. J.; André, P. S.; Rocha, J. F.; Pinto, J. L. *Proc. Conf. Lasers Electro-Optics* 2003, 545.
[36] André, P. S.; Nogueira, R. N.; Teixeira, A. L.; Lima, M. J. N.; Rocha, R. F.; Pinto, J. L. *Laser Physics Lett.* 2004, vol. 1, 1 - 4.

[37] Zhao, C. L.; Yang, X.; Ng, J. H.; Dong, X.; Guo, X.; Wang, X.; Zhou, X.; Lu, C. *Microw. Opt. Technol. Lett.* 2004, vol. 41, 73 – 75.

[38] Nogueira, R. N.; Teixeira, A. L. J.; André, P. S; Rocha, J. F.; Pinto, J. L. *Opt. Commun.* 2006, vol. 262/1, 38-40.

[39] Nogueira, R. N.; Teixeira, A. L. J.; Pinto, J. L.; Rocha, R. F. *IEE Elect. Lett.* 2004, vol. 40, 616–617.

[40] Lee, S.; Khosravant, R.; Peng, J.; Grubsky, V.; Starodubov, D. S.; Willner, A. E.; Feinberg, J., *IEEE Photon. Technol. Lett.* 1999, vol. 11, 1277–1279.

[41] Willner, A. E.; Feng, K.-M.; Cai, J.; Lee, S.; Peng, J.; Sun, H. *IEEE J. Select. Topics Quantum Electron.* 1999, vol. 5, 1298-1311.

[42] Yi, X.; Lu, C.; Yang, X.; Zhong, W.-D.; Wei, F.; Ding, L.; Wang, Y. *IEEE Photon. Technol. Lett.* 2003, vol. 15, 754-756.

[43] Zhang, W.; Williams, J. A. R.; Bennion, I. *IEEE Photon. Technol. Lett.* 2001, vol. 13, 523-525.

[44] Eggleton, B. J.; Mikkelsen, B.; Raybon, G.; Ahuja, Rogers, A.; J. A.; Westbrook, P. S.; Nielsen, T. N.; Stulz, S.; Dreyer, K. *IEEE Photon. Technol. Lett.* 2000, vol. 12, 1022-1024.

[45] Mora, J.; Ortega, B.; Andrés, M. V.; Capmany, J.; Cruz, J. L.; Pastor, D.; Sales, S. *IEEE Photon. Technol. Lett.* 2003, vol. 15, 951-953.

[46] Lauzon, J.; Thibault, S.; Martin J.; Ouellettet, F. *Opt. Lett.* 1994, vol. 19, 2027-2029.

[47] Nogueira, R. N.; Pinto, J. L.; Rocha, J. F. *Microw. Opt. Tech. Lett.* 2006, vol. 48 , 2357-2359.

[48] Gloag, A. J.; Langford, N.; Bennion I.; Zhang, L. *Opt. Commun.* 1996, vol. 123, 553-557.

[49] Inaba, H.; Akimoto, Y.; Tamura, K.; Yoshida, E.; Komukai T.; Nakazawa, M. *Opt. Commun.* 2000, vol. 180, 121–125.

[50] Yamashita S.; Hotate, K. *Electron. Lett.* 1996, vol. 32, 1298-1299.

[51] Wei, D.; Li, T.; Zhao Y.; Jian, S. *Opt. Lett.* 2000, vol. 25, 1150-1152.

[52] Das G.; Lit, J. W. Y. *IEEE Photon. Technol. Lett.* 2002, vol. 14, 606-608.

[53] Graydon, O.; Loh, W. H.; Laming, R. I.; Dong, L. *IEEE Photon. Technol. Lett.* 1996, vol. 8, 63-65.

[54] Nogueira, R. N.; Teixeira, A. L. J.; Pinto, J. L.; Rocha, J. F. *IEEE Photon. Technol. Lett.* 2006, vol. 18, 841 – 843.

[55] Teixeira, A. T.; Nogueira, R. N.; André, P. S.; Lima, M. J. N.; Rocha, J. F., *Electron. Lett.* 2005, vol. 41, 30 - 32.

# INDEX

## A

access, x, 3, 4, 9, 11, 15, 40, 46, 48, 50, 51, 53, 58, 59, 60, 62, 64, 66, 68, 70, 76, 77, 79, 80, 82, 83, 84, 86, 88, 89, 90, 92, 94, 112, 116, 123, 130, 134, 135, 137, 146, 147, 149, 153, 156, 159, 160, 161, 162, 163, 164, 165, 166, 167, 170, 171, 172, 173, 174, 179, 202, 239
accountability, 108
accounting, 10, 53, 60
achievement, 49, 101
acid, 231
acquisitions, ix, 30, 35, 36, 37, 65, 66, 71, 107, 116, 120, 123, 125, 162, 173
adjustment, 85, 186
administration, 15, 54, 55, 56, 128, 131
advertising, 82, 83, 89, 131, 168
aerospace, 73
affect, x, 160
Afghanistan, 68
Africa, 66, 67, 92, 93, 98, 101, 103, 105, 125, 147, 150, 155, 156
age, 5, 9, 174
AGFI, 6
aggregation, 97, 100
agriculture, 40
airports, 76
Alaska, 77
Albania, 27, 33, 37
Algeria, 156
algorithm, 186, 187, 192, 194, 201, 204
alternative(s) 4, 7, 8, 9, 10, 38, 58, 65, 66, 83, 147, 148, 161, 162, 180, 192
ambiguity, 56
amendments, 116
amortization, 78, 81
amplitude, 205, 210, 211

Amsterdam, 175
annealing, 192
annual rate, 3
annual review, 86
antitrust, x, 56, 60, 136, 160, 161, 171
AP, 3
AR, 140, 170
argument, 55, 129
Armenia, 27
ash, 120
Asia, 27, 42, 50, 66, 67, 93, 101, 130, 138, 140, 141
Asian countries, 131
assessment, 156
assets, 33, 46, 53, 59, 61, 62, 63, 65, 67, 70, 81, 84, 112, 117, 122, 153
assignment, 48
assumptions, xi, 80, 178
asymmetric information, 78, 79
asymmetry, 78, 79, 81, 86, 231
AT&T, 52, 82, 87, 139, 160, 161, 172, 173, 175, 176
Athens, 75
Atlantic Ocean, 58
attachment, 84
attention, x, 1, 24, 33, 38, 92, 93, 100, 118, 120, 144, 154, 159, 162
attitudes, 48
Attorney General, 52
auditing, 100
Australia, ix, 69, 127, 128, 130, 131, 132, 133, 134, 135, 137, 139
Austria, 16, 17, 18, 22, 23, 24, 26, 27, 29, 31, 34, 35, 37, 41, 42, 43
authority, 46, 48, 52, 60, 63, 67, 68, 71, 102, 108, 144, 148, 149, 155, 162, 170
availability, 4, 7, 8, 10, 66, 70, 86, 104, 109
average costs, 57, 80, 170
averaging, 57
avoidance, 235

awareness, 93
Azerbaijan, 27

# B

backscattering, 180, 197, 198, 199, 200
balance of payments, 110
balance sheet, 117
Balkans, 38
bandwidth, xi, 79, 85, 87, 114, 130, 177, 180, 189, 190, 191, 195, 196, 199, 201, 203, 204, 210, 241, 242
bank mergers, 125
bankruptcy, 67, 171
banks, 101, 103, 146
barriers, 10, 11, 40, 41, 63, 97, 109, 117, 128
basic services, 58, 136, 144
beams, 50
behavior, 2, 78, 96, 98, 110, 129, 145, 148, 151, 154, 174, 199, 205, 223, 229, 242
Belarus, 27, 30
Belgium, 17, 18, 22, 23, 25, 26, 31, 114, 115
BellSouth, 161, 167, 168, 169, 170, 172
benchmarking, ix, 91, 92, 95
benchmarks, 79
benefits, x, 10, 15, 25, 39, 51, 60, 68, 79, 87, 94, 95, 96, 97, 98, 99, 100, 101, 102, 103, 104, 109, 118, 120, 160, 161, 165, 167, 168, 169, 171, 174
bias, 51, 97
binding, 67
biodiversity, 68
biomechanics, 210
birefringence, xi, 209, 210, 213, 214, 215, 216, 217, 219, 220, 225, 228, 229, 231, 233, 234, 236, 237
birefringent fibers, xi
birth, 154
blog, 84
blue-collar workers, 10
body, 166
Boltzmann constant, 184
borrowing, 65, 103, 110
Bosnia, 27, 29, 30, 37
boundary value problem, 185
bounds, 140
Bragg grating, xi, 178, 180, 209, 210, 211, 218, 224, 225, 228, 231, 233, 243
Brazil, 27, 46
breakdown, 83
Britain, 144, 146
broadband, vii, viii, xi, 1, 2, 3, 11, 70, 75, 76, 77, 78, 79, 80, 81, 82, 86, 88, 89, 90, 139, 140, 165, 172, 174, 178, 180, 190, 191, 206, 211, 241
broadcaster, 133

browsing, 172
budget deficit, 114
Bulgaria, vii, 13, 14, 17, 18, 22, 23, 24, 26, 27, 28, 29, 33, 37, 40, 116
bundling, vi, vii, x, 1, 4, 159, 160, 162, 166, 167, 169, 170, 171, 174, 175, 176
bureaucracy, 108, 110, 145, 146, 149, 155
Bush administration, 128
business model, viii, 3, 4, 75, 76, 77, 82, 84, 89, 127
buyer, 78, 113

# C

cabinets, 179
cable service, 77, 88
cable system, 114
cable television, 128
cables, 50, 56, 58, 80, 81, 172, 179
calibration, 221, 225, 234
California, 172, 173
call centers, 168
canals, 76
candidates, 38, 54
Capacity, 95, 99, 104, 105
capacity building, ix, 91, 92, 96
capital account, 110
capital flows, 108
capital inflow, 110
capital intensive, 123
capital markets, 112, 113
capitalism, 90, 108
Caribbean, 57, 93, 102
carrier, 52, 59, 61, 63, 161, 169, 170, 173, 241, 242
case study, viii, 46, 47, 156
cash flow, 65, 69, 120
cast, 77, 89, 118
catalyst, 51
catalysts, 101
catastrophes, 144
cell, 118
cellulose, 112
cement, 112
Census, 80, 82, 131, 141
Census Bureau, 131, 141
Central Europe, 41
certainty, 84, 86
certification, 65
CFI, 6
changing environment, 97
channels, 109, 120, 123, 129, 131, 135, 168, 180, 183, 189, 202, 203, 204, 210, 242
chemical etching, 225, 231, 233
Chicago, 84

China, 46, 69, 108, 136
Chinese, 69, 131, 136
civil war, 144
cladding, 214, 231, 233
classes, 57
CLECs, 160, 162, 171
clients, 66, 147
Clinton Administration, 59, 62
CNN, 49
code generation, 240
codes, 239, 241
coding, 241
coherence, 215, 236
cohort, 164
Cold War, 51
collaboration, 93, 94, 98, 100
collateral, 36
college campuses, 11
collusion, 136
colonial rule, 131, 133
Columbia University, 71
commerce, 48
commodity, 96
common law, 102
Common Market, 42, 43
communication, vii, viii, xi, 3, 11, 13, 14, 47, 50, 51, 58, 66, 73, 77, 84, 88, 93, 95, 103, 114, 119, 127, 138, 139, 177, 178, 181, 182, 206, 209, 210, 235
communication systems, xi, 177, 178, 181, 182, 206, 235
Communications Act, 65, 67, 144, 149, 154, 156
Communications Act of 1934, 65
Communist Party, 136
community, 43, 56, 76, 81, 82, 83, 84, 110, 128, 129, 135, 137
compatibility, 56, 92, 210
compensation, 123
competition, ix, x, 3, 14, 15, 16, 24, 39, 47, 49, 55, 56, 57, 58, 60, 61, 62, 66, 68, 70, 72, 75, 76, 80, 86, 87, 105, 107, 109, 112, 113, 116, 119, 123, 128, 129, 131, 132, 133, 134, 135, 136, 137, 138, 139, 141, 144, 146, 148, 149, 150, 151, 155, 156, 159, 160, 161, 162, 163, 164, 166, 167, 171, 172, 174, 175, 238
competitive advantage, 169, 170
competitive local exchange carriers, 160
competitive markets, 49, 60
competitiveness, 11, 39, 67, 111, 150
competitor, 31, 148, 160, 167, 172
complement, 33, 65
complementarity, x, 159, 162, 166, 167
complexity, 76
compliance, 48, 65, 66, 94, 95, 148
complications, 117
components, 5, 65, 84, 98, 102, 110, 167, 169, 180, 183, 184, 195, 210, 213, 215, 216, 219, 225, 231, 238
compulsion, 147
computation, xi, 178, 187, 194
computer mouse, 168
computers, 168
concentrates, 28
concentration, 23, 108, 134, 135, 140, 232, 233, 234
concrete, 84
configuration, 53, 183, 192, 199, 202, 204, 205
confinement, 178
conflict, ix, 68, 76, 78, 91, 93, 96, 102, 134, 136
conflict of interest, 96, 102, 134, 136
conflict resolution, 76, 78
conformity, 15
Congestion, 95
Congress, 60, 61, 62, 63, 73
congruence, 48, 137
conjecture, 117
Connecticut, 127, 173
connectivity, viii, 45, 46, 49, 51, 66, 165, 172
consciousness, 9
consensus, 49, 81, 85, 104, 109, 112, 117, 120, 123
consolidation, x, xi, 16, 32, 66, 71, 78, 80, 81, 127, 129, 134, 136, 137, 159, 161, 174, 177
constraints, ix, 91, 93, 155
construct validity, 5, 6
construction, 4, 54, 55, 78, 83
consulting, 59
consumer price index, 163
consumers, ix, x, 2, 4, 7, 8, 9, 10, 11, 62, 127, 128, 130, 132, 134, 135, 136, 137, 140, 160, 161, 162, 164, 167, 169, 171, 173, 174, 175
consumption, 85, 95, 96, 162, 163, 165
context, x, 159, 164
control, 24, 29, 30, 32, 33, 35, 36, 52, 63, 65, 68, 70, 77, 118, 125, 133, 139, 141, 195, 198, 199, 208
convergence, ix, 3, 4, 15, 127, 128, 130, 132, 133, 135, 136, 138, 140, 141, 156, 186, 192, 200
conversion, 87, 117, 235
conviction, 59
cooling, 195, 224, 238
Copenhagen, 40
copper, 118
corporations, 49, 51, 52, 55, 60, 71, 135, 146
correlation, 51
corrosion, 210
corruption, 108
cost saving, 83, 94, 162, 167, 169

costs, 38, 48, 55, 57, 58, 76, 77, 79, 80, 81, 84, 89, 96, 99, 101, 103, 105, 109, 118, 120, 121, 122, 139, 164, 165, 168, 169, 170, 171, 172, 195, 210
Council of Europe, 93
counsel, 53
couples, 181
coupling, 184, 212, 213, 241
coverage, 4, 7, 8, 9, 10, 11, 27, 30, 50, 55, 66, 67, 69, 76, 77, 80, 82, 83, 86, 88, 89, 117, 129, 136
covering, 14, 77
credit, 67, 69, 118, 149, 165
credit rating, 165
Croatia, 27, 29, 33, 34, 37, 116
cross-country, ix, 91, 93, 96
cross-ownership, 128, 138
crystalline, 185
Cuba, 27, 68
cultural practices, 47
culture, 25, 49, 59, 153, 154
current prices, 112
customers, 21, 49, 53, 58, 59, 62, 66, 67, 80, 87, 89, 119, 122, 135, 147, 149, 151, 154, 161, 168, 169, 170
Cyprus, vii, 13, 14, 16, 17, 18, 22, 23, 26, 27, 38
Czech Republic, vii, 13, 14, 17, 18, 23, 24, 25, 27, 28, 29, 30, 34, 38

# D

dailies, 131
data availability, 109
data collection, ix, 91, 93, 100
data communication, 77, 104
data gathering, 99
data transfer, 15, 17
database, 125
debt, 35, 65, 66, 67, 68, 69, 70, 110, 120
debts, 36, 154
decay, 182
decibel, 189
decision makers, 137
decision making, 78, 95, 100, 117
decisions, 38, 47, 53, 60, 164
decoding, 241
decoupling, 235
defense, 61, 63, 73
deficit(s), 111, 114, 121
definition, 25, 79, 93, 95, 96, 97, 205
deformation, 210, 223, 226, 230
degenerate, 198
degradation, 178, 180, 188
Delaware, 63
delivery, 66, 104

demand, 1, 2, 3, 10, 11, 50, 57, 70, 77, 78, 79, 80, 83, 85, 118, 120, 145, 147, 149, 155, 156, 163, 165, 166, 172, 173, 178, 179
democracy, 108, 136, 137, 140
demographic characteristics, 5, 164
Denmark, 17, 18, 22, 23, 26, 27, 34, 35, 36, 41
density, 46, 50, 57, 70, 81, 88, 165, 179, 182, 200
Department of Justice, 69, 161
dependent variable, 186
deposits, 103
depreciation, 80, 112
depression, 154
deregulation, 49, 55, 108, 109, 117, 127, 128, 129, 132, 136, 137, 138, 155, 162
desire, 15, 16, 34, 103
destruction, 179
developed countries, 116
developed nations, 56
developing countries, viii, ix, 45, 46, 49, 50, 51, 56, 57, 58, 68, 70, 71, 91, 92, 93, 98, 102, 104, 107, 108, 109, 112, 116, 120, 127
developing nations, 92, 130
development assistance, 101
development banks, 101, 103
deviation, 195, 203, 204
diffusion, 11, 43, 62
diffusion process, 62
digital divide, 75, 76
diodes, 191
direct investment, 15, 25, 26, 30, 51, 110
discipline, 167
discourse, 47
discrimination, 167
dispersion, 179, 184, 185, 195, 198, 199, 237
dissatisfaction, 147
distilled water, 231
distress, 56, 121
distribution, ix, 10, 46, 51, 80, 107, 108, 112, 113, 168, 185, 196, 197
divergence, 101, 103
diversity, 128, 129, 133, 135, 137, 140, 141
divestiture, 173
dividends, 65, 69
division, 121, 180, 239
domestic markets, 53, 116
domestic policy, 48, 97
dominance, 71, 137, 155
downlink, 40
drinking water, 103
DSL, 77, 80, 88, 165, 172
duopoly, x, 24, 39, 80, 160, 173
duplication, 55, 81
duration, 57, 81, 87, 163

duties, 52, 54

## E

earth, 50, 51, 54, 57, 60, 65, 117
earthquake, 67
East Asia, 27, 93
Eastern Europe, 27, 28, 29, 32, 34, 35, 38, 40, 42, 117
economic crisis, 112, 123
economic development, viii, 51, 75, 76, 83, 92, 95, 136
economic efficiency, 60
economic growth, 76, 127
economic indicator, 118
economic institutions, 90
economic performance, 60
economic problem, 71
economic resources, 102
economic status, 103
economic theory, 173
economics, 115, 140, 164
economies of scale, 55, 133, 136, 169, 170, 174
education, 164
EEA, 206
egalitarianism, 70
Egypt, 93, 105
elaboration, 78
elasticity, 166
electricity, 41, 77, 80, 88, 90, 94
electromagnetic, vii, xi, 47, 209, 210
electromagnetic wave(s) vii
electrons, 181
eligibility criteria, 66
emerging markets, 108
emission, 182, 183, 199, 220, 221, 238
employees, 53, 84, 117, 119, 121, 122, 145, 147
employment, 108, 111, 120, 154
encoding, 241
encouragement, 97
end-users, 165
energy, ix, 40, 78, 91, 92, 93, 94, 181, 189, 204, 213
England, 41, 48, 63
enhanced service, 25, 167
enlargement, 28, 41, 42, 180
environment, ix, 46, 58, 70, 92, 110, 129, 135, 143, 149, 156, 161
environmental issues, 47
equality, 17
equilibrium, 182
equipment, 50, 66, 85, 87, 88, 104, 153, 226, 227, 228, 231, 235
equity, viii, 29, 31, 39, 45, 64, 65, 67, 68, 69, 70

erbium, 220, 238
estates, 112
Estonia, vii, 13, 14, 17, 18, 22, 23, 26, 33, 34, 36
etching, 225, 231, 232, 233
ethnic groups, 145
EU, vii, viii, 13, 14, 15, 16, 17, 21, 22, 23, 24, 25, 27, 28, 30, 31, 33, 34, 35, 38, 39, 40, 41, 42, 43, 95, 115
EU enlargement, 42
Europe, 27, 28, 30, 31, 32, 34, 35, 36, 37, 40, 41, 42, 50, 55, 66, 115, 207
European Commission, 101
European Union, vii, 13, 14, 18, 41, 42, 43, 95, 206
evening, 165
evidence, 47, 48, 56, 57, 69, 101, 108, 124, 129, 162, 163, 166
evolution, vii, viii, 1, 4, 13, 14, 133, 179, 180, 183, 184, 185, 187, 188, 189, 194, 195, 200, 201, 210, 215, 221, 224, 231, 232, 233, 237
exchange markets, x, 159
exchange rate, 110
exclusion, 95
execution, 120
exercise, 30, 109, 133, 167
expenditures, 110, 111, 112, 147
expertise, 79, 92, 116
exploitation, 77, 79
exports, 123, 154
exposure, xi, 31, 209, 211, 232, 233
eyes, 139

## F

facilitators, 104
factor analysis, 5, 6
failure, ix, 51, 87, 88, 94, 107, 116, 117, 144, 146, 211
FDI, 108, 109, 110, 118, 120, 123
FDI inflow, 108, 120
fear, 100, 136, 148
Federal Communications Commission (FCC), 52, 55, 59, 60, 62, 63, 64, 65, 66, 67, 70, 71, 72, 128, 129, 130, 131, 139, 140, 141, 159, 160, 161, 163, 164, 165, 166, 168, 172, 175
federal government, 144
feedback, 180, 182
feet, 88, 156
Fiber Bragg gratings (FBG), xi
fiber optics, 228, 231
fibers, xi, 177, 179, 181, 185, 195, 199, 209, 210, 212, 213, 214, 215, 217, 218, 219, 221, 223, 224, 225, 231, 233, 234, 235, 238
filters, 236

finance, 53, 101, 102, 103, 111, 123
financial crisis (es), 112, 114
financial distress, 121
financial instability, 110
financial institutions, 34
financial intermediaries, 102
financial performance, 78, 118
financial resources, 30, 102, 103
financial stability, 103
financial support, 104
financing, 52, 58, 69, 101, 102, 103, 104, 112, 114, 120
Finland, 17, 18, 22, 23, 24, 26, 34, 36
firms, x, 49, 51, 65, 69, 70, 111, 112, 113, 114, 116, 117, 118, 120, 121, 122, 123, 133, 139, 159, 163, 167, 168, 169, 171, 173
fish, 47
flatness, 181, 196
flexibility, 58, 83, 210, 237
flood, 79
fluctuations, 181
fluid, 138
fluorescence, 182
FMC, 1
foreign direct investment, 15, 110
foreign exchange, 154
foreign firms, 51, 117
foreign investment, 36, 38, 40, 70
foreign policy, 48, 51, 55
forests, 68
fragmentation, 84
France, 17, 18, 22, 23, 24, 25, 26, 31, 32, 33, 34, 41, 42, 114, 115
franchise, viii, 75, 76, 77, 79, 81, 82, 83, 84, 85, 87, 88, 89, 172, 173
franchise area, 173
freedom, 128, 130, 136, 137
friction, 76, 78
FTTH, 76
fuel, 127
function values, 192
funding, 101, 102, 103, 123, 151
funds, 65, 70, 120, 144, 145, 148, 149, 150, 152, 153

generation, 15, 23, 41, 43, 111, 161, 180, 183, 235, 238, 240
Geneva, 70
geography, 59
Georgia, 27
germanium, 219
Germany, 17, 19, 21, 22, 23, 24, 25, 26, 27, 29, 33, 35, 42, 114, 115, 116, 208
GFI, 6
glass, 214, 217
global communications, 50, 125, 210
global trade, 68
globalization, 60, 92, 94, 127, 132
Globalization, 105, 106
GNP, 51, 110, 113
goals, viii, 47, 51, 54, 58, 62, 71, 75, 76, 82, 83, 88, 89, 96, 137, 146, 151
gold, 70
Goodness of Fit, 6
goods and services, 51
governance, 47, 53, 54, 58, 68, 71, 85, 103
government, viii, ix, x, 16, 36, 43, 45, 51, 52, 53, 55, 57, 62, 63, 69, 70, 75, 76, 78, 79, 80, 81, 83, 84, 88, 89, 90, 94, 95, 96, 101, 102, 104, 111, 112, 116, 120, 127, 128, 130, 131, 132, 134, 135, 136, 137, 143, 144, 145, 146, 148, 149, 151, 152, 153, 154, 155
grants, 80, 84
graph, 3, 149, 204, 221
gratings, xi, 180, 199, 209, 210, 217, 218, 225, 228, 229, 231, 235, 236, 241, 243
Greece, 17, 19, 22, 23, 24, 25, 26, 33, 37, 39
greed, 114
grids, 94
grouping, 10, 48
groups, 9, 16, 47, 59, 75, 76, 96, 111, 117, 132, 145, 181
growth, vii, 1, 10, 11, 32, 36, 39, 50, 54, 56, 58, 76, 78, 79, 85, 92, 103, 109, 114, 115, 119, 120, 123, 127, 130, 135, 137, 154, 165, 166, 170, 172
growth rate, 115, 119, 130, 172
growth spurt, 1
guidelines, 98, 132

| G |
|---|

GATS, 48, 60
GATT, 48
Gaussian, 176
GDP, 110, 111, 114, 115, 119, 154
gene, 47
General Agreement on Trade in Services, 60

| H |
|---|

handoff, 87
hands, 131, 149
harm, 56, 136
harmonization, 92
harmony, 136
Hawaii, 65
health, 71

heating, 224
hegemony, 67
heliograph, vii
heterogeneity, 100
high growth potential, 123
higher quality, 137, 169
Homeland Security, 64, 84
homogeneity, 238, 239
Hong Kong, ix, 35, 43, 127, 128, 130, 131, 132, 133, 134, 135, 136, 137, 139, 140
host, 2
hot spots, 2, 3, 7, 11
House, 61, 71, 74, 243
household income, 9
households, 80, 164, 174
human capital, 76
human resources, 102
Hungary, vii, 13, 14, 17, 19, 22, 23, 24, 25, 26, 27, 28, 29, 33, 34, 36, 116
hybrid, 11, 66, 179, 192, 193, 194, 204
hydrofluoric acid, 231

## I

IB, 72
ICT, 75, 105, 207
identity, 29, 135
ideology, 140
imbalances, 120, 121
IMF, 114
immunity, xi, 53, 60, 209, 210
implementation, 15, 102, 109, 111, 114, 116, 117, 123, 140, 181, 183, 187, 193, 194, 239, 240, 242
imports, 154
impurities, 181
in situ, 94, 99
in transition, 131, 140
incentives, 29, 60, 62, 94, 96, 98, 99, 100, 104, 110, 122, 161, 166, 173, 174
inclusion, viii, 75, 76, 83, 89
income, 8, 9, 10, 66, 75, 76, 83, 86, 104, 108, 113, 118
income distribution, 10
increased competition, 15
incumbent local exchange carrier, x, 159
incumbents, 27, 28, 80, 163, 165
independence, 144, 146
independent variable, 4, 5, 6
India, 46, 69, 181
Indian Ocean, 50
indication, 118
indicators, 6, 118
indigenous, 46

indirect measure, 210
indirect taxation, 112
individual character, vii, 1, 4, 7, 10
individual characteristics, vii, 1, 4, 7, 10
Indonesia, 46, 56
industry, x, 1, 14, 40, 43, 47, 51, 61, 62, 63, 66, 71, 78, 79, 85, 124, 125, 129, 134, 136, 139, 151, 156, 159, 160, 161, 163, 167, 171, 173, 174
inefficiency, 154
inelastic, 181
inequality, 76, 90
inflation, 110
influence, 174
information asymmetry, 78, 79, 81, 86
information sharing, 96, 100
information technology, 63, 210
infrastructure, ix, 3, 14, 51, 55, 65, 75, 76, 79, 80, 84, 88, 89, 91, 92, 93, 94, 95, 96, 97, 98, 100, 101, 104, 131, 133, 135, 137, 139, 155
innovation, 4, 6, 7, 11, 52, 161
insertion, xi, 209, 210, 235, 241
insight, 89, 122
instability, 110, 120, 238
institutional change, 48
institutions, 15, 34, 40, 47, 49, 60, 68, 90, 92, 97, 101, 102, 103, 112
instruments, 144
insurance, 79, 90
integration, 4, 40, 84, 92, 125, 170, 174, 186, 187
integrity, 63
intensity, 162, 170, 179, 182, 229
intentions, 31
interaction(s), x, xi, 47, 57, 94, 105, 159, 178, 179, 180, 181, 182, 183, 191, 203, 204
Inter-American Development Bank, 94, 96, 102, 105
interdependence, 49, 68, 92, 100
interest groups, 132
interest rates, 110
interface, 179
interference, 50, 60, 95, 132, 210, 235, 240, 241, 242
intermediaries, 98, 102
internal consistency, 5, 6
internal mechanisms, 59
international communication, 57, 58
International Monetary Fund, 47, 68
international standards, 117
International Telecommunications Satellite Organization (INTELSAT), viii, 45, 46
Internet, vii, x, 1, 2, 3, 8, 9, 10, 11, 12, 21, 49, 62, 64, 70, 71, 73, 76, 77, 79, 82, 90, 100, 103, 114, 115, 118, 129, 130, 131, 133, 135, 140, 159, 161, 163, 165, 168, 172, 175, 207
interpretation, 47, 86

interrelationships, 163
intervention, 102, 110
interview, 62
inversion, 190
investment, vii, viii, 1, 2, 13, 14, 25, 26, 27, 29, 30, 31, 33, 38, 39, 40, 49, 50, 51, 53, 57, 58, 61, 64, 69, 70, 71, 76, 77, 79, 80, 81, 86, 87, 103, 108, 110, 111, 112, 114, 135, 147, 156, 178
investment rate, 110, 114
investment spending, 112, 114
investors, x, 15, 17, 38, 54, 58, 64, 67, 69, 70, 103, 112, 120, 143, 147, 150, 152, 153
IPO, 60, 64, 72, 153
IR, 207
Ireland, 17, 19, 22, 23, 24, 25, 26, 33, 39, 128, 129, 140
iron, 112
Islam, 208
ISPs, 85, 86
Italy, 17, 19, 22, 23, 25, 26, 33, 35, 114, 116, 177, 208
ITC, 208
ITSO, viii, 45, 46, 49, 63, 66, 67, 68, 70, 71, 72, 73
IXCs, 160, 162, 164

## J

Japan, 50
journalists, 136
jurisdiction, 48, 53, 92, 103

## K

Kazakhstan, 27
King, 108, 124, 175
Korea, vii, 1, 2, 3, 4, 9, 12, 68, 140
Kosovo, 37

## L

labor, 109, 113, 122, 123, 152
labor force, 109, 122, 123
Lafayette, 81
LAN, vii, 1, 2, 3, 4, 5, 7, 8, 9, 10, 11
land, 66
landlocked countries, 94
language, 25, 48, 59, 85, 97
language barrier, 97
laser, 179, 180, 182, 191, 197, 198, 199, 201, 202, 208, 235, 238, 239
lasers, xi, 178, 179, 180, 191, 195, 199, 200, 201, 238, 242

lasing effect, 199, 200
lasing threshold, 199
latency, 80
Latin America, 25, 27, 102, 124, 125
Latvia, vii, 13, 14, 17, 22, 23, 24, 26, 31, 33, 34, 35, 36, 43
laws, 15, 60, 63, 69, 71, 95, 98, 102, 117, 123, 136, 139, 140
LEA, 139
lead, 1, 9, 16, 48, 49, 59, 77, 81, 85, 94, 123, 136, 146, 162, 166, 174, 181, 195, 211, 213
leadership, 60, 66, 128, 131, 138, 150
leakage, 179
learning, 125
legislation, viii, ix, 15, 38, 40, 45, 46, 48, 51, 61, 62, 64, 65, 91, 93
Less Developed Countries, 71
liberalisation, 14
liberalism, 47
liberalization, x, 38, 39, 48, 62, 108, 109, 110, 115, 117, 118, 123, 128, 135, 143, 147, 149, 151, 152, 156
licenses, ix, 63, 67, 134, 140, 143, 149, 150, 155, 156
lifestyle, 2, 10, 11
lifetime, viii, 45, 46, 49, 68, 211
likelihood, 8, 10, 59, 99, 134
limitation, 4, 7
linear function, 236
links, 50, 51, 98, 178
liquidate, 154
liquidity, 112, 116, 120
literature, viii, 2, 4, 75, 108, 109, 120, 147, 162, 164, 166, 167, 225
Lithuania, vii, 13, 14, 16, 17, 19, 22, 23, 24, 26, 33, 34, 35, 36, 116
loans, 103, 117
local exchange carriers, x, 159, 160
local government, viii, 75, 79, 81, 88, 89
location, 51, 73, 84, 86
Lockheed Martin, viii, 45, 61, 62, 63, 64, 65, 66, 71, 73, 74
London, 42, 56, 70, 74, 138, 140, 144, 152, 156, 208
long distance, 53, 114, 119, 128, 138, 165, 166, 167, 170
lower prices, x, 160, 162, 167, 169

## M

Macedonia, 27, 29, 37, 41
major cities, 77, 80
major decisions, 53

management, 53, 54, 56, 59, 62, 63, 65, 70, 79, 87, 103, 109, 112, 117, 118, 119, 122, 123, 133, 151, 153, 155, 168, 178
management practices, 63, 122
mandates, 150
manufacturing, 108
marginalization, 71
market, vii, viii, ix, x, 1, 2, 5, 10, 11, 13, 14, 15, 16, 22, 23, 24, 25, 27, 29, 30, 31, 32, 35, 38, 39, 40, 42, 43, 45, 46, 48, 49, 54, 55, 56, 58, 59, 60, 61, 62, 63, 64, 65, 70, 71, 75, 76, 77, 78, 80, 81, 85, 86, 89, 94, 110, 111, 112, 113, 114, 116, 117, 124, 127, 128, 129, 131, 132, 133, 134, 135, 136, 137, 138, 143, 144, 146, 147, 148, 149, 150, 151, 152, 153, 154, 155, 160, 161, 162, 163, 167, 168, 169, 170, 171, 172, 173, 174, 175, 210
market access, 60, 116
market concentration, 23
market economy, 15
market failure, 2, 94
market incentives, 62
market opening, 62
market penetration, 86
market position, 23, 35
market share, ix, x, 23, 30, 61, 143, 151, 152, 170
market structure, x, 43, 124, 128, 160, 163
market value, 30
marketing, 2, 3, 10, 57, 63, 161, 167, 168, 169
marketing strategy, 10
markets, x, 11, 14, 16, 24, 25, 30, 31, 32, 33, 35, 36, 38, 39, 49, 50, 51, 53, 60, 70, 75, 76, 79, 80, 81, 92, 93, 94, 108, 110, 112, 113, 116, 128, 129, 131, 133, 138, 154, 156, 159, 171, 173
marriage, 42
Maryland, 65
mass, 161, 170
mass media, 136
Massachusetts, 208
matrix, 186, 200, 208, 226, 227, 230, 235
maximum price, 86
measurement, 210, 225, 226, 227, 231, 235
measures, x, 5, 82, 89, 94, 159, 187
meat, 112
mechanical stress, 221
media, ix, x, 8, 21, 38, 70, 127, 128, 129, 130, 131, 132, 133, 134, 135, 136, 137, 138, 139, 140, 141, 177
median, 9
medicine, 210
membership, 15, 40, 97
mergers, 65, 107, 116, 124, 125, 136, 140
mesh networks, 3
messages, 168

meta-analysis, 108
Mexico, 116
micrometer, 223
microscope, 231
microwave, 88, 235
Middle East, 66, 69
migration, 174
military, 47, 61, 146, 149, 154
military dictatorship, 146
milk, 112
minerals, 47
minorities, 32
minority, 9, 17, 27, 30, 32, 36, 41, 69
mixing, 181
mobile communication, viii, 3, 11, 13, 14, 15, 43
mobile phone, 8, 11, 115, 118, 122, 166
mobile telecommunication, vii, viii, 13, 14, 39, 43
mobility, 3, 4, 76, 79, 88, 161, 163
mode, 173
modeling, 5, 6, 10, 185, 190, 195
models, 3, 4, 84, 98, 99, 127, 128, 132, 138
moderators, 124
modernization, 94
modules, 178
Moldova, 27, 29
molecules, 181, 182
momentum, 89
money, 25, 101, 102, 150
monitoring, 165
monopoly, ix, x, 52, 55, 59, 77, 80, 112, 113, 114, 115, 116, 119, 134, 136, 143, 146, 148, 150, 151, 153, 154, 159, 175
Montana, 64
Montenegro, 27, 29, 36, 37
Moon, 1, 68, 74
moral hazard, ix, 91, 93, 99, 103, 104, 110, 111
morning, 156
Morocco, 25
motivation, 111, 112
mountains, 51
MTS, 30
multichannel video programming service, 172
multimedia, 3
multinational firms, 49
multiples, 180
multiplexing, 180, 210, 239, 241, 242
multiplicity, 137
mutation, 192, 195

# N

Namibia, 35
nation, 49, 53, 68, 92, 95, 99, 115

nation states, 49, 68, 115
National Aeronautics and Space Administration, 52
national interests, 117
national product, 117
national security, 48, 60
National Security Strategy, 60, 74
NCA, 144, 149, 150, 155, 156
negotiating, 33, 57
negotiation, 15, 79, 80, 88, 153
neoliberalism, 71
neo-liberalism, 47
Netherlands, 16, 17, 20, 22, 23, 24, 26, 27, 29, 31, 33, 34, 60, 114, 115, 116, 175
network, vii, 1, 4, 11, 16, 17, 21, 23, 25, 26, 27, 29, 31, 34, 38, 39, 49, 50, 51, 53, 55, 56, 59, 62, 66, 68, 70, 76, 77, 79, 80, 81, 82, 83, 84, 85, 86, 87, 89, 90, 92, 95, 96, 97, 98, 100, 102, 104, 130, 144, 146, 150, 161, 169, 170, 171, 172, 174, 178
network elements, 171
network members, 96
networking, 4, 77, 90, 241
neural network, 192
neural networks, 192
New Jersey, 12
new media, 129, 130, 136
New York, 52, 63, 70, 71, 90, 105, 106, 124, 138, 139, 140, 147, 155, 172, 175, 207, 243
news coverage, 129
newspapers, 129, 131
Newton, 186
next generation, 161, 235
NFI, 6
Nigeria, ix, 143, 144, 145, 146, 147, 150, 151, 154, 155, 156
Nile, 103, 105
NITEL, ix, x, 143, 144, 145, 146, 147, 148, 149, 150, 151, 152, 153, 154, 155, 156
NNFI, 6
noise, x, 177, 178, 180, 181, 182, 183, 187, 188, 189, 190, 191, 202
North America, 27, 50, 64
North Korea, 68
Norway, 22, 27, 28, 34, 35, 36
NTIA, 52
numerical computations, 192

## O

obligation, 77, 85, 89
observations, 16, 137
OECD, 133, 140, 141
oil, 95, 118, 123
Oklahoma, 52, 107

open markets, 138
openness, 136, 137
operator, 16, 17, 21, 22, 23, 24, 25, 27, 28, 29, 30, 31, 33, 34, 35, 36, 37, 38, 40, 41, 52, 55, 59, 66, 67, 118, 173, 192
Operators, 21, 26, 33
opportunity costs, 164, 165
optical communications, xi, 203, 209, 210, 216, 242
optical fiber, x, xi, 177, 179, 180, 181, 183, 199, 203, 210, 211, 213, 229
optical gain, 182, 183
optical properties, 181, 182, 210
optics, xi, 209, 228, 231, 236, 242
optimization, xi, 178, 187, 191, 192, 194, 195, 196, 203, 204, 209, 242
optimization method, 192
orbit, 50, 53
ORBIT Act, viii, 45, 47, 48, 61, 62, 64, 65, 66, 71, 72, 74
Oregon, 52
organ, 51, 53
organization, viii, 5, 45, 46, 47, 49, 52, 53, 54, 55, 57, 58, 59, 62, 63, 65, 66, 67, 68, 78, 84, 101, 102, 110, 149, 153
Organization for Economic Cooperation and Development, 124
organizational culture, 153
organizations, 47, 49, 51, 52, 53, 65, 71, 101, 102, 104
orientation, 225
OSA, 196, 220, 221
overseas investment, 35
oversight, x, 119, 133, 150, 160
overtime, 95
ownership, viii, ix, x, 13, 14, 15, 23, 27, 29, 33, 38, 42, 45, 52, 54, 57, 59, 60, 61, 64, 65, 68, 69, 76, 77, 78, 79, 81, 87, 112, 113, 116, 117, 118, 122, 123, 127, 128, 129, 130, 131, 132, 133, 134, 135, 136, 137, 138, 139, 140, 141, 146, 155, 159, 174

## P

Pacific, 50, 58, 66, 93, 130
parameter, 198
parents, 167
Pareto, 100, 102
Pareto optimal, 102
Paris, 55, 70, 124
Parliament, 133
partnership, 53, 61, 153
partnerships, 34, 58, 101
passive, xi, 180, 209, 210
PBC, 240

PDAs, 168
penalties, 197
pensions, 153
Pentagon, 63
perception, 7, 8, 10, 49, 78, 79, 81, 85, 132, 145, 147, 155, 162
performance, 11, 54, 60, 78, 79, 81, 82, 86, 88, 109, 110, 118, 123, 124, 128, 136, 189, 190, 192, 240, 242
periodicity, xi, 209, 215
permit, 40, 84, 92
perseverance, 63
personal, 168
perspective, 161, 173
Petroleum, 146
petroleum products, 112
phonons, 181, 182, 190
photoelastic effect, 214, 237
photographs, 217, 231
photons, 181, 182, 183
photosensitivity, 210
plague, 89
planned investment, 114
planning, 55, 83, 84
plants, 112
pluralism, 137
plurality, 133
PM, 196, 217, 219
Poland, vii, 13, 14, 17, 20, 22, 23, 24, 25, 26, 27, 29, 30, 31, 32, 33, 34, 35, 39, 42, 116
polarization, xi, 183, 184, 191, 209, 210, 213, 215, 216, 221, 223, 224, 225, 226, 227, 229, 230, 233, 235, 236, 237, 238, 239, 240, 241, 242
polarized light, 213, 215
policy instruments, 144
policy makers, x, 108, 112, 159, 173
policy making, 117, 132, 137
policy reform, 104
policymakers, 138
political opposition, 112
politics, 48, 110, 117, 123, 140, 156
poor, 51, 63, 118, 122, 123, 145, 146, 147, 151, 155, 156
POPs, 66
population, 5, 32, 79, 81, 131, 132, 144, 165, 190, 192, 193, 195
population density, 81, 165
population size, 193
portability, x, 39, 159
portfolio, 66, 108, 112
portfolio investment, 108
ports, 112

Portugal, 17, 20, 22, 23, 25, 26, 31, 33, 39, 177, 208, 209
positive feedback, 182
post-acquisition, 124
power, viii, x, 6, 46, 47, 48, 49, 50, 51, 62, 71, 75, 78, 80, 81, 84, 86, 88, 89, 94, 95, 96, 134, 135, 136, 160, 162, 163, 167, 169, 170, 174, 178, 179, 180, 182, 183, 184, 185, 186, 187, 189, 190, 191, 192, 195, 196, 197, 198, 199, 200, 201, 202, 203, 204, 205, 206, 215, 238, 241, 242
power relations, 96
prediction, 147
preference, 33, 154
preparation, 172
present value, 119
president, 112
President Clinton, 63
pressure, 32, 62, 64, 77, 78, 89, 171, 195
prevention, 90
price caps, 174
price elasticity, 166
price index, 163
prices, x, 57, 62, 67, 80, 86, 88, 112, 136, 151, 160, 161, 164, 166, 167, 169, 170, 171, 173, 175, 195
pricing flexibility, 58
prisoners, ix, 91, 93, 100, 101, 104
private good, 95, 96
private investment, 79, 81
private ownership, ix, 52, 127, 131, 137
private property, 68
private sector, viii, 46, 49, 52, 68, 89, 104, 113, 149, 154, 156
privation, 121
privatization, viii, ix, 45, 46, 47, 48, 49, 52, 53, 60, 61, 62, 63, 64, 65, 66, 69, 71, 107, 108, 109, 110, 111, 112, 113, 114, 116, 117, 118, 119, 120, 121, 122, 123, 127, 132, 148, 156
probability, 29
probe, 184, 187, 188, 189, 190, 191, 194, 195, 196, 197, 199, 202, 204
product market, 173
production, ix, xi, 47, 51, 91, 93, 94, 95, 96, 97, 98, 99, 100, 101, 102, 103, 104, 112, 122, 209, 210, 216, 235
production costs, 210
production technology, xi, 96, 100, 209, 235
productive efficiency, 60
productivity, 59, 109, 112, 117, 119, 122, 147, 153
profit, 26, 30, 60, 84, 118, 121, 122, 149, 154
profit margin, 122
profitability, 122
profits, 43, 46, 63, 122, 129, 145, 175
program, 67, 83, 84, 85, 110, 112, 135, 206

programming, 129, 172, 192
proliferation, 58, 79, 129
promote, viii, 48, 61, 62, 75, 86, 96, 137, 156, 168
propagation, xi, 178, 187, 192, 201, 202, 209, 210, 211, 213, 215, 235, 236
property rights, 60
proposition, 36, 50
prosperity, 60
protocol, 161
prototype, 181
PTT, 56, 60, 62, 116
public debt, 110
public domain, 94
public enterprises, 60
public finance, 110
public goods, ix, 91, 92, 93, 94, 95, 96, 100, 101, 102, 103, 104
public interest, 47, 52, 65, 70, 136, 137, 138
public policy, 135
public safety, 77, 82, 84, 87
public sector, 110, 111, 112, 117
public service, viii, 45, 46, 49, 67, 68, 138
Public Wireless, vii, 1, 3, 4, 8, 9
publishers, 135
pulse, 235
pumps, xi, 177, 179, 180, 181, 183, 184, 185, 187, 188, 190, 191, 192, 194, 195, 197, 201, 202, 203, 204, 205, 206, 235, 239
P-value, 7

## Q

quadratic programming, 192
quality of service, 151, 169, 174
questionnaire, 5, 9

## R

Radiation, 207
radio, vii, 48, 53, 67, 130, 131, 133, 135, 161, 241
radius, 88
rain, 50
Raman, vi, x, xi, 177, 178, 179, 180, 181, 182, 183, 184, 185, 186, 187, 188, 189, 190, 191, 192, 194, 195, 196, 197, 198, 199, 200, 201, 202, 203, 204, 205, 206, 207, 208
Raman and Brillouin scattering, 181
Raman fiber amplifiers (RFA), x, 177, 178
Raman spectroscopy, 182
range, 10, 31, 78, 88, 130, 165, 182, 184, 191, 192, 203, 210, 229, 235, 237
rate of return, 57

RBOC, x, 160, 162, 166, 167, 172, 173, 174
reading, 136
real estate, 112
reality, 47, 77, 182
reception, 50, 88, 203, 204
recognition, 25, 31, 51, 92, 101, 171
reconcile, 138
reconstruction, 144
reduction, 16, 69, 119, 122, 137, 148, 179, 181, 187
reflection, xi, 165, 197, 199, 209, 211, 216, 217, 219, 220, 221, 223, 224, 225, 228, 229, 230, 232, 233, 235, 240
reflectivity, 198, 211, 212, 213, 216, 241
reforms, 49
refraction index, 211, 214, 215
refractive index, xi, 209, 210, 211, 213, 215
regeneration, 178
regional, ix, x, 40, 56, 59, 91, 92, 93, 94, 95, 96, 97, 98, 100, 101, 102, 103, 104, 135, 149, 159, 163, 173, 175
Regional Bell Operating Companies, x, 159, 160
regional cooperation, 92, 94, 101, 102, 104
regional integration, 92
regional policy, 40, 104
regional public goods (RPGs), ix, 91
Registry, 63
regulation(s), viii, ix, 13, 14, 15, 43, 48, 78, 85, 90, 92, 93, 98, 99, 105, 108, 109, 117, 127, 128, 129, 131, 134, 136, 137, 138, 140, 150, 169, 174
regulators, 21, 90, 92, 95, 98, 132, 160, 173
regulatory framework, ix, 14, 40, 107
regulatory oversight, x, 133, 160
rejection, 2
relationship(s), ix, 7, 10, 11, 36, 47, 78, 79, 81, 82, 89, 93, 101, 111, 127, 128, 130, 132, 149
relaxation, 128, 186, 190
reliability, 5, 6, 51, 70, 86, 94
rent, 110, 145
repair, 153
reproduction, 71
reputation, 70
reserves, 29
resistance, 11, 55, 59, 152, 210
resolution, 51, 55, 59, 62, 76, 78, 92, 187, 210
resources, ix, 30, 38, 68, 72, 91, 92, 93, 97, 98, 99, 100, 102, 103, 104, 107, 116, 119, 121, 122, 123, 129, 149, 167
restructuring, viii, 45, 46, 47, 58, 114, 153, 155
retail, 57, 86, 87, 89, 161, 167, 168, 170, 173, 175
retention, 67
returns, 98
reunification, 136

revenue, 39, 56, 60, 66, 77, 84, 111, 145, 147, 148, 151, 152, 160, 169
revolutionary, 116, 136
rhetoric, 89
rice, 29
rights, 173
rings, xi, 177
risk, 4, 7, 8, 9, 10, 11, 35, 77, 78
RMSEA, 6
Romania, vii, 13, 14, 15, 17, 22, 23, 24, 26, 27, 28, 30, 31, 32, 33, 37, 42, 116
Rome, 208
room temperature, 238
routing, 235
rule of law, 15, 30
rural areas, 50, 70, 120
Russia, 27, 29, 30, 35, 36, 69, 181

## S

SA, 177
safety, 56, 77, 82, 84, 87, 179
sales, 35, 42, 57, 59, 61, 64, 112, 116, 118, 119, 121, 168
sample, 28, 83, 89, 166, 221, 229
sampling, 5, 231
sanctions, 68, 102
Sarin, 32
satellite, viii, 14, 45, 46, 47, 49, 50, 51, 52, 53, 54, 55, 56, 58, 59, 60, 61, 62, 64, 65, 66, 67, 68, 70, 95, 109, 117, 123, 129, 130, 137, 172
satellite service, 14, 46, 53, 55, 61, 62, 64, 66, 70
satellite technology, 46
satisfaction, 86, 147
saturation, 180
Saudi Arabia, 123
savings, 83, 94, 120, 162, 167, 169
scale economies, 77, 80, 81, 95, 129, 170
scarcity, 51, 129, 130
scattered light, 182, 197
scattering, xi, 177, 179, 181, 182, 183, 188, 206
scholarship, 206
school, 60
science, 138
search, 43, 168, 192
searching, 78
Secretary General, 140
securities, 64
Securities and Exchange Commission, 67
security, 4, 7, 11, 48, 60, 64, 117, 118, 164
seed, 101, 181, 201
selecting, 190
self-employed, 8, 9

self-interest, 132
self-promotion, 85
self-regulation, 138
semaphore, vii
semiconductor, 180, 181
Senate, 61, 72, 74, 146
sensing, xi, 209, 211, 225, 231, 242
sensitivity, 220, 221, 223, 224, 225, 241
sensors, xi, 209, 210, 219, 225, 235, 238, 242
separation, 117, 219, 225
Serbia, 27, 30, 32, 33, 36, 37
series, 16, 48, 50, 52, 63, 65, 71, 163, 175
service provider, ix, 2, 25, 49, 53, 54, 86, 108, 109, 116, 118, 123, 127, 128, 131, 132, 134, 136, 149
service quality, 78, 79, 80, 86, 88, 122
services, x, 160, 161, 162, 163, 164, 165, 166, 167, 168, 169, 170, 171, 172, 173, 174
shape, 47, 88, 183, 214
shaping, 235
shareholders, 29, 30, 32, 63, 64, 69, 107, 120, 122
shares, 30, 40, 41, 43, 53, 59, 62, 63, 64, 65, 69, 112, 113, 116, 117, 118, 128, 160, 170
sharing, ix, 84, 91, 92, 93, 94, 95, 96, 99, 100, 102, 104, 105, 168
shock, 150
shortage, 39
sign, 31, 151
signal quality, 88
signals, vii, 50, 88, 161, 179, 180, 183, 184, 185, 187, 188, 189, 190, 191, 194, 195, 196, 197, 200, 204, 225, 241
signal-to-noise ratio, 180
significance level, 6
signs, 184
silica, 213, 219
simulation, 191, 193, 194, 195, 196, 200, 201, 204, 208, 213, 216, 219, 237
Singapore, ix, 43, 127, 128, 130, 131, 132, 133, 134, 136, 137, 141
single market, 14
sites, 178, 198
skills, 48
Slovakia, vii, 13, 14, 16, 17, 20, 22, 23, 25, 26, 27, 29, 31, 32, 36, 37, 40, 41
smoke, vii
SMS, 15
social class, 146
social relations, 71
social responsibility, 130, 133
social security, 118
society, 108, 138, 139, 140, 156
software, 16, 102
Somalia, 68

South Africa, 98, 101, 150
South America, 57, 63, 94
South Asia, 93, 101, 130, 138
South Korea, 140
Soviet Union, 51
space station, 67
Spain, 17, 20, 22, 23, 25, 26, 31, 33, 35, 36, 39, 42, 116
spare capacity, 25
specialization, 78
specificity, 81
spectral component, 183, 184
spectroscopy, 182
spectrum, xi, 15, 16, 40, 47, 68, 85, 129, 130, 161, 182, 184, 192, 196, 197, 198, 199, 201, 202, 203, 204, 209, 216, 219, 220, 221, 225, 228, 229, 230, 233, 239
speech, 137
speed, 2, 4, 15, 17, 50, 62, 64, 79, 83, 86, 88, 112, 172, 192
spillovers, 100
spin, 59
Sprint, 173
SPSS, 5
stability, 4, 49, 69, 103, 186, 195, 200
stabilization, 199, 200, 201, 202
stages, 35, 36, 102
stakeholders, 67, 138
standards, ix, 48, 91, 92, 97, 99, 100, 117, 195
State Department, 52, 69
statistical analysis, 5
statistics, 75, 80
steel, 112
stimulus, 101
stock, 41, 62, 63, 65, 90, 116
stock markets, 116
storage, 108
strain, 103, 211, 219, 220, 221, 223, 224, 225, 226, 227, 228, 229, 230, 231, 234, 235, 242
strategies, vii, 11, 13, 14, 15, 68, 97, 120, 122, 125, 149, 153, 155, 162, 164, 166, 167
stratification, 146
stratified sampling, 5
strength, 47, 98, 131, 150, 184
stress, 130, 213, 214, 217, 220, 221, 231, 233, 234, 236, 237, 242
strictures, 128, 135, 138
structural changes, 108
structural equation modeling, 5, 6, 10
students, 8, 9
subgroups, 2, 9, 10
sub-Saharan Africa, 98, 156

subscribers, x, 2, 3, 16, 17, 18, 21, 23, 24, 27, 28, 29, 31, 32, 33, 34, 35, 36, 37, 114, 115, 118, 130, 148, 151, 153, 159, 161, 162, 163, 164, 165, 166, 167, 168, 169, 170, 171, 172, 173, 174
subscription television, 135
subsidization, 46, 57, 58, 89, 133
subsidy, 56, 57, 148
substitutes, 162, 163, 164, 165, 166
substitution, x, 159, 165, 166, 174
Sudan, 68
Sun, 139, 244
superiority, 61
supervision, 109
suppliers, 173
supply, 92, 93, 95, 97, 100, 104, 118, 148, 149
support services, 63
survivability, 152
survival, 150
susceptibility, 183
sustainable growth, 92
Sweden, 17, 21, 22, 23, 26, 27, 32, 33, 34, 35, 36, 39, 43, 116
switching, vii, 1, 110, 146, 235
Switzerland, 31, 34, 35
systems, x, xi, 49, 50, 54, 55, 56, 58, 61, 66, 68, 86, 96, 102, 114, 130, 164, 166, 177, 178, 179, 180, 181, 182, 189, 190, 196, 197, 202, 203, 204, 205, 206, 209, 210, 235, 241

## T

tactics, 149, 151
Taiwan, 4
takeover, 16, 36
Tanzania, 156
targets, 112
tariff, 54, 58, 100
tariff rates, 58
tax collection, 109, 123
tax rates, 120
taxation, 60, 112
technical assistance, 57
technological advancement, 166
technological change, 68, 71, 134, 137
technological developments, 58, 156
technological progress, 85
technology, vii, ix, xi, 1, 2, 3, 4, 7, 8, 9, 10, 11, 15, 16, 21, 38, 46, 51, 57, 63, 76, 77, 78, 79, 80, 81, 82, 83, 84, 85, 88, 89, 92, 96, 97, 98, 100, 102, 112, 116, 127, 132, 137, 138, 139, 140, 144, 153, 161, 177, 178, 181, 195, 209, 210, 235
technology transfer, 57
teens, 9

telecommunications, vii, viii, ix, x, 2, 13, 14, 15, 35, 38, 39, 40, 41, 43, 46, 48, 50, 51, 52, 53, 55, 56, 59, 60, 61, 62, 63, 64, 66, 68, 70, 71, 72, 91, 92, 93, 94, 98, 107, 108, 109, 114, 115, 116, 117, 118, 119, 120, 122, 123, 124, 125, 128, 135, 138, 140, 146, 148, 149, 150, 152, 153, 154, 155, 156, 159, 160, 161, 162, 163, 164, 165, 171, 172, 173, 174, 175, 179, 182, 238
Telecommunications Act, viii, x, 45, 47, 48, 72, 141, 159, 160
telecommunications policy, 140
telecommunications services, x, 2, 46, 53, 56, 60, 61, 63, 66, 68, 70, 150, 159, 161, 163, 171, 172, 173, 175
telephone, vii, ix, x, 4, 51, 53, 56, 100, 114, 115, 116, 118, 128, 136, 139, 143, 144, 146, 147, 149, 154, 155, 156, 160, 163, 164, 172
telephony, x, 14, 15, 40, 50, 52, 55, 138, 160, 161, 162, 163, 164, 165, 166, 167, 171, 172, 173, 174
television, vii, 53, 128, 129, 130, 132, 133, 135, 138, 139, 172
temperature, 183, 184, 210, 219, 220, 224, 225, 226, 227, 228, 229, 230, 231, 234, 235, 237, 238, 242
temperature dependence, 183, 224, 229
temperature gradient, 237
tenants, 84
Tennessee, 52
terminals, 50
territory, 66, 172
Texas, 173
text messaging, 163, 168
textbooks, 115
The Economist, 41, 42, 154
theft, 146
theory, viii, 75, 77, 130, 136, 137, 139, 156, 173, 211, 213
thermal equilibrium, 182
thermal expansion, 219
thinking, 36, 104
Third World, 72
threat, 61, 68, 129, 140, 153
threats, 79
threshold, 97, 100, 179, 182, 199, 201
thresholds, 6, 199
time, vii, x, xi, 1, 3, 8, 14, 17, 26, 27, 28, 29, 35, 48, 52, 53, 55, 57, 59, 61, 64, 71, 78, 80, 83, 85, 86, 100, 103, 104, 109, 111, 118, 119, 130, 135, 137, 144, 146, 147, 149, 151, 152, 153, 159, 161, 162, 166, 168, 170, 171, 174, 178, 184, 185, 187, 191, 192, 193, 199, 210, 232, 233, 235, 239, 241, 242
Title III, 73
top management, 122
topology, 195

tourism, 112
trade, 48, 63, 68, 110
trade policies, 68
trading, 28
traffic, 11, 46, 50, 56, 60, 66, 68, 77, 79, 87, 161, 164, 180
training, ix, 57, 83, 91, 92, 147
transaction costs, 48, 76, 78
transactions, 34, 59, 69, 169
Transcorp. Ltd., x, 143
transference, 208
transformation, viii, 14, 41, 45, 49, 62, 129
transition, 4, 110, 131, 140
translation, 219, 223
transmission, vii, x, 2, 3, 4, 50, 51, 53, 54, 58, 130, 137, 146, 160, 161, 177, 178, 179, 180, 181, 183, 184, 190, 197, 198, 203, 204, 205, 210, 241, 242
transmission path, 205
transnational corporations, 49
transport, 94, 160
transportation, ix, 91, 93, 94, 95, 103, 108
transverse section, 217, 221, 232
trend, 3, 62, 161, 163, 166, 233
triangulation, 47
triggers, 182
Turkey, 32, 35, 108, 109, 110, 111, 112, 114, 115, 116, 117, 118, 120, 123, 124

## U

UK, 16, 17, 21, 22, 23, 24, 26, 27, 29, 31, 32, 33, 40, 74, 116, 208
Ukraine, 27, 30, 34
ultraviolet light, xi, 209, 211
uncertainty, 34, 78, 79, 85, 109
unemployment, 113, 120
unemployment rate, 113
UNESCO, 49
unforeseen circumstances, 144
uniform, 57, 58, 86, 95, 195, 211, 212, 213, 237
United Kingdom, 63, 69
United Nations, 51, 52, 61, 71, 105, 125, 149
United States, 46, 51, 53, 62, 71, 72, 74, 129, 130, 131, 136, 140, 141, 147
updating, 168
uplink, 40
urban areas, 3, 5, 50
urban centers, 2, 3, 94
Uruguay, 48
Uruguay Round, 48
users, 2, 9, 10, 11, 53, 55, 57, 58, 66, 71, 83, 85, 86, 95, 114, 115, 165, 172, 173, 174, 239, 240, 241
Uzbekistan, 30

## V

vacuum, 58
validity, x, 5, 6, 48, 56, 100, 159
values, 49, 59, 68, 186, 187, 192, 194, 195, 197, 203, 214, 219, 223, 224, 226, 227, 228, 231, 235
variable(s), 4, 5, 6, 163, 164, 186
variation, ix, 75, 86, 100, 219, 220, 221, 225
VAT, 119
vehicles, 211
velocity, 184, 213, 231
Verizon, 80, 160, 168, 170, 172, 173, 175
video programming, 172
voice, x, 14, 15, 50, 66, 98, 104, 130, 137, 160, 161, 162, 165, 168, 170, 171, 172, 173, 174, 175
voice mail, 161, 168, 170
volatility, 110
voting, 53, 70
vulnerability, 49, 68

## W

Wales, 63
Wall Street Journal, 90, 139, 140
war, 35, 67, 69, 144
Washington, 45, 71, 72, 73, 74, 105, 112, 124, 127, 128, 139, 159, 160, 175
water resources, 68, 103
watershed, ix, 143
wavelengths, xi, 177, 179, 180, 181, 190, 191, 195, 198, 199, 202, 203, 204, 213, 216, 223, 225, 233, 236, 239, 240, 241
wealth, 46, 71, 131
web, 10, 106, 136, 172
websites, 21, 29, 31, 33, 34, 35, 36, 37, 88, 130
welfare, x, 2, 160, 162, 166, 167, 169
well-being, 61, 65
Western countries, 53
Western Europe, 31, 35, 38, 50
White House, 60, 74
wholesale, 32, 57, 86, 87
Wi-Fi, 77, 80, 85, 88, 173
WiMAX, vii, 1, 2
windows, 180
Wireless LAN, vii, 1
wireless networks, 80, 81, 84, 87, 162, 169
wireline/wireless bundling, x, 160, 162, 169, 171
wires, 80, 136, 141
withdrawal, 32, 36
workers, 8, 9, 10, 52, 119, 145, 147, 153
World Bank, 47, 66, 68, 101, 105, 112, 114, 124, 125, 154, 156
World Trade Center, 63
World Trade Organization, 47, 48, 58, 60
Worldcom, 139
writing, 35, 85
WTO, 47, 48, 60, 68, 70, 71

## X

X-axis, 222, 223

## Y

Y-axis, 222, 223, 229
yield, 98

## Z

Zimbabwe, 156